International Conference on Mechanical, Manufacturing, Modeling and Mechatronics 2016 (IC4M 2016)

MATEC Web of Conferences Volume 51 (2016)

Kuala Lumpur, Malaysia
27 - 29 February 2016

Editors:

Mohd Nasir Tamin
Du Ruxu

ISBN: 978-1-5108-2270-2

Printed from e-media with permission by:

Curran Associates, Inc.
57 Morehouse Lane
Red Hook, NY 12571

Some format issues inherent in the e-media version may also appear in this print version.

Printed by Curran Associates, Inc. (2016)

For additional information, please contact EDP Sciences – Web of Conferences at the address below.

EDP Sciences – Web of Conferences
17, Avenue du Hoggar
Parc d'Activité de Courtabœuf
BP 112
F-91944 Les Ulis Cedex A
France

Phone: +33 (0) 1 69 18 75 75
Fax: +33 (0) 1 69 28 84 91

contact-edps@webofconferences.org

Additional copies of this publication are available from:

Curran Associates, Inc.
57 Morehouse Lane
Red Hook, NY 12571 USA
Phone: 845-758-0400
Fax: 845-758-2633
Email: curran@proceedings.com
Web: www.proceedings.com

TABLE OF CONTENTS

PREFACE

2016 International Conference on Mechanical, Manufacturing, Modeling And Mechatronics (IC4M 2016) was held in Kuala Lumpur, Malaysia during February 27-29, 2016. The proceeding is a compilation of the accepted papers and represent an interesting outcome of the conference.

IC4M 2016 was organized by Hong Kong Society of Mechanical Engineers (HKSME). The conference provides a useful platform both for showing the latest research directions and for exchange of research results and ideas in broad Mechanical, Manufacturing, Modeling and Mechatronics areas.

We would like to acknowledge all of those who supported IC4M 2016. Each individual and institutional help were very important for the success of this conference. Especially we would like to thank the organizing committee for their valuable advices in the organization and helpful peer review of the papers.

We hope that IC4M 2016 will be a forum for fruitful discussions that will shed light on new ideas and foster collaborative researches. We are sure that the proceedings will serve as an important research source of references and the knowledge, which will lead to not only scientific and engineering progress but also new products and processes.

Prof. Mohd Nasir Tamin
Universiti Teknologi Malaysia, Malaysia

Prof. Du Ruxu
The Chinese University of Hong Kong, Hong Kong

COMMITTEE

Advisory Committee Chair
Prof. Mohd Nasir Tamin,Universiti Teknologi Malaysia, Malaysia

Conference Committee Co-Chairs
Prof. Dr. Mohd Hamdi Bin Abd Shukor, University of Malaya, Malaysia
Prof. Du Ruxu, The Chinese University of Hong Kong, Hong Kong
Prof. Amaresh Chakrabarti, Indian Institute of Science, Bangalore, India

Program Committee
Prof. Yong Zeng Concordia University, Canada
Dr. Musa bin Mailah, Universiti Teknologi Malaysia, Malaysia
Prof. Shamsuddin b. Sulaiman, Universiti Putra Malaysia, Malaysia

Steering Committee
Prof. Yuliang Chen, National Defense University, Taiwan
Prof. Raj Das, University of Aukland, New Zealand
Prof. Wong Shaw Voon, Universiti Putra Malaysia, Malaysia
Prof. Sim Kok Swee, Multimedia University, Malaysia.

Technical Committee
Prof. Zengtao Chen, University of Alberta, Canada
Prof. Wang, Wei-Chung , National Tsing Hua University, Taiwan
Prof. Norhamidi Muhamad, Universiti Kebangsaan Malaysia, Malaysia
Prof. Mohd. Zaidi Omar, Universiti Kebangsaan Malaysia, Malaysia
Prof. Hwang, Yean-ren, National Central University, Taiwan
Prof. Huh, Hoon, Korea Advanced Institute of Science and Technology, South Korea
Dr. YING LIU, Cardiff University, UK
Dr. Sr Andri Andriyana, University of Malaya, Malaysia
Dr. Muhammad Zahir Hassan, Universiti Teknikal Malaysia Melaka
Dr. Mohd Khairol Anuar b. Mohd Ariffin, Universiti Putra Malaysia, Malaysia
Dr. Khalid Hasnan, Universiti Tun Hussein Onn Malaysia, Malaysia
Dr. ANG Marcelo Jr Huibonhoa, National University of Singapore, Singapore
Prof. Ping-Hei Chen, National Taiwan University, Taiwan

Local Committee
D. MOHD SAYUTI BIN AB KARIM, University of Malaya, Malaysia

Statement of Peer review

In submitting conference proceedings to MATEC Web of Conferences, I certify to the Publisher that I adhere to the **Policy on Publishing Integrity** of the journal in order to safeguard good scientific practice in publishing.

1. All articles have been subjected to peer review administered by the proceedings editors.
2. Reviews have been conducted by expert referees, who have been requested to provide unbiased and constructive comments aimed, whenever possible, at improving the work.
3. Proceedings editors have taken all reasonable steps to ensure the quality of the materials they publish and their decision to accept or reject a paper for publication has been based only on the merits of the work and the relevance to the journal.

Title, date and place of the conference

2016 International Conference on Mechanical, Manufacturing, Modeling and Mechatronics (IC4M 2016)

February 27-29, 2016 Kuala Lumpur, Malaysia

Proceedings editor(s):

Prof. Mohd Nasir Tamin, Prof. Du Ruxu

Date and editor's signature

MATEC Web of Conferences **51**, 01001 (2016)
DOI: 10.1051/matecconf/20165101001

Decoupling Kinematic Loops for Real-Time Multibody Dynamic Simulations

Mohamed Omar[1,a]

[1]*Taibah University, Mechanical Engineering Department, Almadinah Almonawarah, Saudi Arabia*

Abstract. Earth moving equipment are typically equipped with hydraulic cylinder actuators to perform the designated tasks. Multibody modelling of such systems results in models with kinematic loops that couples the motion variables of the loop bodies. Iterative solutions will be needed to satisfy the loop constraints and the applied constraints, which require evaluation of the constraint Jacobean matrix. The size of the Jacobean matrix and the associated projections depends on the number of motion variables in each kinematic loop. Consequently, the computational cost significantly increases as the number of variables in the kinematic loop increases. Real-time control and hybrid hardware-in-the-loop systems both require efficient and fast computational algorithms. Eliminating the kinematic loops can improve the computational efficiency and effectiveness of the control algorithms. This paper presents an efficient approach to eliminate the coupling due to the cylinder-rod connections and consequently the kinematic loops in the multibody models leading to efficient simulation. The proposed approach calculates the spatial accelerations and inertia forces of the actuator bodies and the interaction forces with other components. The actuator forces are then projected onto the connecting bodies leading to exact dynamics of the system.

1 Introduction

Most earth-moving machineries carry main attachments that perform the designated functions of the machine e.g. loader bucket, dozer blade, excavator bucket, dozer ripper, etc. As shown in Figure 1, the motions of these attachments are controlled by linkages and mechanisms. Depending on the machine function and design, the control linkages topology could contain four-bar linkages (FBL), four-bar slider linkages (FBSL), or compound linkage. In complex compound linkages, FBL and FBSL could be connected in series or parallel combinations. Some product lines could have different model capacities while performing the same function. Due to the workload capacity, significant difference in the design and the function requirements could lead to different design configurations and kinematic topology of the linkages. In many designs, the buckets are attached to the rocker link of a four-bar linkage that controls the bucket rotational motion. Also, blades and rippers could be attached to the coupler link of a four-bar linkage to control the translational motion. In both cases, the four-bar mechanism is actuated by an inverted FBSL mechanism with a translational joint to model the cylinder-rod motion. In most earth-moving equipment, the translational motion is controlled by a hydraulic actuator, as shown in Figure 1. The kinematic topology of the FBL and FBSL lead to

[a] Corresponding author: momar@taibahu.edu.sa, momar23@gmail.com

kinematic closed-loops [1, 2, 3]. Real-time simulation for such systems poses a challenge for software developers and for analysts [4, 5, 6, 7].

This paper presents an approach for replacing the linear actuator and rigid link bodies in the multibody system model with passive force and constraint elements whose kinematic and dynamic equations are expressed in terms of the attached body state variables [8, 9]. The spatial displacement and velocity of any number of markers fixed in the element are computed, and this information is used to compute interaction between the element and other parts of a model. Spatial contact forces between the system bodies and element bodies are computed and applied to their respective markers. This process could significantly reduce model size and computer simulation time by eliminating excess bodies, state variables, and kinematic constraints from the equations of motion. The actuator is attached to the main driven bodies through primary and secondary ends through universal and spherical joints respectively. No other bodies may be kinematically coupled to the passive element, and no kinematic constraints, other than a distance constraint between the two attachment points, may be imposed upon it. In order to implement the passive element, the kinematic motion need to be expressed, the constraint equations need to be formulated, the inertia forces need to be calculated and the reaction forces need to be evaluated.

This paper is arranged as follows, in section 2 the passive element's spatial displacement, velocity, acceleration, and constraints are expressed. In section 3, the inertia and spatial forces acting on the element are derived and the reaction forces are evaluated. An example will be presented in Section 4.

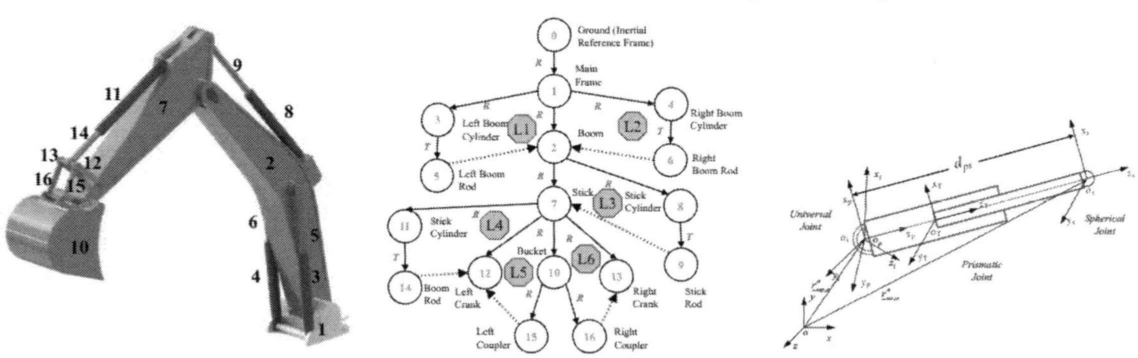

Figure 1. A) CAD model of a hydraulic excavator. B) Model kinematic tree. C) Kinematic representation

2 Kinematic relations of displacement, Velocity and acceleration

The algorithms in this paper refer to an actuator with two common attachment points that must not be allowed to coincide. The linear actuator element contains a single internal translational degree of freedom (DoF) parallel to a line passing through its two attachment points, and this DoF could be driven by force or kinematically constrained. A distance constraint between the primary and secondary attachment points may kinematically control this DoF. The constraint can be a function of time and if the constraint is constant, the passive actuator will act as a rigid link. The universal joint with intersecting orthogonal rotation axes is assumed. The universal joint connects two marker, the first fixed in a driven body and the second is fixed in the passive element, serving as its primary reference. The universal joint's first DoF is aligned with one of the body marker axes and the second with one of the primary reference marker axes, as shown in Figure 1. The spherical joint at the other end is considered the secondary attachment point, and it connects two marker frames, one fixed in a driven body and the other in the passive element. The translational joint between the element's two components constrains the primary and secondary markers to always have the same orientation. Thus there is no need to compute the spherical joint's rotation angles. Bodies of the passive elements are not included in the model's kinematic tree, so the interaction forces through the connecting joints are applied directly to the respective bodies. Thus an element's mass and inertia terms and the associated kinematic constraint equations aren't included in the equations of motion. Instead, all inertial and applied forces are added to the right hand side of the equations of motion.

Let m_p and m_s denote the primary and secondary markers each fixed in a respective driven body. The absolute position vector of these markers are represented by $\underline{t}^0_{0mp}$ and $\underline{t}^0_{0ms}$. Since the primary and secondary attachment points coincide with the respective body marker origins, it follows that

$$\underline{t}^0_{0p} = \underline{t}^0_{0mp} \qquad \text{and} \qquad \underline{t}^0_{0s} = \underline{t}^0_{0ms} \tag{1}$$

equate the global position of the mating marker pair origins. Let $\underline{t}^0_{ps} = \underline{t}^0_{0s} - \underline{t}^0_{0p}$ represent coordinates of a vector from the primary to secondary attachment point. Let d_{ps} represent the distance between the attachment points and u_{ps} represent coordinates of a unit vector parallel to the centerline and oriented from primary toward secondary attachment points as follows

$$d_{ps} = \left|\underline{t}^0_{ps}\right| > 0 \qquad \text{and} \qquad \underline{u}^0_{ps} = \underline{t}^0_{ps} / d^0_{ps} \qquad \text{leading to} \quad d_{ps} = \underline{u}^{0T}_{ps}\underline{t}^0_{ps} \tag{2}$$

Equation (2) may be written in constraint form where $d_{ps}(t)$ is a function of time or constant as:

$$\psi = \underline{u}^{0T}_{ps}\underline{t}^0_{ps} - d_{ps}(t) \tag{3}$$

Assuming that the axes of a right-handed Cartesian coordinate system are numbered (1,2,3), and let k denote the common centerline axis of the primary and secondary markers. The identifier k may be assigned a value of 1, 2, or 3, depending on desired definition. The (i, j) pair defines the universal joint's orthogonal rotation axes and the angles θ and ϕ represent the rotation angles about these axes. Let $(\underline{i}, \underline{j}, \underline{k})$ denote a triad of 3x1 vectors, each containing a 1 in the row associated with the numeric value assigned to the respective identifier in (i, j, k). Coordinates of the unit vector along the centerline, as defined in (2) and referenced to the primary frame, are $\underline{u}^p_{ps} = \underline{k}$. This unit vector defined by $\underline{k}$ points from the primary toward the secondary attachment point. Since $\underline{k}$ contains a single unit entry in the *kth* row, the matrix product

$$\underline{u}^0_{ps} = \underline{R}^{0p}\underline{k} \tag{4}$$

indicates that $\underline{u}^0_{ps}$ defines the *kth* column of $\underline{R}^{0p}$. Considering t^p_{kps} is the *kth* and only nonzero entry of $\underline{t}^p_{ps}$ and substituting $\underline{u}^0_{ps}$ into (3) yields a more compact local constraint equation as follows

$$\psi = \underline{k}^T\underline{t}^p_{ps} - d_{ps}(t) = t^p_{kps} - d_{ps}(t) \qquad \text{where} \qquad \underline{t}^p_{ps} = \underline{R}^{0pT}\underline{t}^0_{ps} \tag{5}$$

At this point, the remaining two columns of $\underline{R}^{0p}$ need to be determined. Let $\underline{R}^{0mp}$ represent the known direction cosine matrix for marker m_p. Then

$$\underline{R}^{0p} = \underline{R}^{0mp}\underline{R}^i\underline{R}^j \quad \text{where} \quad \underline{R}^i = I + \sin\theta\,\underline{\tilde{i}} + (1-\cos\theta)\,\underline{\tilde{i}}^2 \quad \text{and} \quad \underline{R}^j = I + \sin\phi\,\underline{\tilde{j}} + (1-\cos\phi)\,\underline{\tilde{j}}^2 \tag{6}$$

$\underline{R}^i$ and $\underline{R}^j$ define rotational transformations about the respective universal joint axes identified by i and j. To determine the angles θ and ϕ, express (6) as follows

$$\underline{R}^i\underline{R}^j = \underline{R}^{0mpT}\underline{R}^{0p} \tag{7}$$

From the known column in $\underline{R}^{0p}$, namely $\underline{u}^0_{ps}$, and the known $\underline{R}^{0mp}$ matrix, sufficient information is available to evaluate the angles and the remaining entries in $\underline{R}^{0p}$. Manipulating the equations to get

$$\underline{u}_{ps} = \sin\phi\,\underline{i} - \cos\phi\sin\theta\,\underline{j} + \cos\phi\cos\theta\,\underline{k} \tag{8}$$

Let the scalar quantities (u_1, u_2, u_3) represent the coordinates of $\underline{u}_{ps}$, the joint angles are identified as

$$\theta = \operatorname{atan2}(-u_2, u_3) \qquad \text{and} \qquad \phi = \operatorname{asin}(u_1) \tag{9}$$

With all displacements known, the element's joint and spatial velocities may be computed. Let $v^0_{0mp,p}$ and $v^0_{0ms,s}$ represent the known body marker spatial velocities. Similar to (1), the vectors

$$\underline{\dot{t}}^0_{0p,p} = \underline{\dot{t}}^0_{0mp,p} \qquad \text{and} \qquad \underline{\dot{t}}^0_{0s,s} = \underline{\dot{t}}^0_{0ms,s} \tag{10}$$

represent the element's translational velocity components at the respective connecting points. The secondary marker's velocity relative to the primary marker is obtained from the general equation

$$\underline{\dot{t}}^0_{0s,s} = \underline{\dot{t}}^0_{0p,p} - \underline{\tilde{t}}^0_{ps}\underline{\omega}^0_{0p} + \underline{u}^0_{ps}\dot{d}_{ps} \tag{11}$$

where $\dot{d}_{ps}$ is the time derivative of d_{ps}. From (11) and noting that $\underline{u}^0_{ps}$ represents a unit vector, is parallel to $\underline{t}^0_{ps}$, and is orthogonal to $\underline{\tilde{t}}^0_{ps}\underline{\omega}^0_{0p}$, we obtain

$$\dot{d}_{ps} = \underline{u}^{0T}_{ps}\underline{\dot{t}}^0_{ps} \quad \text{where} \quad \underline{\dot{t}}^0_{ps} = \underline{\dot{t}}^0_{0s,s} - \underline{\dot{t}}^0_{0p,p} \tag{12}$$

Equation (12) may also be expressed locally which shows that $\dot{d}_{ps}$ is just the *kth* entry of $\underline{\dot{t}}^p_{ps}$ .as

$$\dot{d}_{ps} = \underline{k}^T\underline{\dot{t}}^p_{ps} = \dot{t}^p_{kps} \tag{13}$$

Equation (13) also represents the first time derivative of the constraint equation in (3)

$$\dot{\psi} = \underline{u}^{0T}_{ps}\underline{\dot{t}}^0_{ps} - \dot{d}_{ps}(t) \quad \text{or expressed locally as} \quad \dot{\psi} = \underline{k}^T\underline{\dot{t}}^p_{ps} - \dot{d}_{ps}(t) = \dot{t}^p_{kps} - \dot{d}_{ps}(t) \tag{14}$$

The element's absolute rotational velocity is determined from

$$\underline{\omega}^0_{0p} = \underline{\omega}^0_{0mp} + \underline{\omega}^0_{mp} \tag{15}$$

where $\underline{\omega}^0_{0mp}$ is the primary body marker's known absolute rotational velocity, and $\underline{\omega}^0_{mp}$ is the element's rotational velocity relative to its parent marker. Because a translational joint connects the element's two components, $\underline{\omega}^0_{0p}$ represents the absolute rotational velocity of both parts.

Manipulating the kinematic equations, equation for $\underline{\omega}^p_{mp}$ could be expressed as

$$\underline{\omega}^p_{mp} = \cos\phi\dot{\theta}\underline{i} + \dot{\phi}\underline{j} + \sin\phi\dot{\theta}\underline{k} \tag{16}$$

Let the scalar quantities (ω_1, ω_2, ω_3) represent the coordinates of $\underline{\omega}^p_{mp}$ in (16), we can have

$$\dot{\theta} = \omega_1 / \cos\phi \quad \text{and} \quad \dot{\phi} = \omega_2 \tag{17}$$

Now that the element's absolute spatial velocities are known, the spatial accelerations must be determined, so the actuator inertial forces can be computed and applied to the parent body attachment markers. Since high accuracy is not required for inertial forces and because an iterative method is not used in solving the equations of motion, the parent marker accelerations from the previous time step will be used in these calculations. Also needed is the relative rotational acceleration $\underline{\dot{\omega}}^p_{mp}$ for calculating the element's spatial accelerations. It could be obtained by differentiating the equation used to solve for $\underline{\omega}^0_{mp}$. Coordinates of the translational accelerations of markers m_p and m_s are $\underline{\ddot{t}}^0_{0mp,p}$ and $\underline{\ddot{t}}^0_{0ms,s}$, and their corresponding spatial accelerations are given as

$$a^0_{0mp,p} = \begin{bmatrix} \underline{\dot{\omega}}^0_{0mp} \\ \underline{\ddot{t}}^0_{0mp,p} + \underline{\tilde{t}}^0_{0mp,p}\underline{\omega}^0_{0mp} \end{bmatrix} \quad \text{and} \quad a^0_{0ms,s} = \begin{bmatrix} \underline{\dot{\omega}}^0_{0ms} \\ \underline{\ddot{t}}^0_{0ms,s} + \underline{\tilde{t}}^0_{0ms,s}\underline{\omega}^0_{0ms} \end{bmatrix} \tag{18}$$

From (18) the marker accelerations $\underline{\ddot{t}}^0_{0mp,p}$ and $\underline{\ddot{t}}^0_{0ms,s}$ may be extracted. Since the element frame origins coincide with the respective marker origins, it follows that

$$\underline{\ddot{t}}^0_{0p,p} = \underline{\ddot{t}}^0_{0mp,p} \quad \text{and} \quad \underline{\ddot{t}}^0_{0s,s} = \underline{\ddot{t}}^0_{0ms,s} \tag{19}$$

Differentiating (2) to obtain

$$\ddot{d}_{ps} = \underline{u}^{0T}_{ps}\underline{\ddot{t}}^0_{ps} + \underline{\dot{u}}^{0T}_{ps}\underline{\dot{t}}^0_{ps} \quad \text{where} \quad \underline{\ddot{t}}^0_{ps} = \underline{\ddot{t}}^0_{0s,s} - \underline{\ddot{t}}^0_{0p,p} \quad \text{and} \quad \underline{\dot{u}}^0_{ps} = \underline{\tilde{\omega}}^0_{0p}\underline{u}^0_{ps} \tag{20}$$

Equations (20) represent the second time derivative of the constraint equation that could be written as

$$\ddot{\psi} = \underline{u}^{0T}_{ps}\underline{\ddot{t}}^0_{ps} + \underline{\dot{u}}^{0T}_{ps}\underline{\dot{t}}^0_{ps} - \ddot{d}_{ps}(t) \quad \text{or} \quad \ddot{\psi} = \underline{k}^T\underline{\ddot{t}}^p_{ps} - \underline{k}^T\left(\underline{\tilde{\omega}}^p_{0p}\right)^2 \underline{k}d_{ps}(t) - \ddot{d}_{ps}(t) = \ddot{t}^p_{kps} + \left(\omega^{p2}_{i0p} + \omega^{p2}_{j0p}\right)d_{ps}(t) - \ddot{d}_{ps}(t) \tag{21}$$

It could be shown that the second derivative of angular velocities could be expressed as:

$$\underline{\dot{\omega}}^p_{mp} = \cos\phi\left(\ddot{\theta} + \dot{\theta}\underline{\tilde{\omega}}^p_{0mp}\right)\underline{i} + \left(\ddot{\phi} + \dot{\phi}\underline{\tilde{\omega}}^p_{0p}\right)\underline{j} + \sin\phi\left(\ddot{\theta} + \dot{\theta}_i\underline{\tilde{\omega}}^p_{0mp}\right)\underline{k} \tag{22}$$

Associating the scalar terms (α_1, α_2, α_3) in $\underline{\dot{\omega}}^p_{mp}$ with the corresponding components of ($\underline{i}$, $\underline{j}$, $\underline{k}$) are

$$\ddot{\theta} = \left(\alpha_1 + \dot{\phi}\underline{k}^T\underline{\omega}^p_{0p} - \sin\phi\dot{\theta}\underline{j}^T\underline{\omega}^p_{0mp}\right)/\cos\phi \quad \text{and} \quad \ddot{\phi} = \alpha_2 - \cos\phi\dot{\theta}\underline{k}^T\underline{\omega}^p_{0mp} + \sin\phi\dot{\theta}\underline{i}^T\underline{\omega}^p_{0mp} \tag{23}$$

3 Spatial and inertia forces

Once the element's spatial displacements, velocities, and accelerations are known, the spatial forces acting on the respective element bodies are computed and applied to their parent markers. Separate free-body diagrams of the passive element's components yield

$$M^{pp}_{p,pp}a^{p}_{0p,p} = \tilde{v}^{pT}_{0p,p}M^{pp}_{p,pp}v^{p}_{0p,p} + g^{p}_{p,p} + f^{p}_{u,p} - f^{p}_{t,p} \quad \text{and} \quad M^{pp}_{s,ss}a^{p}_{0s,s} = \tilde{v}^{pT}_{0s,s}M^{pp}_{s,ss}v^{p}_{0s,s} + g^{p}_{s,s} + f^{p}_{s,s} + f^{p}_{t,s} \tag{24}$$

where $g^{p}_{p,p}$ and $g^{p}_{s,s}$ are the net spatial forces applied to the respective primary and secondary parts. The spatial forces $f^{p}_{u,p}$ and $f^{p}_{s,s}$ are coordinates of the corresponding universal and spherical joint reaction forces applied to the elements, and the forces $f^{p}_{t,p}$ and $f^{p}_{t,s}$ are the internal reaction forces developed between the mating elements at their common translational joint interface.

The forces $f^{p}_{u,p}$ and $f^{p}_{s,s}$ must be computed and subtracted from the respective parent body attachment marker forces to complete the equations of motion. To simplify the development let:

$$s^{p}_{p,p} = -M^{pp}_{p,pp}a^{p}_{0p,p} + \tilde{v}^{pT}_{0p,p}M^{pp}_{p,pp}v^{p}_{0p,p} + g^{p}_{p,p}, \quad s^{p}_{s,s} = -M^{pp}_{s,ss}a^{p}_{0s,s} + \tilde{v}^{pT}_{0s,s}M^{pp}_{s,ss}v^{p}_{0s,s} + g^{p}_{s,s} \quad \text{and}$$

$$s^{p}_{ps,s} = T^{pT}_{ps}s^{p}_{p,p} + s^{p}_{s,s} \tag{25}$$

represent the known spatial force quantities. Solving for $f^{p}_{u,p}$ and $f^{p}_{s,s}$, gives

$$f^{p}_{u,p} = -s^{p}_{p,p} + f^{p}_{t,p} \quad \text{and} \quad f^{p}_{s,p} = -s^{p}_{s,s} - f^{p}_{t,s} \tag{26}$$

A force-balance equation based on free-body diagram of the combined element eliminates the internal reaction forces $f^{p}_{t,p}$ and $f^{p}_{t,s}$. Using the secondary frame origin as reference and applying (25) gives

$$s^{p}_{ps,s} + f^{p}_{u,s} + f^{p}_{s,s} = 0 \quad \text{or} \quad f^{p}_{u,s} + f^{p}_{s,s} = -s^{p}_{ps,s} \tag{27}$$

Equation (27) indicates that the sum of spatial joint reaction forces $f^{p}_{u,s} + f^{p}_{s,s}$ may be determined here, but not the separate components $f^{p}_{u,s}$ and $f^{p}_{s,s}$ as needed in the equations of motion. To determine these forces, we take advantage of additional information about the system by first expanding the spatial forces to show their rotational and translational components

$$f^{p}_{u,p} = \begin{bmatrix} \underline{\tau}^{p}_{u,p} \\ \underline{f}^{p}_{u} \end{bmatrix} \quad \text{and} \quad f^{p}_{s,s} = \begin{bmatrix} \underline{\tau}^{p}_{s,s} \\ \underline{f}^{p}_{s} \end{bmatrix} \tag{28}$$

Since the secondary connection is a spherical joint, it can't support any internal torque, thus $\underline{\tau}^{p}_{s,s} = 0$. Using the direction cosine matrix in (6) to transform $\underline{\tau}^{u}_{u,p}$ to the primary coordinates gives

$$\underline{\tau}^{p}_{u,p} = \underline{R}^{jT}\underline{\tau}^{u}_{u,p} = \left(I - \sin\phi\tilde{\underline{j}} + (1-\cos\phi)\tilde{\underline{j}}^{2}\right)\underline{\tau}^{u}_{u,p} \quad \text{leading to} \quad \underline{\tau}^{p}_{u,p} = \left(-\sin\phi\underline{i} + \cos\phi\underline{k}\right)\tau^{u}_{ku,p} \tag{29}$$

The desired force equation is extracted as follows

$$\underline{f}^{p}_{u} = -\tilde{\underline{k}}\left(\underline{\tau}^{p}_{u,p} + \underline{\tau}^{p}_{ps,s}\right)/d_{ps} + \underline{k}Q_{t} - \underline{k}\underline{k}^{T}\underline{s}^{p}_{p} \tag{30}$$

Q_t is the element's known internal driving force. The components of $\underline{f}^{p}_{u}$ are obtained by premultiplying (30) by respective elements of (i^T, j^T, k^T). The secondary spatial forces are computed as follows

$$f^{p}_{s,s} = -T^{pT}_{ps}f^{p}_{u,p} - s^{p}_{ps,s} \tag{31}$$

And extracting the translational component

$$\underline{f}^{p}_{s} = -\underline{f}^{p}_{p} - \underline{s}^{p}_{ps} \tag{32}$$

The resulting spatial forces must be transformed and subtracted from the appropriate body marker frames to complete the equations. The passive actuator equations may appear to involve many multiplications, but the overall count when compared to implementing a regular actuator model is small. Also, eliminating the associated constraint Jacobian and mass matrix operations provides a substantial computational savings.

4 Descriptive example

Figure 1 shows CAD model and the kinematic tree diagram of a partial hydraulic excavator. In this example, the machine's undercarriage and the lower structure is not taken into consideration. The kinematic model consists of 16 dynamic bodies with 6-DoF. This model will result in 96 second order differential equations in the Cartesian space. Since each dynamic body is connected to its parent with single DoF joint (translational or rotational), the kinematic tree has total of 16 joint variables. According to the system topology 6 closed kinematic loops are formed.

In the joint-based multibody dynamics formulation, chord-joints could be modeled by introducing a massless chord-body that is rigidly attached to a marker on the parent body of the chord-joint. The chord-body is then connected to the child body of the chord joint by kinematic joint, as shown in Figure 2, [1, 2, 9]. In order to maintain the integrity of the chord joint, 6 kinematic constraints are used to represent the rigid connection between the chord-body and chord-joint parent body at the joint locations. For example, in the kinematic loop L1, the massless chord-body 17 is rigidly attached to the joint parent body 2 and connected to the joint child body 5 with a spherical joint. The other loops are modeled similarly as shown in Figure 2 Figure 2. using the loop description in Table 1.

Table 1. Kinematic loop structure.

Loop #	1	2	3	4	5	6
Loop bodies	1, 3, 5, 2	1, 4, 6, 2	2, 8, 9, 7	7, 11, 14, 12	7, 12, 15, 10	7, 13, 16, 10
Chord body	17	18	19	20	21	22
Number of DoF	6	6	6	6	6	6

In joint-coordinates multibody dynamics formulation, the equation of motion of system with closed loop could be written as follows [2, 9]:

$$\begin{bmatrix} \mathbf{M} & \mathbf{T}_a & 0 & \mathbf{T}_a^T\bar{\mathbf{H}}_c^* \\ \mathbf{T}_a^T & 0 & -\mathbf{H}_a & 0 \\ 0 & -\mathbf{H}_a^T & 0 & 0 \\ \bar{\bar{\mathbf{H}}}_c^{*T}\mathbf{T}_a^T & 0 & 0 & 0 \end{bmatrix} \begin{bmatrix} \mathbf{a}_a \\ -\mathbf{F}_a \\ \ddot{\mathbf{q}}_a \\ -\bar{\bar{\mathbf{Q}}}_c \end{bmatrix} = \begin{bmatrix} \mathbf{T}_a^T\mathbf{H}_c^*\mathbf{Q}_c + \mathbf{G} + \tilde{\mathbf{v}}^T\mathbf{P} \\ \gamma_a \\ \mathbf{Q}_a \\ \bar{\bar{\mathbf{H}}}_c^{*T}\gamma_c \end{bmatrix} \tag{33}$$

where $\mathbf{M}$ is the system inertia matrix, $\mathbf{T}_a$ is the system topology matrix, $\mathbf{H}_a$ is the influence coefficient matrix of the system kinematic joints, $\mathbf{H}_c$ is the influence coefficient matrix of the system chord joints, $\mathbf{a}_a$ is the vector of Cartesian accelerations of all bodies in the system, $\mathbf{F}_a$ is the vector of spatial reaction forces at all the kinematic joints, $\mathbf{q}_a$ is the vector of kinematic joint variables and the corresponding derivatives, $\mathbf{Q}_c$ is the vector of chord-joint forces , $\mathbf{G}$ is the vector of externally applied forces to bodies in the system, $\tilde{\mathbf{v}}^T\mathbf{P}$ is the vector of quadratic inertia forces, γ_a is the vector of quadratic velocity terms associated with all kinematic joints, $\mathbf{Q}_a$ is the vector of kinematic joint forces, and γ_c is the vector of quadratic velocity terms associated with all chord joints.

The reduced matrix in (33) is shown partitioned where the block 3 by 3 block matrix corresponds to the augmented open-loop system equations. To obtain the joint variable accelerations, the Cartesian variables $\mathbf{a}_a$ and $\mathbf{F}_a$ can be systematically eliminated. Eliminating the Cartesian accelerations and reaction forces will lead to the system equations of motion as follows:

$$M_q\ddot{q}_a = -J_{ca}^T\bar{\bar{Q}}_c + Q_a - J_{ca}^T Q_c + H_a^T\left(T_c T_a^{-1}\right)^T\left(G + \tilde{v}^T P - MT_c T_a^{-1}\gamma_a\right) \tag{34}$$

The equation of motion of the multibody system with appended constraints could be rearranged as:

$$\begin{bmatrix} M_q & J^T \\ J & 0 \end{bmatrix}\begin{bmatrix} \ddot{q}_q \\ \lambda_q \end{bmatrix} = \begin{bmatrix} Q \\ \gamma_q \end{bmatrix} \tag{35}$$

where $\mathbf{M}_q$ is the generalized mass matrix as defined in (9), J is the Jacobian matrix, and λ_q is the vector of Lagrange multipliers associated with the constraints.

The problem dimensionality of the system under consideration has total of 16 spatial bodies where each body has 6 spatial accelerations. The dimensions of the inertia matrix $\mathbf{M}$ is (96x96), the connectivity matrix $\mathbf{T}_a$ is (96x96), and the joint influence matrix $\mathbf{H}_a$ is (96x16). The amount of expensive calculations required to transform the equation of motion from the augmented form in (33) to the joint-subspace form in (34) should be clear to the reader. Any saving or reduction in this projection step will significantly impact the simulation speed. To this end, the resulting equation of motion in (34) will be highly condensed and requires the evaluation of the Jacobean matrix J at every time step. The Jacobean matrix could be calculated from the derivatives of the chord joint constraint equations and from the other motion constraints imposed on the system as follows:

$$\begin{bmatrix} \Phi_q^1 & \cdots & 0 \\ \cdots & \cdots & \cdots \\ 0 & \cdots & \Phi_q^6 \end{bmatrix} \begin{bmatrix} \ddot{q}^1 \\ \vdots \\ \ddot{q}^6 \end{bmatrix} = \begin{bmatrix} \gamma^1 \\ \vdots \\ \gamma^6 \end{bmatrix} \tag{36}$$

where Φ_q^i is the Jacobean matrix associated with the kinematic constraints of the closed loop i.

Each kinematic loop requires 6 constraint equations. Iteratively satisfying the loop constraints require computing the Jacobean matrices. Decoupling the loop variables requires additional transformation and projections that will consume extra processing time. The Jacobean matrix needs to be analyzed to determine the dependent and independent variables that will be used by the integrator.

Figure 2. Schematic description of the passive element kinematic connection

Using the proposed approach leads to eliminating the kinematic closed-loops and the associated passive bodies from the kinematic tree. As a result, the differential equations of the system will significantly decrease. Figure 2. shows the kinematic tree of the resulting system after eliminating the kinematic loops. The first four kinematic loops are replaced with primary and secondary bodies. A time-dependent constraint or a joint force could be used to drive the translational joint. While the last two loops, two constant-distance constraints are imposed between the primary and secondary markers to represent the length of the coupler links. The remaining number of kinematic bodies reduced to 5, the number of joint variables becomes 5, and the kinematic loops are eliminated as shown in Figure 2. This led to significant reduction in the computational time of the whole model. After implementation, the dynamics simulation of the system runs more than 10-times faster than real-time with no accuracy loss. This achieved reduction enables the integration of multibody solver for more advanced simulations applications: Virtual reality, Hardware in the loop, real-time control etc.

References

1. R. Featherstone, Body Dynamics Algorithms, Springer, New York (2008)
2. M Omar, The Scientific World Journal, Vol 2014, 409402.
3. M Omar, Journal of Advances in Mechanical Engineering, Vol 2014,
4. P. Nikravesh, Product Engineering (2005).
5. G Jalón, E. Bayo, Kinematic and Dynamic Simulation of Multibody Systems,. 1994
6. G. Rodriguez, A Jain, and K Kreutz-Delgado, J of Astronautical Sciences, vol. 40, (1992).
7. S Saha , WSchiehlen, Int. J. of Mechanics of Structures and Machines, V. 29, N. 2, (2001).
8. R. Featherstone, D. Orin, Proc. IEEE Int. Conf. Robotics & Automation, (2000)
9. M Omar, Studies in Engineering and Technology, Vol 1, No. 2, (2014).

MATEC Web of Conferences **51**, 01002 (2016)
DOI: 10.1051/matecconf/20165101002

Rollover Dynamics of a Narrow Tilting Three-Wheeled Vehicle

Jubin Antony J[1,a] and Jayabal K[1,b]

[1]*Indian Institute of Information Technology Design and Manufacturing, Kancheepuram, Tamil Nadu, India*

Abstract. A narrow-track vehicle possesses many advantages like fuel efficiency and reduced road footprint among others. But the main drawback of this model stems from the fact that they tend to rollover more easily due to their narrow base area. In order to overcome this rollover instability, these vehicles are made to tilt to obtain increased virtual base area. The focus of this paper is to analyse the minimum lean angle necessary for a narrow three-wheeled vehicle under lateral acceleration to negotiate a curve safely. In addition, the influence of the horizontal positions of the centre of mass of the vehicle over the rollover speeds is studied.

1 Introduction

Due to sharp increase in the cost of fuel and environmental pollution during the past ten years, there is an increasing demand for producing an energy efficient passenger vehicle. In the existing four wheeled designs, the scope for further fuel efficiency is limited. On the other hand, the two wheeled vehicles have a better fuel efficiency for single users but in this model the passenger is exposed to bad weather and generally has a very less safety factor. Therefore, the three-wheeled vehicle design would be a better scope for producing a fuel efficient model, without sacrificing the safety and comfort of the passenger.

Since the three-wheeled vehicle has a reduced base as compared to an equivalent four-wheeled vehicle, it has a higher probability of tipping over while cornering a turn. Huston [1] has proved it by comparing a non-tilting three wheeled vehicle with a standard four-wheeled vehicle. In order to prevent the rollover, it was suggested by Karnopp [2] that the vehicle should lean like a motorcycle.Subsequently, Hibbard and Karnoppdiscussed the dynamics of a narrow tilting vehicle [3]and the method to control the leaning angle of tilting vehicles [4].

Among tilting three-wheeled vehicles, Palash [5] carried on with the analysis of a tilting tadpole model and their advantages.Berote [6] described about the dynamics of a tilting delta model and the effect ofhydraulics have on the system. Despite the future potential of this kind of vehicles, not too many research articles are found in the literature on the rollover dynamics of tilting of three-wheeled vehicles.

In this paper, the effect of leaning on the rollover speed is analysed and comparisons of rollover speeds are discussed between two types of three-wheeled vehicles, the tadpole and delta configurations; the former one will have two wheels at the front axle and one at the rear whereas the later one will have vice-versa.

[a]Correspondingauthor : mdm12b011@iiitdm.ac.in
[b]jayabal@iiitdm.ac.in

2 Nomenclature

l_1	Longitudinal distance from centre of mass to front axle
l_2	Longitudinal distance from centre of mass to rear axle
L	Longitudinal distance from rear axle to front axle
W	Weight of the vehicle
h	Height of centre of mass of the vehicle without leaning
R	Radius of curvature of the turn
V_y	Longitudinal velocity along y-direction
a_y	Longitudinal acceleration along y-direction
E	Inertia force vector due to lateral acceleration
A	Inertia force vector due to longitudinal acceleration
D	Inertia force vector due to longitudinal deceleration
Z	Resultant force acting on the centre of mass
θ	Angle of leaning of the vehicle
$\varnothing$	Angle between **Z** and x-axis
γ	Angle between **E** and x-axis
$\alpha, \beta, \alpha_d, \beta_d$	Angle between **Z** and the normal to TT axis passing through centre of mass
δ	Angle between the normal to TT axis passing through centre of mass and x-axis

3 Rollover Stability

Rollover stability is the ability of the vehicle to resist tipping over under the action of lateral body force. The lateral force may be due to the centrifugal force while cornering or due to the air resistance. For a good vehicle design, rollover stability should be as high as possible.

In all figures appearing in the article, the axis of tipping is denoted by TT. To ensure rollover stability, the clockwise moment about the TT axis, as viewed from the rear end, should be less than zero.

$$\sum M_{TT} < 0 \tag{1}$$

3.1 Three-wheeled Tadpole Model

In this section, a three-wheeled tadpole configuration is considered. It is assumed that when the vehicle is leaned, the centre of mass of the vehicle is displaced by an angle θ with respect to the point P in the ground plane GG as shown in Figure 1. Let X be the perpendicular distance between the TT axis and the centre of mass as shown in Figure 2. The magnitude of **Z**, **A** and **E** are given by

$$Z = |\mathbf{Z}| = |E\ cos\gamma\ \hat{\imath} + (E\ sin\gamma - A\ )\hat{\jmath}| = [A^2 + E^2 - 2AE\ sin\gamma]^{0.5} \tag{2}$$

$$A = |\mathbf{A}| = \frac{W a_y}{g} \tag{3}$$

$$E = |\mathbf{E}| = \frac{W}{g}\frac{V_y{}^2}{R} \tag{4}$$

The moment about the axis TT will be

$$\sum M_{TT} = -WX + Zh\ \cos\alpha\ cos\theta < 0 \tag{5}$$

From Figure 4, it can be shown using trigonometric rules that

$$X = (c\ tan\delta)\ cos\delta = \left(l_2 + \frac{2hL\ sin\theta}{b}\right) sin\delta \tag{6}$$

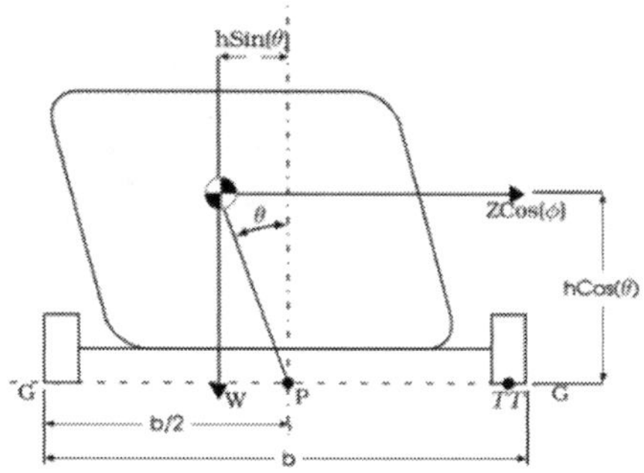

Figure 1. Front view of three wheeled tadpole model

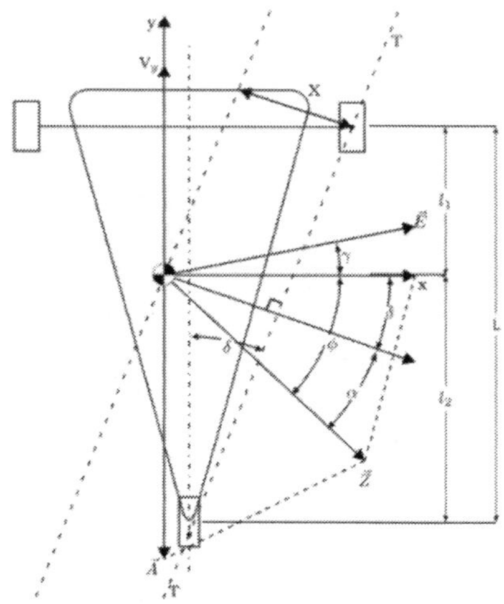

Figure 2. Top view of three wheeled tadpole model while accelerating

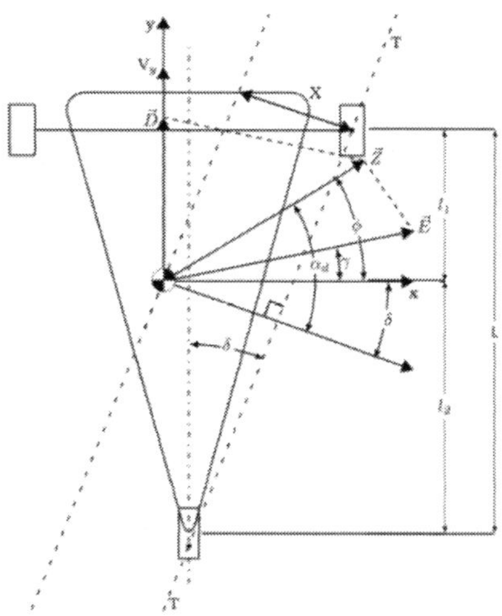

Figure 3. Top view of three wheeled tadpole model while braking

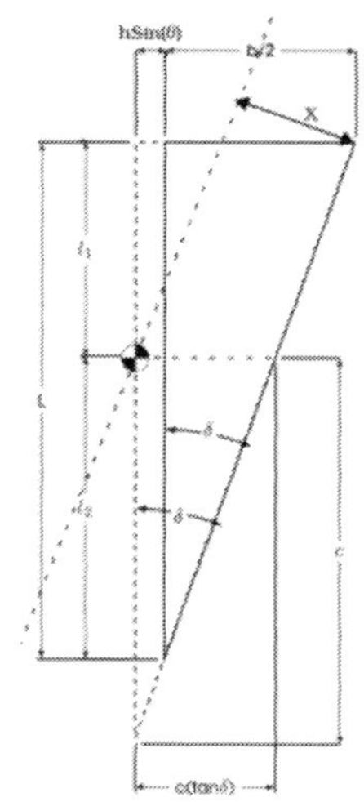

Figure 4. Top view of the centre of mass displaced area

3.1.1 Cornering a curve with a constant velocity

If the vehicle is negotiating a turn with a constant velocity, then the longitudinal acceleration will be zero, a_y=0 (i.e. **Z**=**E**) and $\alpha = \gamma + \delta$. Thus, eq. (5) will be

$$\sum M_{TT} = -W\left(l_2 + \frac{2hL\ sin\theta}{b}\right) sin\delta + \frac{W}{g}\frac{{V_y}^2}{R} h\ cos\alpha\ cos\theta < 0 \tag{7}$$

When eq. (7) is equated to zero, the rollover speed is determined as

$$V_{ro} = \sqrt{\frac{Rgb\left(l_2 + \frac{2hL\ sin\theta}{b}\right)}{2hL\ cos\alpha\ cos\theta}} \tag{8}$$

3.1.2 Cornering a curve with an acceleration

In this case, the vehicle is cornering a turn with a longitudinal acceleration (i.e. A≠0) and also**Z** will have component of **A**as shown in eq. (2). Thus, the rollover stability condition will be

$$\sum M_{TT} = -W\left(l_2 + \frac{2hL\,sin\theta}{b}\right) sin\delta + Zh\,cos\alpha\,cos\theta < 0 \tag{9}$$

$$Z\,cos\alpha < \frac{W(l_2 b + 2hL\,sin\theta)\,sin\delta}{bh\,cos\theta} \tag{10}$$

where

$$Sin\delta = \frac{b}{2\left[L^2 + \left(b/2\right)^2\right]^{0.5}} \tag{11}$$

$$\alpha = \phi - \delta = tan^{-1}\left(\frac{E\,sin\gamma - A}{E\,cos\gamma}\right) - tan^{-1}\left(\frac{b}{2L}\right) \tag{12}$$

Eq. (11) and (12) can be geometrically calculated from Figure 2.

3.1.3 Cornering a curve with braking

While braking on a turn, **D**will be acting along the positive y-axis as shown in Figure 3. Similarly like the above condition, the rollover stability condition can be represented as

$$Z\,cos\alpha_d < \frac{W(l_2 b + 2hL\,sin\theta)\,sin\delta}{bh\,cos\theta} \tag{13}$$

In eq. (13), the value of α_d, ($\alpha_d = \phi + \delta$) can be geometrically found out from Figure 3. Eq. (10) and (13) should be solved numerically in order to obtain the rollover speeds.

3.2 Three-wheeled DeltaModel

Similarly like the tadpole model, the three wheeled delta model, as shown in Figures 5 and 6, can also be solved to find rollover speeds.

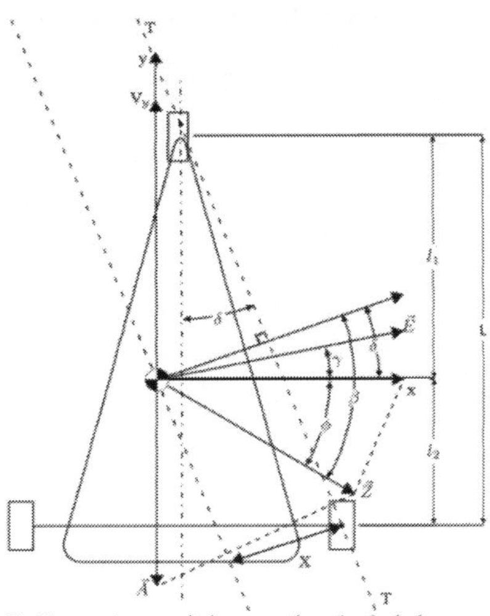

Figure 5. Top view of three wheeled delta model while accelerating.

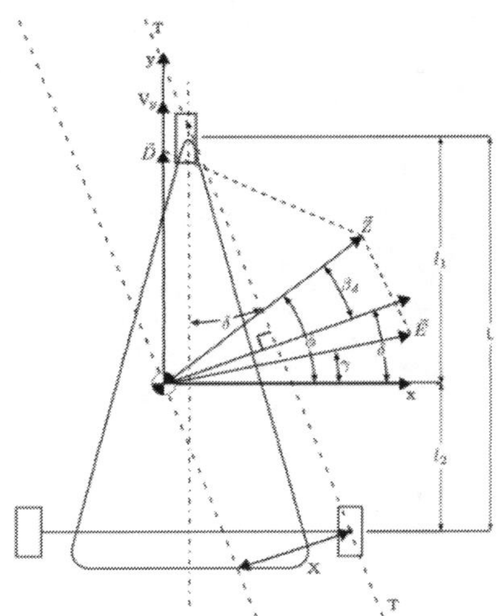

Figure 6. Top view of three wheeled delta model while braking.

The rollover speeds for various cases are given as

$$V_{ro} = \sqrt{\frac{Rgb\left(l_1 + \frac{2hL\, sin\theta}{b}\right)}{2hL\, cos\alpha\, cos\theta}} \tag{14}$$

$$ZCos(\beta = \phi + \delta) = \frac{W(l_1 b + 2hL\, sin\theta)\, sin\delta}{bh\, cos\theta} \tag{15}$$

$$ZCos(\beta_d = \phi - \delta) = \frac{W(l_1 b + 2hL\, sin\theta)\, sin\delta}{bh\, cos\theta} \tag{16}$$

Eq. (14), (15) and (16) provide the rollover speed equations, respectively, when the vehicle is moving at a constant speed, during acceleration and during deceleration. Eq. (15) and (16) should be solved numerically to find the rollover speeds.

4 Simulation

For simulation, the vehicle model taken had a mass of m=400 Kg, length of L=2 m, a base length of b=0.6 m and a centre of mass height of h=0.6 m. It is assumed to negotiate a curve of radius, R=100 m with the acceleration magnitude of (0.25g) m/s^2. To illustrate the importance of the horizontal position of the centre of mass, various positions of centre of mass is taken. Also, the graphs are plotted for various lean angles to know the significance of leaning; refer Figures 7 to 12.

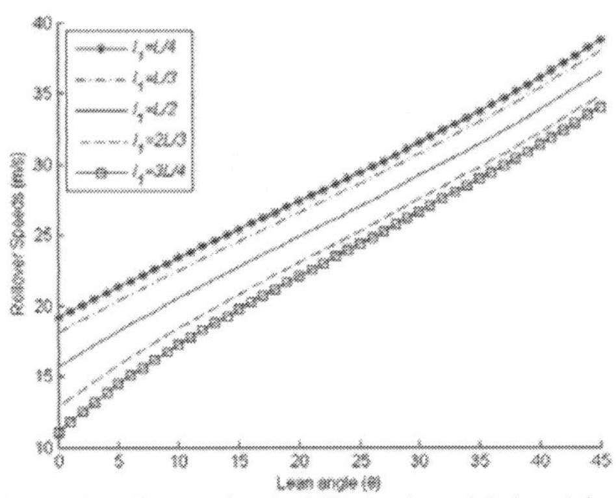

Figure 7. Three-wheeled Tadpole vehicle with no acceleration

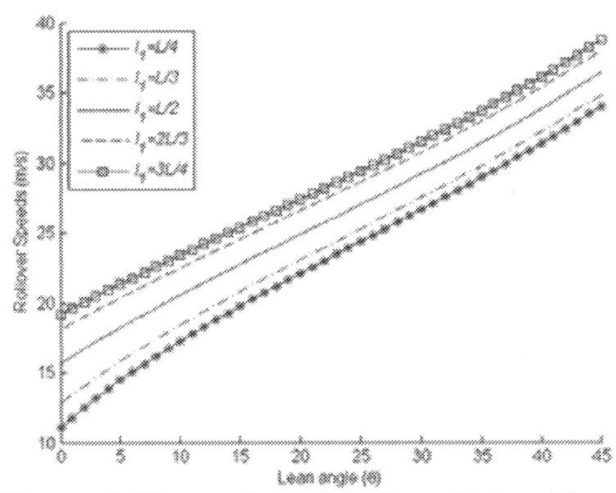

Figure 8. Three-wheeled Delta vehicle with no acceleration

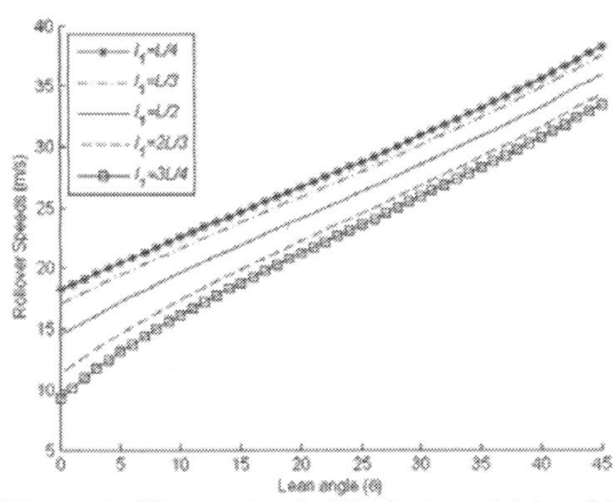

Figure 9. Three-wheeled Tadpole vehicle with acceleration

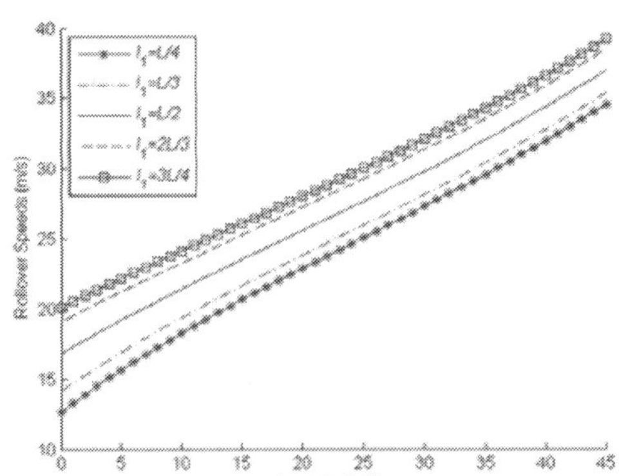

Figure 10. Three-wheeled Delta vehicle with acceleration

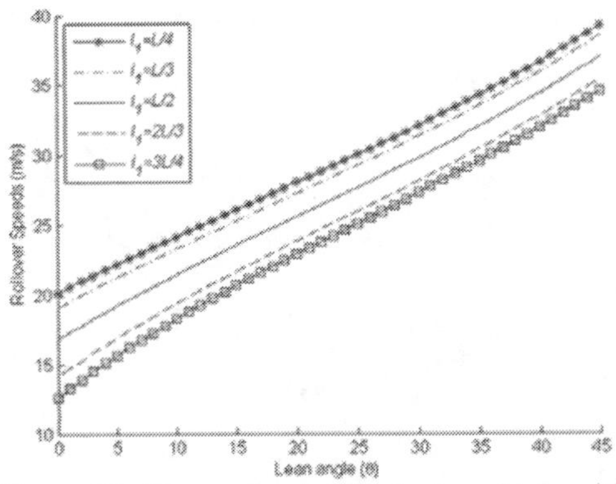

Figure 11. Three-wheeled Tadpole vehicle with deceleration

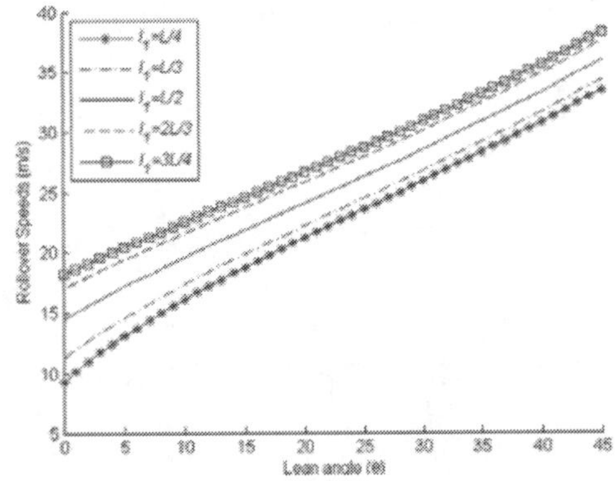

Figure 12. Three-wheeled Delta vehicle with deceleration

5 Conclusion and discussion

In this paper, the rollover stability conditions governing the tadpole and delta configurations of leaning three-wheeled vehicles have been presented and compared. From the above graphs, it is clear that leaning of the vehicle in a turn does increase the rollover stability for both the configuration of three wheeled vehicle. But, the maximum leaning angle will be limited by the vehicle design and, the friction between tire and road.

Furthermore from the graphs, it can be inferred that the vehicle has a better stability if the horizontal position of the centre of mass is close to the end with two wheels. In other words, the tadpole model is more stable if the centre of mass position is nearer to the front end and for the delta model, it should be nearer to the rear end. The mass of the passenger or the vehicle does not directly affect the rollover stability, but it affects the position of the centre of mass, which is indirectly going to change the stability in case of a three-wheeled vehicles. It can also be noted that the delta model is more stable while accelerating in a turn and the tadpole model is more stable while braking in a turn.

Thus, it can be concluded that by leaning the vehicle, the stability of a narrow three-wheeled vehicle could be matched with that of a standard four-wheeled vehicle. The outcome of this work can be useful for defining the optimal lean angle for a three-wheeled vehicle. Other factors like comfort of the passenger should also be considered while designing the optimal lean angle for a three-wheeled vehicle.

References

1. J.C. Huston, B.J. Graves and D.B. Johnson, "Three wheeled vehicle dynamics", SAE, Paper no. 820139 (1982)
2. Karnopp, D., Fang, C., "Simple model of steering-controlledbankingvehicles",ASME Dynamics Systems and Control Division (DSC), **44**, Pg. 15-28 (1992).
3. Hibbard, R., Karnopp, D., "The dynamics of small, relativelytall and narrowtiltinggroundvehicles", ASME Dynamics Systems and Control Division (DSC), **52**, Pg. 397-417 (1993).
4. Hibbard, R., Karnopp, D., "Methods of controlling the lean angle of tilting vehicles," ASME Dynamics Systems and Control Division (DSC), **52**, Pg. 311-320(1993).
5. PalashPatodi, Vinay Saxena, YogeshRathore, "Review on tadpole design – Issues & challenges", International Journal of Researchin Aeronautical and Mechanical Engineering, **2**, Pg. 122-128 (2014).
6. Berote, Johan, Van Poelgeest, Auguste, Darling, Jos, Edge, Kevin, Plummer, Andrew, "The dynamics of a three-wheeled narrow-track tilting vehicle," University of Bath, F2008SC032.

MATEC Web of Conferences **51**, 01003 (2016)
DOI: 10.1051/matecconf/20165101003

Multibody Dynamics Formulation for Modeling and Simulation of Roller Chain Using Spatial Operator

Mohamed Omar[1,a]

[1]*Taibah University, Mechanical Engineering Department, Almadinah Almonawarah, Saudi Arabia*

Abstract. This paper proposes a new approach for modeling and simulating roller-chain drives using multibody formulations. The presented approach models the chain links as spatial decoupled dynamic bodies. The chain links are modeled using two different dynamic representations for the pin-link and the bushing-link. The bushing-link is modeled with two descendent dynamic bodies to represent the rollers. The adjacent pin and bushing links are connected by compliant bushing force elements. An efficient search algorithm is used to detect the contact between the roller and the sprocket teeth while a nonlinear force module is used to predict the contact force. A generalized sprocket representation is used to model the sprocket. The spatial motion of the chain links allows the out-of-plane vibrational motion of the links as well as simulating sprockets misalignment. Using the compliant connection between links avoids using the iterative calculation needed to satisfy the joint constraint equations leading to more efficient calculation scheme. The nonlinear dynamic equations of motion are solved using recursive approach. Complex roller chains drives, bicycle chain and conveyance systems can be easily modeled and analyzed using the proposed approach.

1 Introduction

Recent design advances in roller chain opened new areas of simulation-based-optimization to improve the product quality. Xi [1] utilized multibody dynamic simulation of roller chains to identify the noise sources in diesel engines. The effect of the chain impact on the noise was also investigated by Zheng [2] where the noise level was correlated to the rigid cylinder acceleration. The finite element approach was used and results were compared to experimental results. Ryabov [3] presented an optimization approach for selecting the sprocket teeth based on chain drive cost and service life.

Purdue [4] studied the effect of varying the bike configuration on improving the transmitted pedaling force using commercial program in his investigation. Assuming the chain motion is planar, the simulation software was used to study the effect of non-circular sprocket on minimizing the force exerted by the rider. Approximate method of modeling the chain as a travelling heavy link was used by Rivola [5] to investigate how the transmitted power is fluctuated as a result of the sprocket eccentricity and polygonal effect. Altered conjugated sprocket tooth profile was presented by Wang [6] with the objective of reducing the polygonal action and impact in high speed drive chain systems. Another investigation utilized multibody numerical results to study the impact of using asymmetric sprocket tooth. The effect of using new composite materials for links in conveyer chains was

[a] Corresponding author: momar@taibahu.edu.sa, momar23@gmail.com

investigated by Kadam [7]. The contact models in cylindrical objects were reviewed by Pereira [8] and the application range of the nonlinear contact forces as function of the indentation were examined. Choi [9] examined the roller chains with tensioners and the chain polygonal effect across the chain span length. Veikos [10] modeled the chain as lumped masses connected with simple spring-damper element to study the chain vibration. Modeling the link-sprocket contact was investigated by Liu [11] using impulsive function. Sheu [12] proposed a vector approach for analyzing the friction kinetostatics and the transmission efficiency. The forces in the chain links were measured experimentally by Kidd et al [13] and showed that sprocket misalignment leads to localized bending of the chain links. Marshek et al [14] showed that the misalignment leads to rubbing of the links with the sprockets.

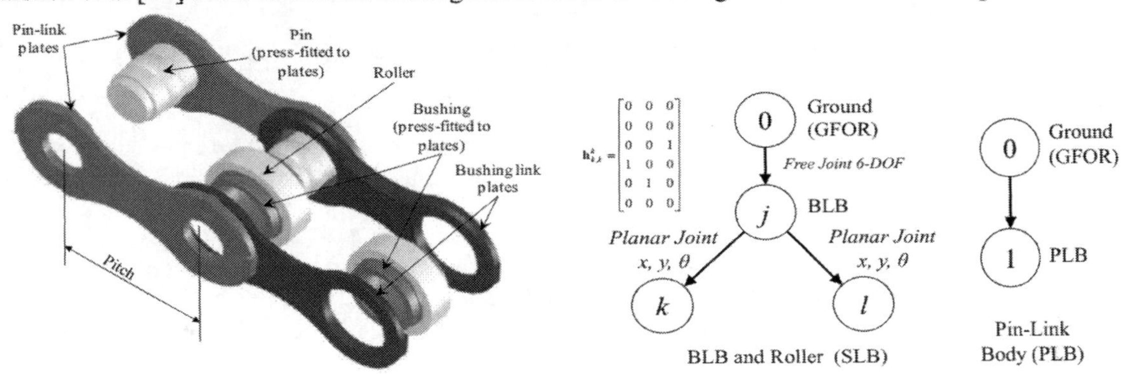

Figure 1. Roller chain links: pin-link, bushing-link and rollers.

This review shows the importance and need to develop an efficient multibody simulation capability that should be able to address the out-of-plane link motion, sprocket misalignment, comprehensive algorithms for contact detection between links and sprockets, and generalized description of the sprocket tooth and profile. In earlier investigation, the author [15] presented a novel approach for modeling chain drives without taking the roller into account. This could lead to unrealistic high friction and contact forces between the sprocket and the links. This paper presents a modified formulation that accounts for the roller dynamics. The system nonlinear dynamic equations are modified accordingly and presented in details.

This paper is organized as follows. In the following section, the link kinematic description and the equations of motion are derived. In section 3, the compliant bushing force element is presented, the roller-bushing contact force module is also discussed. A brief discussion of the nonlinear link-sprocket contact force model is presented. Finally section 4 presents an example of the proposed approach.

2 Link equations of motion

The chain links will be divided into two distinct dynamic body types; the pin-link body (PLB) and the bushing-link body (BLB). The PLB is composed of two side plates press-fitted with two pins to form one rigid body. Each PLB is considered to be kinematically independent from other links and will be modeled as a rigid body with 6-DOF relative to the global frame of reference (GFOR). Similarly, the BLB is composed of two side plates press-fitted with two bushings and form one rigid body. The rollers are considered as dynamic bodies connected to BLB with planar joint with 3 DOF (two translational and one rotational DOF). In order to increase the computational efficiency, the two rollers and the BLB will be treated as on super body link (SBL). The BLB will be modeled with 6-DOF relative to the GFOR while the two rollers are modeled with 3-DOF each relative to the BLB. The SBL will have a total of 12-DOF.

The vector of spatial momentum of the PLB expressed in the GFOR could be differentiated to obtain the body equation motion. The spatial momentum vector of pin-link-body j, could be calculated as follows:

$$\mathbf{P}_j^0 = M_j^0 \mathbf{v}_{0j}^0 \qquad \text{where } M_j^0 = \begin{bmatrix} \underline{J}_j^0 & 0 \\ 0 & m_j \underline{I} \end{bmatrix} \text{ and } \mathbf{v}_{0j}^0 = \begin{bmatrix} \underline{\omega}_{0j}^0 \\ \dot{\underline{r}}_{0j}^0 \end{bmatrix} \tag{1}$$

where $\mathbf{p}_{j,0}^0$ is the spatial momentum vector of body j, M_j^0 is the inertia matrix transformed to GFOR, m_j is the body mass, $\underline{I}$ is 3x3 identity matrix, $\underline{J}_j^0$ is the 3x3 matrix representing the moment of inertia defined at a triad located at the body center of mass and transformed to GFOR, and $\mathbf{v}_{0j}^0$ is the spatial velocity vector in GFOR, and $\dot{\underline{r}}_{0j}^0$ and $\underline{\omega}_{0j}^0$ are the translational and angular velocity vectors in GFOR. Differentiating (1) leads to the BLB equation of motion as follows:

$$M_j^0 \mathbf{a}_{0j}^0 - \tilde{\mathbf{v}}_{0j}^{0^T} M_j^0 \mathbf{v}_{0j}^0 = \mathbf{G}_j^0 , \qquad \text{where} \qquad \mathbf{G}_j^0 = [\underline{\tau}_j^0 \quad \underline{g}_j^0]^T \tag{2}$$

where $\mathbf{G}_j^0$ is the spatial force vector including gravity defined in GFOR, τ_j^0 and g_j^0 represents the external moments and forces acting on the body. The mass matrix in (2) could be inverted to calculate the body absolute accelerations which can be directly integrated to obtain the velocity and displacement of the PLB as follows:

$$\mathbf{a}_{0j}^0 = \left(M_j^0\right)^{-1} \left(\mathbf{G}_j^0 + \tilde{\mathbf{v}}_{0j}^{0T} \mathbf{P}_j^0\right) \tag{3}$$

In the case of SLB, the rollers are considered as descendants of the BLB. The relative velocity of child bodies relative to the parent could be expressed in local and global coordinate systems using the joint variable velocity as follows [15]:

$$\mathbf{v}_{jk}^k = \mathbf{h}_{k,k}^k \dot{\mathbf{q}}_k \qquad \text{and} \qquad \mathbf{v}_{0k}^0 = X^{0k} \mathbf{h}_{k,k}^k \dot{\mathbf{q}}_k \tag{4}$$

where $\mathbf{h}_{k,k}^k$ is a 6x3 joint influence coefficient matrix, as shown in Figure 1, $\dot{\mathbf{q}}_k$ is 3x1 vectors of joint relative velocities, and X_{0k}^{0k} is the spatial transformation matrix between the body k and GFOR. The joint influence coefficient matrices could also be used to project the spatial force vectors, F_k, into the joint subspace force vector, Q_k, as follows:

$$Q^k = \left(\mathbf{h}_k^k\right)^T F_k \tag{5}$$

Using the recursive formulation, the absolute velocity, of the roller bodies, k and l, could be calculated using the BLB parent, j, and the relative joint velocities as follows:

$$\mathbf{v}_{0k}^0 = \mathbf{v}_{0j}^0 + X_{0k}^{0k} \mathbf{v}_{jk,k}^k = \mathbf{v}_{0k}^0 + X_{0k}^{0k} \mathbf{h}_{k,k}^k \dot{\mathbf{q}}_k \qquad \text{or} \qquad \mathbf{v}_{0k}^0 - \mathbf{v}_{0j}^0 = X_{0k}^{0k} \mathbf{h}_{k,k}^k \dot{\mathbf{q}}_k \tag{6}$$

Differentiating velocity equations, the acceleration of the child bodies across the joint DOF could be obtained as follows:

$$\mathbf{a}_{0k,0}^0 = \mathbf{a}_{0j,0}^0 + \mathbf{h}_{k,0}^0 \ddot{\mathbf{q}}_k + \tilde{\mathbf{v}}_{0k,0}^0 \mathbf{h}_{k,0}^0 \dot{\mathbf{q}}_k = \mathbf{a}_{0j,0}^0 + \mathbf{h}_{k,0}^0 \ddot{\mathbf{q}}_k + \boldsymbol{\gamma}_{0k}^0 \qquad \text{or} \qquad \mathbf{a}_{0k,0}^0 - \mathbf{a}_{0j,0}^0 = X_{0k}^{0k} \mathbf{h}_{k,k}^k \ddot{\mathbf{q}}_k + \boldsymbol{\gamma}_{0k}^0 \tag{7}$$

where $\ddot{\mathbf{q}}_k$ is 3x1 vectors of joint relative velocity and acceleration, and $\boldsymbol{\gamma}_{0k}^0$ is the vector of quadratic velocities.

The velocity and accelerations of the SLB could be written in a compact form as follows:

$$\mathbf{T}_a^0 \mathbf{v}_a^0 = \mathbf{H}_a^0 \dot{\mathbf{q}}_a \qquad \text{and} \qquad \mathbf{T}_a^0 \mathbf{a}_a^0 = \mathbf{H}_a^0 \ddot{\mathbf{q}}_a + \boldsymbol{\gamma}_a^0 \tag{8}$$

where $\mathbf{T}_a$ is called the system topology matrix and $\mathbf{H}_a$ is the system influence coefficient matrix.

The equation of motion of the connected bodies j, k, and l could be driven from the free body diagram. For body j, the equation of motion in its local coordinate system could be written as:

$$M_j^0 \mathbf{a}_{0j}^0 - F_j^0 + F_k^0 + F_l^0 = \mathbf{G}_j^0 + \tilde{\mathbf{v}}_{0j}^{0^T} \mathbf{P}_j^0 , \quad M_k^0 \mathbf{a}_{0k}^0 - F_k^0 = \mathbf{G}_k^0 + \tilde{\mathbf{v}}_{0k}^{0^T} \mathbf{P}_k^0 , \text{ and } M_l^0 \mathbf{a}_{0l}^0 - F_l^0 = \mathbf{G}_l^0 + \tilde{\mathbf{v}}_{0l}^{0^T} \mathbf{P}_l^0 \tag{9}$$

These equations could be arranged in a matrix form as follows:

$$\begin{bmatrix} M_j^0 & & \\ & M_k^0 & \\ & & M_l^0 \end{bmatrix} \begin{bmatrix} \mathbf{a}_{0j}^0 \\ \mathbf{a}_{0k}^0 \\ \mathbf{a}_{0l}^0 \end{bmatrix} - \begin{bmatrix} I & -I & -I \\ & I & \\ & & I \end{bmatrix} \begin{bmatrix} F_j^0 \\ F_k^0 \\ F_l^0 \end{bmatrix} = \begin{bmatrix} \mathbf{G}_j^0 + \tilde{\mathbf{v}}_{0j}^{0^T} \mathbf{P}_j^0 \\ \mathbf{G}_k^0 + \tilde{\mathbf{v}}_{0k}^{0^T} \mathbf{P}_k^0 \\ \mathbf{G}_l^0 + \tilde{\mathbf{v}}_{0l}^{0^T} \mathbf{P}_l^0 \end{bmatrix} \tag{10}$$

Equation (10) represent the SLB equation of motion that contains the bodies Cartesian accelerations and reaction forces between the rollers and the BLB. It could be written in a compact matrix form as:

$$\mathbf{M}_a \mathbf{a}_a - \left(\mathbf{T}_a\right)^T \mathbf{F}_a = \mathbf{G}_k^0 + \tilde{\mathbf{v}}_a^T \mathbf{P}_a \tag{11}$$

Using Equations (5) , (8) and (11) , the system equations could be rearranged as follows:

$$\begin{bmatrix} \mathbf{M}_a & 0 & \left(\mathbf{T}_a\right)^T \\ 0 & 0 & -\mathbf{H}_a \\ \mathbf{T}_a & -\left(\mathbf{H}_a\right)^T & 0 \end{bmatrix} \begin{bmatrix} \mathbf{a}_a \\ \ddot{\mathbf{q}}_a \\ -\mathbf{F}_a \end{bmatrix} = \begin{bmatrix} \mathbf{G}_a^0 + \left(\tilde{\mathbf{v}}_a^0\right)^T \mathbf{P}_a \\ \mathbf{Q}_a \\ \gamma_a \end{bmatrix} \tag{12}$$

The system coefficient matrix in the left hand side of equation (12) could be inverted to solve for the unknown joint variables accelerations as follows:

$$\left(\mathbf{H}_a^T \mathbf{T}_a^{-T} \mathbf{M}_a \mathbf{T}_a^{-1} \mathbf{H}_a\right) \ddot{\mathbf{q}}_a = \mathbf{Q}_a + \mathbf{H}_a^T \mathbf{T}_a^{-T} \left(\mathbf{G}_a^0 + \tilde{\mathbf{v}}_a^T \mathbf{P}_a\right) \quad \rightarrow \quad \mathbf{M}_{qq} \ddot{\mathbf{q}}_a = \mathbf{Q} \tag{13}$$

The detailed block-entries of the inertia matrix and generalized force vector are written as follows:

$$\ddot{\mathbf{q}}_a = \begin{bmatrix} \mathbf{a}_j^0 \\ \ddot{\mathbf{q}}_k \\ \ddot{\mathbf{q}}_l \end{bmatrix} \quad \mathbf{M}_{qq} = \begin{bmatrix} & 6 & 3 & 3 \\ 6 & M_j & M_j H_k & M_j H_l \\ 3 & H_k^T M_j & H_k^T \left(M_j + M_k\right) H_k & H_k^T M_j H_l \\ 3 & H_l^T M_j & H_l^T M_j H_k & H_l^T \left(M_j + M_l\right) H_l \end{bmatrix} \quad \mathbf{Q} = \begin{bmatrix} Q_j + \mathbf{G}_j^0 + \tilde{\mathbf{v}}_{0j}^{0^T} \mathbf{P}_j^0 + \cdots \\ \mathbf{G}_k^0 + \tilde{\mathbf{v}}_{0k}^{0^T} \mathbf{P}_k^0 + \mathbf{G}_l^0 + \tilde{\mathbf{v}}_{0l}^{0^T} \mathbf{P}_l^0 \\ H_k^T \left(\mathbf{G}_k^0 + \tilde{\mathbf{v}}_{0k}^{0^T} \mathbf{P}_k^0\right) \\ H_l^T \left(\mathbf{G}_l^0 + \tilde{\mathbf{v}}_{0l}^{0^T} \mathbf{P}_l^0\right) \end{bmatrix} \tag{14}$$

Noting that constructing the H matrices doesn't involve computations, rather, it is picking the right column from the spatial transformation matrix. This 12x12 matrix could be easily evaluated and inverted to compute the motion derivatives. Using Equation (3) and Equation (14) all the accelerations of the links in the chain could be calculated and then integrated very efficiently.

3 Bushing, sprocket and contact force elements

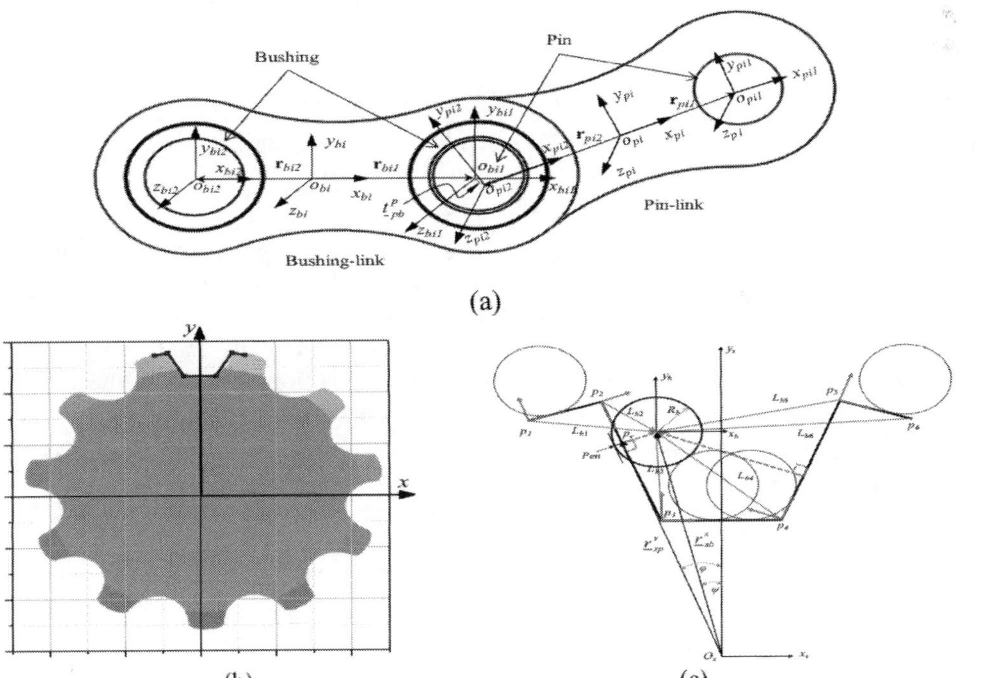

Figure 2. (a) Bushing element kinematic diagram. (b) Approximated sprocket tooth profile (c) Roller sprocket contact detection kinematic model

The connections between the pin-links and the bushing-links are modeled using bushings force elements. The bushing stiffness accounts for the pin-link flexibility and the bushing-link flexibility. The repeated pattern of pin-link and bushing-link connection allows using the same bushing parameters over the chain. As shown in Figure 2, two markers are located in each link O_{pi1} and O_{pi2}, to define the two pin-centers relative to the body reference-frame, O_{pi}. Similarly, two markers, O_{bi1} and O_{bi2}, are used to define the bushing center relative to the bushing-link reference frame, O_{bi}. The bushing force element connects the pin center marker to the corresponding bushing center marker. At the beginning of the simulation when there is no deflection or relative rotation between the pin-link and bushing-link, the pin-center and bushing-center markers are considered to be coincident and parallel. To calculate the elastic forces of the bushing element, the translational deflection and relative orientation between the pin-center marker and the bushing-center marker need to be calculated every time step.

Due to the bushing elastic deformation and structural material damping, the bushing force element develop a linear elastic forces and torques proportional to the bushing deformation. Also, calculating the approximated viscous damping forces and torques proportional to the relative deformation velocity could be done. The elastic torque component about the bushing axis is assumed to be zero in order to allow relative rotation. The bushing forces and torques can be computed as follows:

$$F^{p}_{p,p} = \begin{bmatrix} M \\ F \end{bmatrix} \quad \text{where} \quad M = \begin{bmatrix} -(\theta K_{tx})u^{r}_{1} - \omega_x C_{tx} \\ -(\theta K_{ty})u^{r}_{2} - \omega_y C_{ty} \\ -(\theta K_{tz})u^{r}_{3} - \omega_z C_{tz} \end{bmatrix} \quad \text{and} \quad F = \begin{bmatrix} -t_x K_x - \dot{t}_x C_x \\ -t_y K_y - \dot{t}_y C_y \\ -t_z K_z - \dot{t}_z C_z \end{bmatrix} \tag{15}$$

where K_x, K_y, K_z, K_{tx}, K_{ty}, and K_{tz}, the stiffness coefficients associated with translational and rotational displacement respectively and C_x, C_y, C_z, C_{tx}, C_{ty}, and C_{tz} the damping coefficients associated with translational and rotational velocities respectively.

Estimating the contact force between the roller and the bushing is crucial in studying the durability and wear of the two parts. The contact between the bushing and the roller is modelled using the Hertz contact model. The roller has relative planar displacement and rotation with respect to the BL. The magnitude of penetration, δ, could be calculated from the components of relative displacements. Employing Johnson's formula based on Hertz theory [16], the normal contact force, P, could be calculated as function of material properties $E^{*} = E/2(1-\nu^{2})$ and difference in radii $\Delta R = R_i - R_j$, as:

$$\delta = P\left(\ln\left(4\pi E^{*}\Delta R/P\right) - 1\right)/\pi E^{*} \tag{16}$$

Using the relative velocities the penetration rate could be calculated and used to determine the damping force. The magnitude of the friction force and torque could be calculated from the normal force, coefficient of friction, and the relative velocity.

The contact between the sprocket teeth and chain links is the main motion transmission mechanism in chain drive system. The contact forces are modeled to prevent penetration of the chain-rollers into the sprocket teeth. The contact force magnitude depends on tooth curvature at the contact point and on the localized mutual deformation of the roller and the tooth. Predicting the contact location requires accurate calculation of the roller center position relative to the tooth profile. The friction force due to roller sliding over the tooth requires accurate calculation of the relative velocity between the roller and sprocket tooth at the point of contact. In the current proposed approach, the sprocket tooth is modeled using five line segments to approximate the tangent lines in plane, shown in Figure 2 (b). More accurate representation of the tooth profile could utilize spline to smooth the tooth geometry. The kinematic model for contact point detection is shown in Figure 2 (c).

The relative position and relative velocity vectors between the roller center and the sprocket center, $\underline{r}^{s}_{sb,s}$ and $\dot{\underline{r}}^{s}_{sb,s}$, are calculated. Knowing the tangent line segment and the roller radius, the magnitude of penetration could be calculated and the contact forces are estimated. In order to calculate the normal and sliding velocity components, $\dot{\underline{r}}^{s}_{sb,s}$ the relative velocity vector needs to be transformed using the inclination angle φ as follows:

$$\begin{bmatrix} v_{x-slid} \\ v_{norm} \\ v_{z-slid} \end{bmatrix} = \underline{R}\dot{\underline{r}}^{s}_{sb,s} = \begin{bmatrix} \cos\varphi & \sin\varphi & 0 \\ -\sin\varphi & \cos\varphi & 0 \\ 0 & 0 & 1 \end{bmatrix} \begin{bmatrix} v_{rx} \\ v_{ry} \\ v_{rz} \end{bmatrix} \tag{17}$$

The sliding velocity could be computed as:

$$\underline{v}_s = |\underline{v}_s| \underline{u}_{vs} \quad \text{where} \quad |\underline{v}_s| = \sqrt{v^2_{x-slid} + v^2_{z-slid}} \quad \text{and} \quad \underline{u}_{vs} = \underline{v}_s / |\underline{v}_s| \tag{18}$$

The normal contact force composed of elastic spring and damping force can be evaluated as follows:

$$F_n = K\delta + C v_{norm} \tag{19}$$

The nonlinear sliding friction force is a function of the normal elastic force and the friction coefficient. The friction coefficient can be considered as smooth function of the sliding velocity as follows:

$$\underline{F}_f = \mu S_\mu (K\delta) \underline{u}_{vs}, \quad \text{where} \quad S_\mu = \min(1, 2\xi - \xi^2) \quad \text{and} \quad \xi = |\underline{v}_s| / v_\mu \tag{20}$$

where v_μ is a predefined maximum sliding velocity at which the friction force reaches maximum value, S_μ is the friction force softening.

Then contact forces are now applied in the sprocket reference frame and the link reference frame as:

$$F^s_{s,s} = \begin{bmatrix} I & 0 \\ \tilde{\underline{r}}^s_{sp} & I \end{bmatrix}^{-T} \begin{bmatrix} 0 \\ \underline{F}^c \end{bmatrix} \quad \text{and} \quad F^l_{l,l} = \left(X^{sl}_{sl}\right)^{-T} F^s_{s,s} \tag{21}$$

4 Demonstrating example

This example represents a bicycle chain that contains 56 PLB and 56 SLB. The multibody model of the bike is composed of: front drive sprocket, wheel sprocket, the lower idler sprocket, and the upper idler sprocket. The bicycle chassis is fixed to the GFOR and all sprockets are attached to the chassis with revolute joints. Every PLB and SLB body is referenced to the ground with full 6-DOF joints. The drive sprocket attachment position is offset from the wheel sprocket attachment position by 30mm in the lateral direction. The chain link-pitch is 12.7mm, the bushing outer radius is 3mm, the pin radius is 2.1mm, and roller radius is 4mm. The sprocket tooth thickness is 3.2mm. A kinematic driver is applied to the driving sprocket ramping up the speed from zero to 1 rev/s in 3 seconds, as shown in Figure 3.

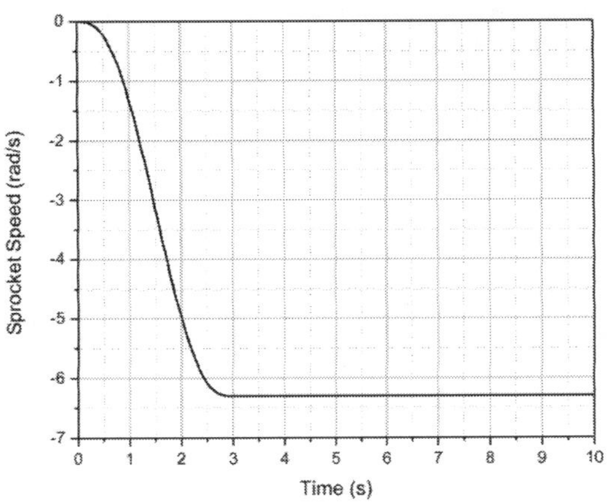

(a)

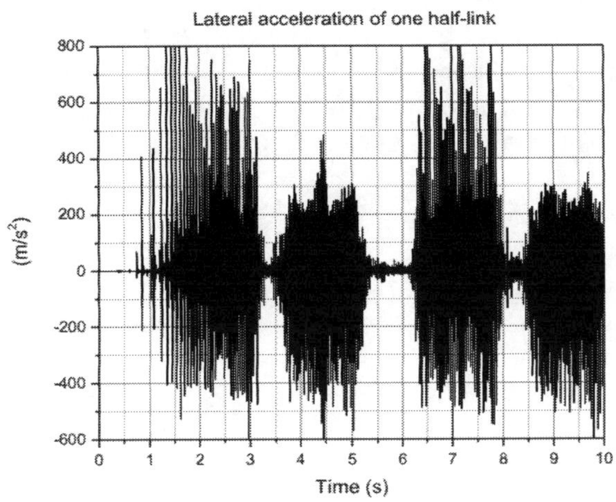

(b)

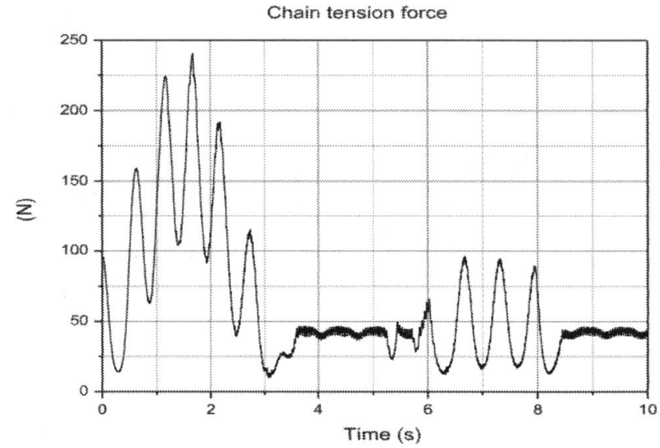

(c)

Figure 3 (a) Drive sprocket speed (b) Lateral acceleration of chain link (c) Bushing tension force

Each link travels a distance of 1.422m around the sprockets to complete one cycle. Sample simulation results are shown in Figure 3. Figure 3 (b) shows the links' lateral accelerations. The axial tension force in link 1 during the simulation is shown in Figure 3 (c).

References

1. J Xi, Z Feng, G Wang, F Wang, J of Mech Science & Technology, Vol: 29-1, pp: 181~189, 2015
2. H. Zheng, Y. Y.Wang, G. R. Liu And K. Y. Lam, K. P. Quek, T. Ito And Y. Noguchi, JSV 245(1), 133-150, 2001
3. G. K. Ryabov and M. A. Seledzhieva, Chemical and Petroleum Engineering, **Vol. 49**, Nos. 2013.
4. A. I. Purdue, A. I. J. Forrester, M. Taylor, M. J Stokes, E. A. Hansen, J. Rasmussen, 8th ISEA, Procedia Engineering 2 pp.: 2543–2548. 2010
5. A Rivola, M Troncossi, Mechanical Systems and Signal Processing, 48, 325–338, 2014
6. Y Wang, D Ji, K Zhan, Mechanism and Machine Theory **70**, 380–393, 2013
7. A J. Kadam, S V. Deshpande, Recent Trends In Mechanical engineering, 2015
8. C M. Pereira, A L. Ramalho ·J A. Ambrósio, Nonlinear Dyn , **63**: 681–697, 2011
9. W Choi, G E Johonson, MEAM Technical Report Number 92-90, 1990.
10. NM Veikos and F Freudenstein, Mechanism design and Synthesis, DE **Vol 46**, pp. 431-438, 1992
11. X Liu, T Wu , Z Q Xiang, W Q Jia, Applied Mech and Materials, **Vol. 387** pp 237-240, 2013
12. K B Sheu, C W Chien, S T Chiou, T S Lai, Mechnsm and Machine Theory, 39 (8):819-837, 2004
13. M. D. Kidd, N. E. Loch, R. L. Reuben, Experimental Mechanics, **Vol. 39**, No. 4, 1999.
14. K. M Marshek, Mechanism and Machine Theory, **Vol 14**, pp 135-139, 1978.
15. M. A. Omar, AIME, **2014**, Article ID 378030, dx.doi.org/10.1155/2014/378030.
16. Johnson, K.L.: One hundred years of Hertz contact. Proc. Inst. Mech. Eng. **196**, 363–378 (1982)

MATEC Web of Conferences **51**, 01004 (2016)
DOI: 10.1051/matecconf/20165101004

Vibration Characteristics of Tapered Roller Bearings with Roller Diameter error

Van-Canh Tong[1] and Seong-Wook Hong[1,a]

[1]*Department of Mechanical System Engineering, Kumoh National Institute of Technology, 61 Daehak-ro, Gumi, Gyeongbuk, Republic of Korea 730-701*

Abstract. This study investigates the vibration characteristics of tapered roller bearings with the presence of error in roller diameter. A general model for tapered roller bearing with five degree-of-freedom was adopted with taking into account the error in roller diameter. Then, the effect of one single errored roller and multiple errored rollers on the vibration behaviour of tapered roller bearing was investigated. Simulation results confirmed that the bearing vibration frequencies are equal to cage rotational frequency in the case of axial preloading, and multiple of cage rotational frequency in the case of combined loadings, irrespective of the error magnitude, the number or position of errored rollers, which, however, significantly affect the vibration amplitude of bearing.

1 Introduction

Rolling element bearings have been extensively used in rotating machineries to support load and ensure the rotational motion of the shaft. Vibration of rolling bearings is an important issue, which has a great attention in design and application of bearings. This is because the bearing vibration can considerably influence the overall performance of not only bearings but also the whole rotating systems containing bearings. Recently, the requirements for rotational vibration of bearings have become strict. Numerous research works have been implemented for detecting, monitoring, and controlling bearing vibration. A large number of studies have concerned about bearings with an assumption that no geometric error exists in the bearings. That assumption is, however, impractical since the geometric imperfections are often occurred in bearings during manufacturing process. This study deals with the vibration behaviour of tapered roller bearings (TRBs) with error in roller diameter.

The geometric error in rolling bearing can be classified into three groups, i.e. high frequency error as roughness, intermediate frequency error as waviness, and low frequency error as form error [1]. The most typical error source of roller is the form error, such as out of roundness and off-sized roller. The effect of off-sized roller, in particular, error in roller diameter on bearing vibration has been investigated by several researchers. However, most of them carried out investigation with regard to ball or cylindrical roller bearings [2-7]. Although TRBs are widely used in heavy load applications, the effect of geometric error on the performance of TRBs has not been sufficiently considered.

In this study, the effect of roller diameter error on the vibration characteristics of TRBs is investigated. A general five DOFs TRB model is employed with taking into account the error in roller

[a] Corresponding author : swhong@kumoh.ac.kr

diameter. Then the effect of single and multiple errored rollers on the vibration of TRBs are investigated. The vibration of TRBs is evaluated using time and frequency domains, along with the demonstration of bearing centre orbits. Simulation is performed for a TRB subjected to axial preload and combined radial and axial loadings. The obtained results would be beneficial for tolerance design, manufacturing and loading of TRBs.

2 Tapered roller bearing model

Figure 1(a) shows the geometry of TRB along with external load and displacement vectors as

$$\{F\} = \{F_x \quad F_y \quad F_z \quad M_y \quad M_z\}^T \tag{1}$$

$$\{\delta\} = \{\delta_x \quad \delta_y \quad \delta_z \quad \gamma_y \quad \gamma_z\}^T \tag{2}$$

The 5-DOF bearing model outlined by de Mul et al. [8] was used in this study. This model, however, is a simple quasi-static model, which does not take into account the time-varying position of the rolling elements. In order to establish a time-varying bearing model, the instantaneous angular position of a roller at time t is defined as

$$\phi_j(t) = \frac{2\pi(j-1)}{Z} + \Omega_m t \tag{3}$$

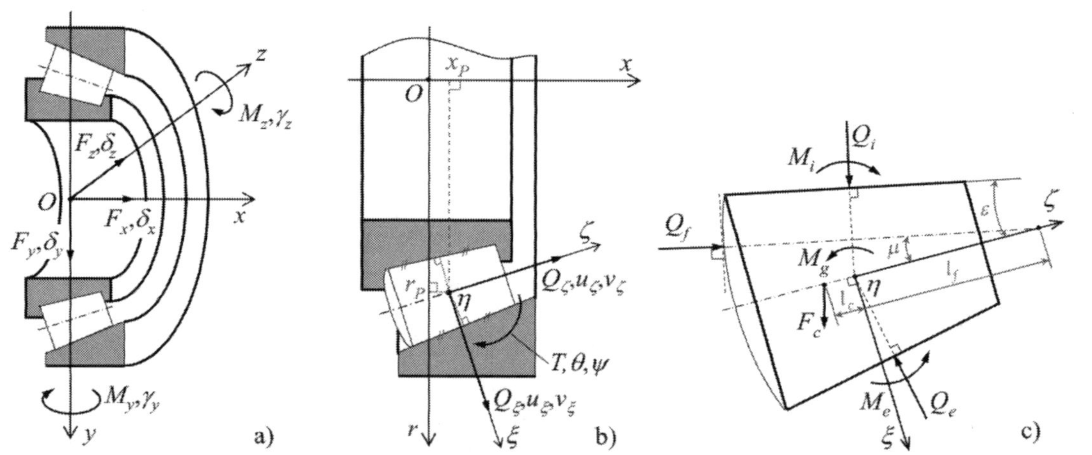

Figure 1. TRB geometry: a) TRB loading and displacements, b) TRB cross-section geometry at location angle $\phi_j(t)$ with roller and inner ring displacements c) roller free-body diagram

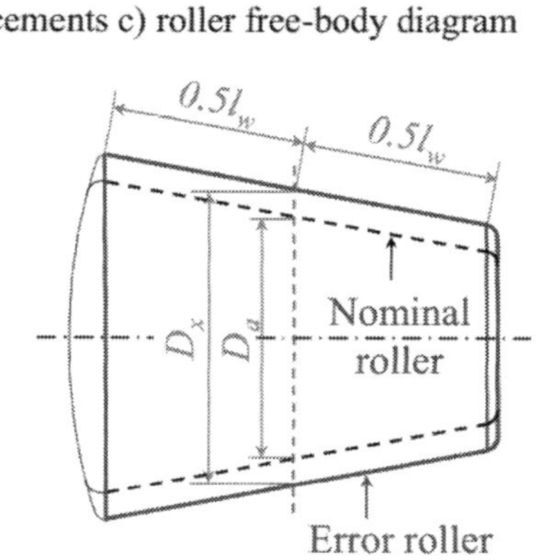

Figure 2. Nominal and errored roller geometry

where j and Z represent the roller index and the total number of total rollers. Ω_m is the orbital speed of roller [9]. Figure 1(b) indicates the geometry of TRB cross-section at a particular roller location angle $\phi_j(t)$. The displacements of inner ring cross-section $\{u_\kappa\} = \{u_r, u_z, \theta\}^T$, can be calculated using transformation matrices $[K]$ and $[R_\phi]$ (see [8])

$$\{u_\kappa\} = [K]([R_\phi]\{\delta\}) \tag{4}$$

Because the roller is displaced from its original position by a displacement vector $\{v_\kappa\} = \{v_r, v_z, \psi\}^T$, the roller and raceway contact loads (Q_i, Q_e, M_i and M_e) are generated as shown in Figure 1(c). These contact loads can be estimated using the slicing technique based on the contact compression between the roller and raceways [8]. The roller-flange contact force Q_f is determined using the classical Herzian contact theory based on the contact between the spherical roller end and flat flange. The inertial loads acting on the roller including the centrifugal force F_c and gyroscopic moment M_g [9], are also considered in this study. Then the roller equilibrium equations can be obtained by considering two force equations about ξ and ζ axes and a moment equation about η axis as shown in Figure 1(c). By solving the roller equilibrium equations, the displacement of roller $\{v_\kappa\}$ can be determined, as well as the final value of roller-races contact loads. The roller contact loads must reflect the error in diameter. Figure 2 shows the geometry of the roller with and without error. The error magnitude is calculated as

$$\delta_D = D_x - D_a \tag{5}$$

where D_x and D_a are the errored and nominal diameters of roller (Figure 2).
Having obtained the inner ring contact loads at all rollers $\{Q_\kappa\} = \{Q_r, Q_z, T\}^T$, the inner ring equilibrium equations can be derived by summation of all inner ring contact loads and an external load $\{F\}$ as

$$\{F\} + \sum_{j=1}^{z} \{[R_\phi]^T([K]^T\{Q_\kappa\})\}_j = \{0\} \tag{6}$$

Equation (6) is subsequently solved for the inner ring displacements $\{\delta\}$. All the dynamic equations of TRB including the roller equations and global equations of inner ring have been previously described in [6-7]. Because the roller and global equations are non-linear, the iterative technique should be used as a solution, for example Newton-Raphson method.

3 Numerical results

This section presents numerical results for the effect of roller diameter error on the TRB vibration behaviour. A sample TRB (30212-A) with its basic parameters given in Table 1 was used for investigation. The rotational speed of TRB is selected at 1000 rpm throughout the paper.

Table 1. Basic parameters of TRB 30212-A.

Parameter	Symbol	Value	Parameter	Symbol	Value
Bore diameter	d [mm]	60	Number of rollers	z	19
Outside diameter	D [mm]	110	Pitch diameter	d_m [mm]	84.12
Total width	B [mm]	23.75	Nominal roller mean diameter	D_a [mm]	13

3.1 Single errored roller

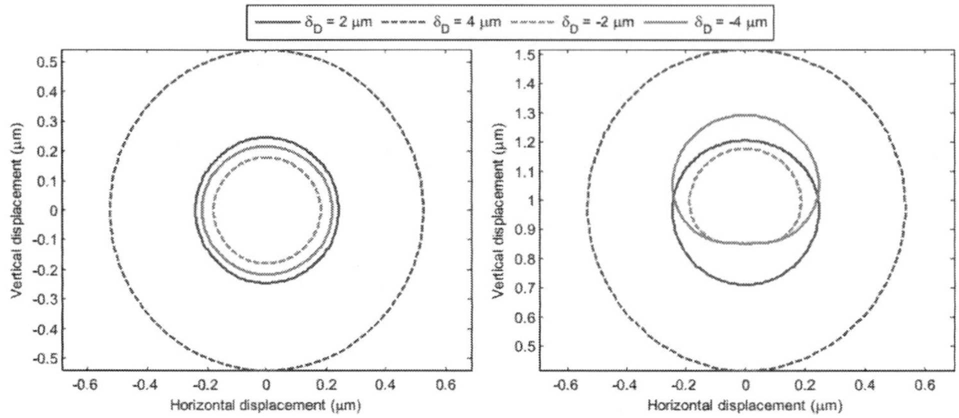

Figure 3. TRB centre orbits with single errored roller: a) $F_x = 1000N$, b) $F_x = F_y = 1000N$

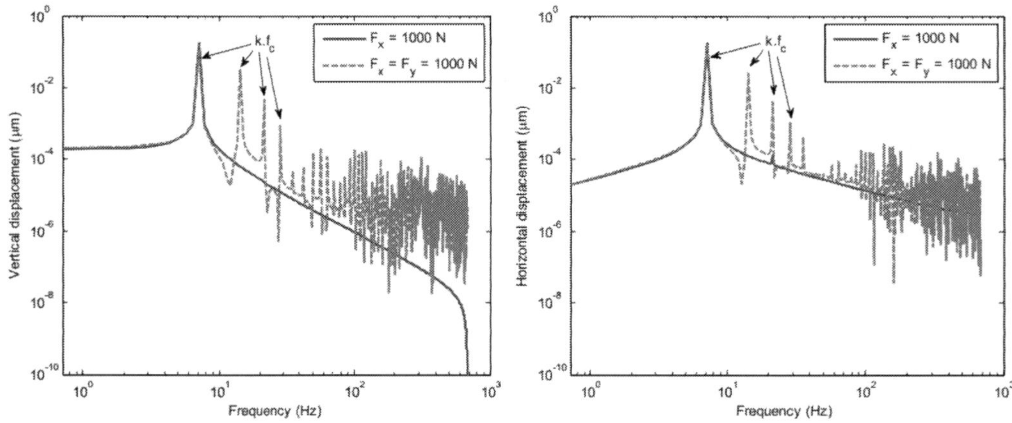

Figure 4. Spectrums of TRB displacements with errored roller with the error magnitude of $\delta_D = 2\mu m$

Figure 3 illustrates the centre orbits of the TRB with one errored roller. It is seen that the magnitude of the centre orbits is increased with increasing the error magnitude. In the case of axial preload TRB, the bearing centre trajectories are remained circular regardless of error magnitude. However, the trajectories are no longer circular when the TRB is loaded with combined radial and axial loads. Figure 4 shows the frequency spectrums of vertical and horizontal displacements of TRB. The TRB under only axial preload shows only single dominant peak frequency equal to the cage rotational frequency (f_c). However, under combined loads, the roller error excites many vibration frequencies corresponding to multiples of the cage frequency ($k.f_c$)

3.2 Multiple errored rollers

3.2.1 Identical error magnitude

Figure 5 indicates the centre orbits of TRB with different combination of errored rollers. The TRB has two errored rollers whose error magnitude is selected 2 μm. From Figure 5, the set of errored rollers at #1 and #2 shows the highest radial vibration, and the next orders are the sets of errored roller at #1 and #4, #1 and #8, and #1 and #10. Here the number represents the consecutive order among the rollers in the TRB. This is attributed by the increase of the distance angle between the errored rollers. Therefore, the vibration of TRB can be reduced when errored rollers are distributed around the bearing circumference.

Figure 6 demonstrates the frequency spectrums of TRB with errored rollers at #1 and #2. The frequency components are quite similar to those in the case of single errored roller shown in Figure 4 with its dominant frequencies at multiple harmonics of f_c. It means that the number, position and magnitude of error rollers do not significantly affect the frequency component of TRB vibration.

3.2.2 Different error magnitude

Vibration of TRB is evaluated for the case when there exist two errored rollers with different error magnitude of 2 and -2 μm. Figure 7 illustrates the centre orbits of TRB with different combinations of errored rollers. From Figure 7, the vibration of TRB shows a behaviour contrary to that for the case of identical errored rollers shown in Figure 5. The largest centre orbit is found when the errored rollers are at #1 and #10, whereas the smallest one is for the case of errored rollers at #1 and #2.

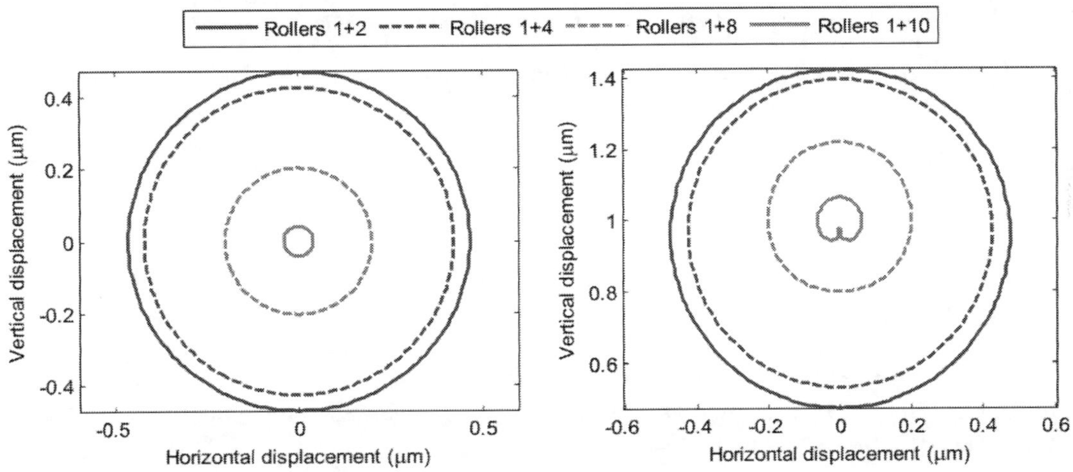

Figure 5. TRB centre orbits with two errored rollers with error magnitude of 2 μm: a) $F_x = 1000N$, b) $F_x = F_y = 1000N$

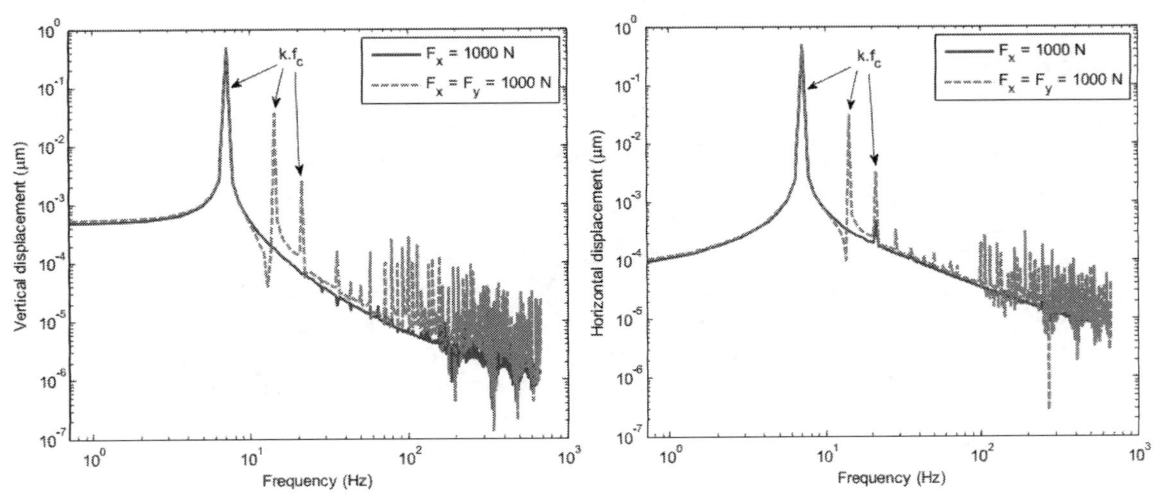

Figure 6. Spectrums of TRB displacements with errored rollers at #1 and #2 with error magnitude of 2 μm

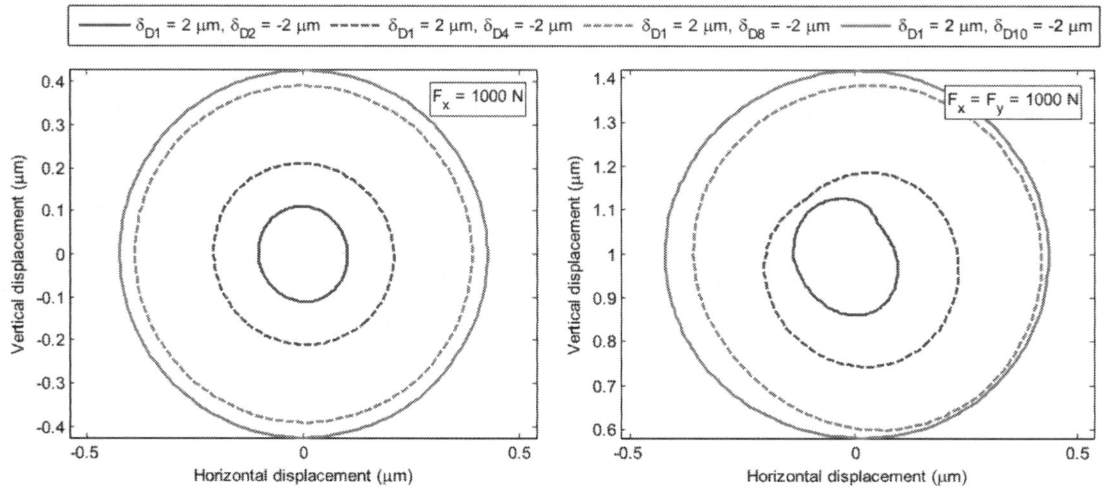

Figure 7. TRB centre orbits with two errored rollers with different error magnitudes, 2 μm and -2 μm

4 Conclusion

From the simulation results for vibration response of TRB with errored rollers, following conclusions are obtained:

1. The roller diameter error excites the TRB at the cage rotational frequency in the case of axial preload, at multiple harmonics of cage rotational frequency in the case of combined axial and radial loads.
2. Increasing roller error magnitude significantly increases the vibration amplitude of TRB. However, roller error magnitude does not affect the frequency components of TRB, likewise the number and position of errored rollers.

Acknowledgement

This research was financially supported by Daegyeong Institute of Regional Program Evaluation.

References

1. X. D. Zhang, C. Zhang, B. Wang. American Soc. Precis. Eng. (1999)
2. L. D. Meyer, F. F. Ahlgren, B. Weichbrodt. J. Mech. Des **102**, 2 (1980)
3. P. K. Gupta. ASME J. Tribol. **110**, 1 (1988)
4. S. Noguchi, K. Ono. Precis. Eng. **28**, 2004
5. G. Chen, F. Mao, B. Wang. Proc. IMechE. Part J: Part J: J. Eng. Tribol. **226**, 8 (2012)
6. S. Yu, D. Wang, H. Dong, B. Wang. Proc. IMechE. Part C: J Mech. Eng. Sci. **227**, 11 (2013)
7. P. Ji, Y. Gao, F. Ma and Q. An. Proc. IMechE. Part J: J. Eng. Tribol. **229**, 6 (2015)
8. J. M. de Mul, J. M. Vree, D. A. Maas. ASME J. Tribol. **111**, (1989)
9. V. C. Tong, S. W. Hong. Proc. IMechE. Part C: J Mech. Eng. Sci. **230**, 2 (2016)

MATEC Web of Conferences **51**, 01005 (2016)
DOI: 10.1051/matecconf/20165101005

Electro-Mechanical Response and Engineering Properties of Piezocomposite with Imperfect Interface

Rattanan Tippayaphalapholgul[1] and Yasothorn Sapsathiarn[1,a]

1Department of Civil and Environmental Engineering, Faculty of Engineering, MahidolUniversity, Thailand

Abstract. Composites of piezoelectric materials are widely use in practical applications such as nondestructive testing devices, smart adaptive structures and medical devices. A thorough understanding of coupled electro-elastic response and properties of piezocomposite are crucial for the development and design of piezoelectric composite materials used in advanced applications. The micromechanics analysis is employed in this paper to determine the response and engineering properties of the piezocomposite. A mechanical imperfect interface bonding between piezoelectric inclusion and polymer matrix is taken into consideration in the analysis. The micromechanics analysis is based on the Boundary Element Method (BEM) together with the periodic micro-field micromechanics theory. A selected set of numerical results is presented to investigate the influence of volume ratio and interface bonding condition on effective piezocomposite material coefficients and portray basic features of coupled electroelastic response within the domain of piezocomposite unit cell.

1 Introduction

Piezoelectric materials have been used in the development of a variety of electro-mechanical devices due to their inherent electro–mechanical coupling phenomenon; that is, they produce an electric field when deformed under a mechanical stress (direct piezoelectric effect) and, conversely, undergo deformation when subjected to an electric field (converse piezoelectric effect) [1, 2]. Several issues related to properties of monolithic piezoelectric elements, for example, low fracture strength, high stiffness and difficulty in producing complex shapes, lead to the development of composites of piezoelectric materials to improve overall properties which are suitable for practical applications. Figure 1(a) shows a composite of piezoelectric material and polymer matrix. The soft polymer matrix supports the brittle piezoelectric fibers and also offers more conformability of piezocomposite.

An understanding of coupled electroelastic behavior of piezoelectric composite materials and their properties is critical for an effective and reliable design of electro-mechanical devices based on piezocomposites. The response of piezoelectric composite materials is governed by several parameters, such as the fiber-to-matrix volume ratio, the properties of the fiber, the surrounding matrix material, and the interphase layer between piezoelectric fibers and the matrix [3-5]. Properties of piezocomposites can be globally represented by a homogeneous medium with anisotropic properties.Micromechanics model can provide the relations between the microstructure and the overall

[a]Corresponding author : Yasothorn.sap@mahidol.ac.th

or effective properties of composites based on the known properties of each material phases in the composites.

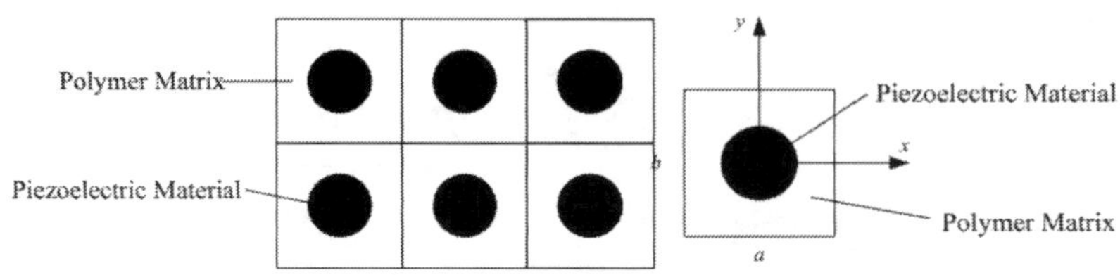

(a) (b)

Figure 1.(a) Schematic of piezoelectric composite material; (b)Representative Volume Element (RVE)

Several researchers have employed a variety of analytical and numerical techniques to predict the effective electromechanical moduli of piezocomposites [6-9]. Most of studies on responses and properties of composite materials appeared in the literature assume that the bonding between the fiber/inclusion and the matrix phase are perfect. In fact, there is a thin interphase layer between phases leading to imperfect bonding along the interface between the fiber/inclusion and matrix phases. Assumptions are usually employed in the modeling of the interphasial layer due to the structure of interphase layer is very complicated. Among others, a spring–factor model is the most widely used in the analysis. The model assumes that the displacement jump across the interface to be proportional to the corresponding interfacial stresses [10]. The spring–factor model for imperfect interface has been employed by [11] and [12] in the analysis of elastic fiber–reinforced composites.

The micromechanics model for composites of piezoelectric materials and polymer matrix is developed in this paper. The model is based on the periodic microfield micromechanics theory and the boundary element method (BEM). A selected set of numerical results are presented to portray the essential characteristics of electroelastic fields and effective properties of a piezocomposite

2 Theoretical consideration for BEM-based micromechanics analysis of piezocomposite

The composite of piezoelectric materials (fiber/inclusion) embedded in a polymeric matrix material is considered in this paper (Figure 1). An efficient methodology of micromechanics analysis for determining the effective properties of piezoelectric composite materials based on the boundary element method (BEM) and the periodic microfield approach is developed and explained in the subsequent sections. The static electroelastic response of a piezoelectric material is governed by the following equations expressed using the standard indicial notation [13]; $\sigma_{ij,j} + F_i = 0$ and $D_{i,i} = \rho_e$,where $\sigma_{ij}(i, j = x, y, z)$, D_i , F_i and ρ_e denote the stress tensor, electric displacement and body force in the $i-$direction, and a body charge, respectively.

The constitutive equations, strain–displacement and electric field–potential relations can be expressed as $\sigma_{ij} = c_{ijkl}\epsilon_{kl} - e_{kij}E_k$, $D_i = e_{ikl}\epsilon_{kl} + \varepsilon_{ik}E_k$, $\epsilon_{ij} = \frac{1}{2}(u_{i,j} + u_{j,i})$ and $E_i = -\phi_{,i}$, where u_i , ϵ_{ij} , E_i and ϕ denote the displacement in the $i-$direction, strain tensor, electric field in the $i-$direction and electric potential, respectively; and c_{ijkl} , e_{kij} and ε_{ik} denote elastic, piezoelectric and dielectric permittivity tensors, respectively. In the case of ideal elastic materials and rigid dielectrics, $e_{kij} = 0$. It can be also proved that $e_{kij} = 0$ in the case of an isotropic material and the piezoelectric effect does not exist.

The number of independent coefficients depends on the symmetry of the material. Piezoelectric solids used in most practical applications are either crystal with hexagonal symmetry or polarized ceramics. Constitutive equations for a piezoelectric solid having hexagonal symmetry about the z-axis undergoing plane strain deformations ($\epsilon_{yy}, \epsilon_{xy}, \epsilon_{zy} = 0,\ E_y = 0$) can be expressed as

$$\sigma_{xx} = c_{11}\epsilon_{xx} + c_{13}\epsilon_{zz} - e_{31}E_z ; \sigma_{zz} = c_{13}\epsilon_{xx} + c_{33}\epsilon_{zz} - e_{33}E_z ; \sigma_{zx} = 2c_{44}\epsilon_{xz} - e_{15}E_x \quad (1a)$$

$$D_x = 2e_{15}\epsilon_{xz} + \varepsilon_{11}E_x ; D_z = e_{31}\epsilon_{xx} + e_{33}\epsilon_{zz} - \varepsilon_{33}E_z \quad (1b)$$

where c_{11}, c_{13}, c_{33} and c_{44} denote elastic moduli; e_{31}, e_{33} and e_{15} denote piezoelectric constants and ε_{11}, ε_{33} denote the dielectric permittivity.

Based on the micromechanics theory, the piezoelectric composite is a homogeneous orthotropic material. The micromechanical analysis of composite material based on the periodic micro-field approach has a key assumption that the composite material consists of periodically repeated microstructures.

The behavior and properties of the microstructure can be identified by a Representative Volume Element (RVE) which or unit cell which can be isolated and is in a state of equilibrium. The concept of periodic RVE of a piezoelectric composite is schematically represented in Figure 1(b). The RVE has the overall dimensions $a \times b$. If the ratio $a/b = 1$, then the effective properties in both transverse direction of the composite are identical. Effective material coefficients of piezocomposites can be determined by analysis of a periodic RVE with appropriate boundary conditions. The macroscopic constitutive relation, which defines the effective material properties c_{ij}^{eff}, e_{ij}^{eff} and ε_{ij}^{eff} of piezocomposites, is represented in the form given by (1) which can be determined by consideringthe macro-stress and macro-strain $\bar{\sigma}_{ij}$ and $\bar{\varepsilon}_{ij}$ of the composite [8]. According to the micromechanics theory, the macro-stress ($\bar{\sigma}_{ij}$) and macro-strain ($\bar{\epsilon}_{ij}$) of composites can be defined as the volume average in an RVE as $\bar{\sigma}_{ij} = (1/V)\int_{\Omega} \sigma_{ij}\, d\Omega$; $\bar{\epsilon}_{ij} = (1/V)\int_{\Omega} \epsilon_{ij}\, d\Omega$ where V is the volume of an RVE in the domain Ω. Similarly, the average electric displacement and electric field can be defined as $\bar{D}_i = (1/V)\int_{\Omega} D_i\, d\Omega$; $\bar{E}_i = (1/V)\int_{\Omega} E_i\, d\Omega$.

The volume average of field variables can be related to the boundary field variables of the RVE by using Gauss theorem.Therefore, the boundary field variables of the RVE can be used in the calculation of the effective composite material properties and, with the boundary element analysis, the calculation involves discretization of the boundary of a microstructure only. Therefore, the BEM-based micromechanics model is very suitable for performing the calculation of effective composite material properties.

In this study, the mechanically imperfect fiber–matrix bonding is represented by a spring–factor model, which assumes that the displacement jumps along the interface are related to the interface stresses through a spring–factor parameter. The interface compatibility and continuity conditions between the fiber and matrix are given by $\tau_i^f + \tau_i^m = 0$ and $u_i^f - u_i^m = m_i \tau_i^f$ where m_i denotes the spring–factor parameter and superscript "f" and "m" are used to identify the variables associated with the fiber and matrix phases respectively. It is noted that for the limiting case of perfect fiber-matrix bonding: $m \to 0$ and $u_i^f = u_i^m$.

3 Numerical results and discussion

A computer code based on the boundary element method (BEM) and micromechanics theory outlined in the preceding sections has been developed. The accuracy of the present formulation is first verified through the comparison with existing solutions reported in the literature. The results of comparison are not presented for brevity. Excellent agreement between the present solutions and existing solutions is noted confirming the accuracy and validation of the present formulation.

A selected set of numerical solutions is presented in Figure 2–4 to demonstrate the basic features of electroelastic responses within a domain of piezoelectric composite material and the influence of various parameters, e.g., piezoelectric volume fraction and interface bonding condition, to the effective properties of piezoelectric composite material piezocomposite. Piezocomposite in which the elastic polymer matrix is periodically embedded with $BaTiO_3$ piezoelectric material is considered in the numerical example. The material properties of $BaTiO_3$ are: c_{11} = 15.0×10^{10} N/m^2, c_{33} = 14.6×10^{10} N/m^2, c_{13} = 6.6×10^{10} N/m^2, c_{44} = 4.4×10^{10} N/m^2, e_{15} = 11.4 C/m^2, e_{31} = -4.35 C/m^2, e_{33} = 17.5 C/m^2, d_{11} = 9.87×10^{-9} F/m, d_{33} = 11.15×10^{-9} F/m ;and the polymer material properties are: c_{11} = 1.49×10^{10} N/m^2, c_{33} = 4.72×10^{10} N/m^2, c_{13} = 0.52×10^{10} N/m^2, c_{44} = 0.47×10^{10} N/m^2.

The boundary conditions along the outer surfaces of the RVE of piezocomposite are given by

Surface x/a = -0.5: u_x = 0.0 m, t_z = 0.0 Pa, q = 0.0 C/m^2

Surface x/a = 0.5: u_x = 0.0 m, t_z = 0.0 Pa, q = 0.0 C/m^2

Surface z/a = -0.5: u_z = 0.0 m, t_x = 0.0 Pa, ϕ = 0.0

Surface z/a = 0.5: $u_z = \delta$, t_x = 0.0 Pa, q = 0.0 C/m^2

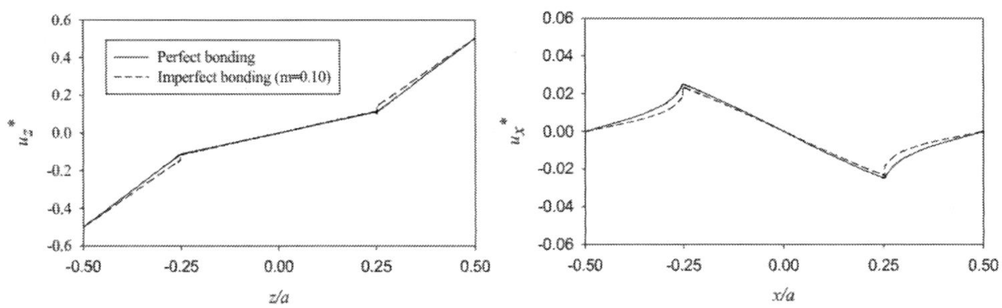

(a) Profile of u_z^* along the z-axis (x = 0.0)(b)Profile of u_x^* along the x-axis (z = 0.0)

Figure 2. Nondimensionaldisplacements (a) u_z^* along the z-axis (x = 0.0); (b) u_x^* along the x-axis (z = 0.0)

Figure2 present profiles of the nondimensional displacements, $u_z^* = u_z/\delta$ along the z-axis (x = 0.0) in Figure 2(a) and $u_x^* = u_x/\delta$ along the x-axis (z = 0.0) in Figure2(b) respectively, where δ denotes the prescribed displacement of the piezocomposite unit cell in z-direction. The fiber-matrix volume ratio considered in the present numerical example is 0.2 ($V_f = 0.2$). The results are presented in Figure2 for both perfect (solid line) and imperfect (dashed lined) interface bonding conditions between piezoelectric fiber/inclusion and polymer matrix. It can be seen in Figure 2(a) that all of the displacement curves are increasing functions (the slope is positive) indicating that the unit cell is extended with tensile strain along the z-axis under prescribed tension displacement in the z direction. It should be observed from the material properties that polymer matrix is less stiff than those of piezoelectric inclusion leading to a higher tensile strain as shown in Figure 2(a).

The corresponding contraction of piezoelectric material along the transverse direction (x-axis) for piezocomposite is presented in Figure 2(b). Due to the prescribed zero displacement in transverse direction of the RVE, the polymer matrix expands while the piezoelectric phase contracts. The influence of imperfect interface bonding condition on displacement profiles of piezocomposite unit cell can also be observed in Figure 2(a) and 2(b). The presence of imperfect interface bonding generally causes a lower strain in the unit cell for both piezoelectric and polymer phases. Moreover, a jump in displacements at the inclusion-matrix interface is noted in both figures.

Figure 3 shows the distribution of nondimensional stress ($\sigma_{zz}^* = \sigma_{zz} a / c_{33}\delta$) and electric displacement ($D_z^* = D_z a / e_{33}\delta$) within the unit cell of piezoelectric composite material in Figure 3(a) and Figure 3(b) respectively. It can be observed in Figure 3(a) and 3(b) that the stress and electric displacement contour are continuous over the domain of piezocomposite. Also, it can be seen in Figure 3(a) that the entire domain of piezoelectric composite material is under tensile stress with high concentration of σ_{zz}^* stress induced over the central region over the piezoelectric inclusion. This is due to the presence of high stiffness piezoelectric inclusion at the center of the unit cell. In addition, due to symmetric loading and geometry of the unit cell, the symmetry of both stress σ^*_{zz} (Figure 3(a)) and electric displacement D^*_z distribution (Figure 3(b)) is noted.

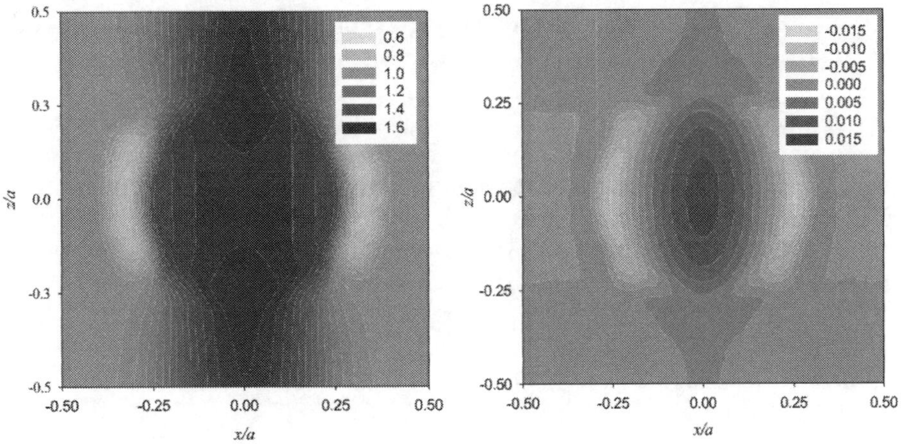

(a) Distribution of σ^*_{zz} (b) Distribution of D^*_z

Figure 3. Distribution of (a) stress σ^*_{zz} and (b) electric displacement D^*_z within the unit cell of piezocomposite

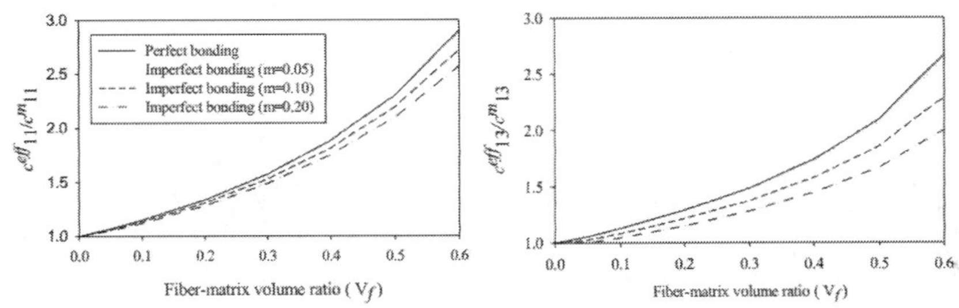

(a) $c_{11}^{eff} / c_{11}^{m}$ (b) $c_{13}^{eff} / c_{13}^{m}$

Figure 4. Variation of effective coefficients of piezoelectric composite (a) $c_{11}^{eff} / c_{11}^{m}$; (b) $c_{13}^{eff} / c_{13}^{m}$

The variation of overall or effective material coefficients of the piezoelectric composite material with the volume fraction of the fiber (V_f) are presented in Figure 4. Effective coefficients of material are calculated for the piezoelectric volume fraction in a range from 0.0 to 0.6 and the spring parameters m is set to be 0.01, 0.1 and 0.2. As expected, the effective coefficients of piezocomposites depend on the fiber volume fraction and the interface bonding condition as can be seen in Figure 4(a) and 4(b). The nondimensional effective material coefficients dimensional $c_{11}^{eff} / c_{11}^{m}$ effective material coefficients (Figure 4(a)) and $c_{13}^{eff} / c_{13}^{m}$ (Figure4(b)) has a unit magnitude at $V_f = 0$. For the given material properties used in the numerical example, $BaTiO_3$ piezoelectric material is stiffer than the polymer matrix. Therefore, when the fiber volume fraction increases, it leads to increasing values of effective composite material coefficients (see Figure4(a) and 4(b)). Nevertheless, it should be noted that the effective coefficients of the composite materials are not linearly increase with increasing piezoelectric volume fraction. Moreover, the numerical results in Figure 4 show that effective composite material properties do not only depend on the material properties of each phase in the composite but also depend on the bonding condition at the interface which can be interpreted from the value of the spring–factor parameter (m). In particular, it can be seen in Figure4(a) and 4(b) that the effective composite material coefficients $c_{11}^{eff} / c_{11}^{m}$ and $c_{13}^{eff} / c_{13}^{m}$ decrease for the case of a weaker interface bonding (higher value of m) between piezoelectric material and polymer matrix.

4 Conclusion

In this paper, a micromechanics model is developed and employed to study the influence of imperfect interface on effective piezoelectric composite material properties. The micromechanics analysis is based on the Boundary Element Method (BEM) together with the periodic micro-field micromechanics theory. The coupled electroelastic responses within the domain of piezoelectric composite material are also presented to portray basic features of piezocomposite unit cell. A mechanical imperfect interface bonding between piezoelectric inclusion and polymer matrix is also taken into consideration in the present analysis. The results from the numerical study show that the effective piezocomposite properties depend on the fiber-matrix volume fraction and interface bonding condition. The variation of the effective coefficient are non-linear with piezoelectric volume fraction.

The entire unit cell is stretched in the direction of prescribed displacement and stress concentration of stress is found over the central region. The distributions of nondimensional stress and electric displacement are continuous and symmetric due to symmetric loading and geometry of the unit cell considered. The numerical results presented in this paper provides a thorough understanding of coupled electro-elastic response within a unit cell and effective materials properties of piezocomposite which are useful for the design of piezocomposite-based electro-mechanical devices for advanced applications.

Acknowledgements

The work presented in this paper is supported by Mahidol University and partially supported by Graduate Studies of Mahidol University Alumni Association.

References

1. N.W.Hagood,A. A.Bent,Proc 43th AIAA/ASME, paper 93–1717(1993).
2. L. J. Nelson,Mater. Sci. Technol., **18**, 1245(2002).
3. Y.Sapsathiarn, T.Senjuntichai,R.K.N.D. Rajapakse,Composites Part B: Engineering, **39**, 1114(2008).
4. Sapsathiarn Y., Senjuntichai T. and Rajapakse R.K.N.D., Composites Part B: Engineering,**43(5)**, 2257 (2012).
5. Sapsathiarn Y., Singh Y. ,Rajapakse R.K.N.D. and Mumford D. ,ActaMechanica, **225**, 2943 (2014)
6. Kar-Gupta R, Venkatesh T. Acta Mater, **55,** 1093 (2007)
7. Berger, H., Kari, S., Gabbert, U., Rodriguez-Ramos, R., Bravo-Castillero, J., Guinovart-Diaz, R., Sabina, F. J., Maugin, G. A., Smart Materials and Structures, **15(2)**, 451 (2006).
8. Poizat, C., and Sester, M.,Comput. Mater. Sci.,**16**, 89–97(1999).
9. Rajapakse, R. K. N. D., and Chen, Y., IEEE Trans. Ultrason. Ferroelect. Freq. Contr., **55**, 1847 (2008).
10. Hammamia, H., Arousb, M., Lagachea, M., Kallel, A.Composites A, **37**, 1 (2006).
11. Nairn, J.A. and Liu, Y.C., Int. J. Solids Structures,**34**, 1255 (1996).
12. Lenci, S. and Menditto, G., Int. J. Solids Structures.,**37**, 4239 (2000).
13. Parton, V.Z., &Kudryavtsev, B.A. *Electromagnetoelasticity: Piezoelectrics and Electrically Conductive Solids*(Gordon and Breach Science,1988).

MATEC Web of Conferences **51**, 01006 (2016)
DOI: 10.1051/matecconf/20165101006

Methodology for Structural Calculation of Gear Teeth with Unconventional Profile

Andrea Chiaramonte Radicella [1], Giulio Di Francesco[2,a], Vincenzo La Battaglia[2] and Stefano Marini[2]

[1] *Toyota Motor Europe, Avenue Du Bourget 60, 1140 Ville de Bruxelles, Belgium*
[2] *Università degli Studi Roma Tre - Dipartimento di Ingegneria, Via della Vasca Navale 79, 00146 Roma, Italy*

Abstract. After having made reference to the structural analysis used in the study of gear wheel teeth, we then move on to the state of the art on the topic. We proceed to identify the boundary conditions used in the structural analysis of unconventional teeth with sides having a profile of an involute of a circle but with different pressure angles in each of the two sides. A procedure for the discretization of traditional teeth and of innovative teeth is presented and compared with the discretization obtained using current software.

1 General details

Gear wheels with teeth having sides with different pressure angles are explored further given their advantages, especially in some applications, over traditional teeth [1] [2]. With the center distance being the same, they are stronger which means that, for a given torque, smaller gears can be used [3] [4] [5].

Hence the importance of analyzing further the structural characteristics of said tooth and, in particular, of identifying a method for automatically subdividing the structure into elements beforehand [6] [7].

We now deal with the issue of the Finite Elements Method mesh of the structure of the tooth having two sides with different involutes.

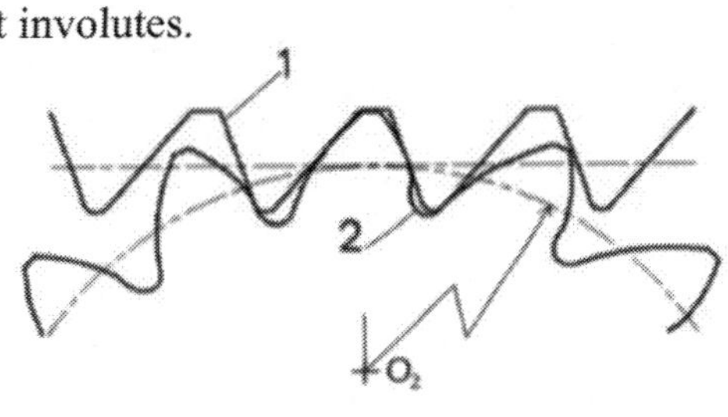

Figure 1. Asymmetric gear engaged.

The Figure 1 shows a gear wheel (2) with a tooth having pressure angles α_{01} and α_{02} which engages with an infinite radius wheel (1), also having the same corresponding angles α_{01} and α_{02}. Contact conditions between active flanks (right for wheel 2 in Fig. 1) are the same we have in traditional teeth.

[a] Corresponding author : difrance@uniroma3.it

2 Current state of the art

It has been pointed out that the use of advanced and current structure modelling software traditional teeth does not generally allow to obtain a mesh that can be used for subsequent in-depth FEM structural calculations.

These generic software applications fail to achieve the necessary accuracy for discretization and for the subsequent structural calculation stage. This drawback occurs in some areas in which a denser mesh would in fact be required; generally, the choice of a suitable tooth mesh density must take into account the stresses occurring during operation.

The meshing system that we refer to is based on the subdivision of the tooth through a network of "meridians" and "parallels": the "meridians" are roughly composed of circular arcs, the "parallels", on the other hand, are curves with a trend similar to that of the tooth profile.

3 Delimiting the tooth area to be discretized

Based on the foregoing considerations, it is therefore necessary to propose a type of discretization that can be used in the subsequent stage of structural analysis of the tooth; to that end, the area of the tooth to be discretized needs to be delimited. We then conduct a preliminary analysis of the entire structure (gear) using a non-dense subdivision and identifying beforehand the areas of the tooth under significant stresses; in this way, we can delimit the area of the tooth to be processed with a more dense discretization in order to achieve a more in-depth structural analysis [8] [9].

The area thus defined, delimited by its external profile, is subdivided into elements using meridians and parallels branching off, perpendicularly to the external boundary profile, from one side of the tooth and extending perpendicularly to the external boundary profile of the opposite side.

Said meridians and parallels subdivide the structure into quadrilateral elements with the four sides being as parallel as possible between them, two by two, and not too dissimilar from one another in size.

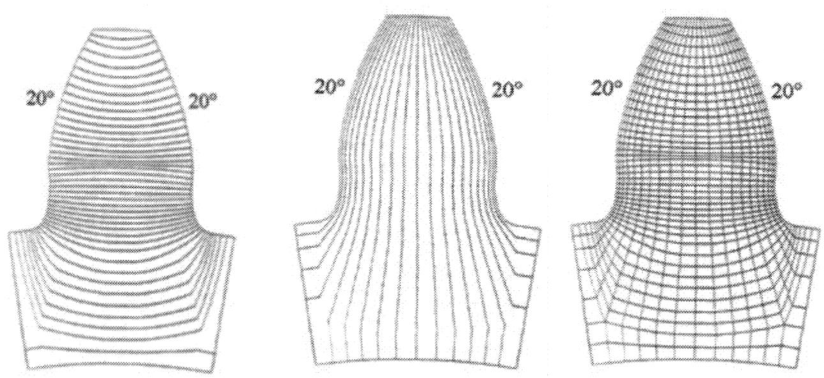

Figure 2. Parallels and meridians discretization

The lower part of the tooth model digs below the base root circle reaching a depth at which the stress is practically insignificant.

Fig. 2 shows a conceptual diagram of what is set out above.

In order to trace the tooth profile, we need to know its defining parameters including those of the tool that we plan to use for cutting the gear tooth.

4 Discretization of traditional teeth

Fig. 2 shows the discretization of a traditional tooth using the parallels and meridians referred to above and, lastly, the fully discretized tooth.

As regards the discretization of the tooth structure, we noted that the use of advanced and current structure modelling software applications for traditional teeth does not generally allow (Fig.3, software Comsol 5.1) to obtain a mesh that can be used for the subsequent in-depth FEM structural analysis.

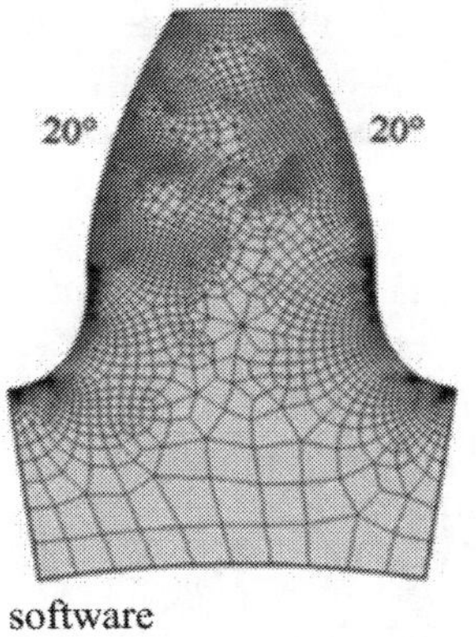

Figure 3. Mesh obtained using commercial software

As these figures clearly show, said generic applications fail to achieve the necessary precision for the discretization and the subsequent structural analysis stage; this drawback occurs especially in some areas in which a more dense mesh would in fact be required; generally, the choice of a suitable tooth mesh density must take into account the stresses occurring during operation.

As stated above, the discretization example obtained using current excellent modelling software applications (software Comsol 5.1) is not satisfactory in terms of the accuracy required in the structural analysis of this specific case of innovative teeth; the areas with a denser and more homogeneous mesh may indeed not fully correspond with those subject to higher tensile stresses.

5 Discretization of innovative teeth

The figure below shows a tooth characterized by α_{01} = 14° and α_{02} = 20° (Fig. 4 left), α_{01} = 20° and α_{02} = 26° (Fig. 4 center) , obtained through the procedure proposed in this study and a corresponding tooth α_{01} = 20° and α_{02} = 26° (Fig.4 right), obtained using the current commercial software.

As the images clearly show, the drawbacks referred to above are all the more evident when we move from the traditional structure (20° – 20°) to the innovative one (20° - 26°), object of this in-depth study.

As stated above, meridians and parallels subdivide the structure into four-sided elements with the four sides being as parallel as possible between them and not too dissimilar from one another in size.

The area to be analyzed more in depth is obtained by joining the nodes whose shift is greater than a value that is considered non-significant.

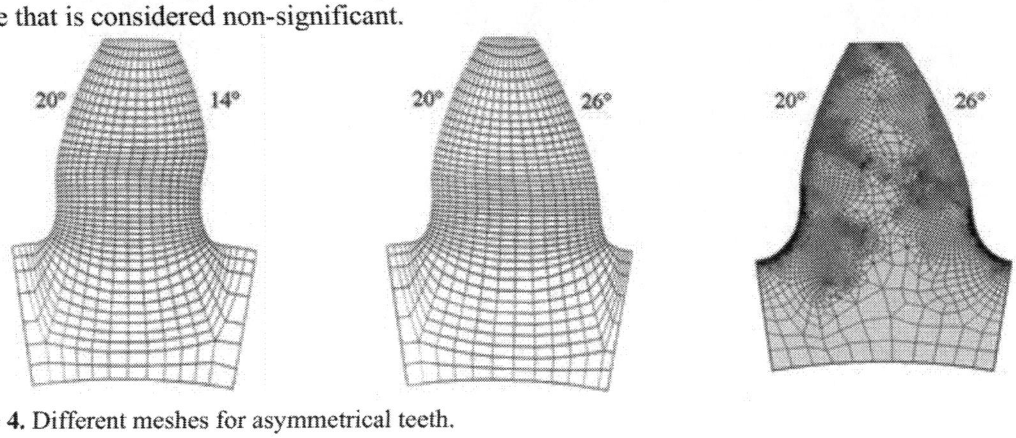

Figure 4. Different meshes for asymmetrical teeth.

The structure of the lower part of the tooth protrudes below the root circle up to a depth at which the stresses are practically irrelevant.

The software under analysis must first allow to obtain an automatic discretization of the structure; once the innovative tooth, using the traditional method, has been identified from a geometrical perspective, (base circle, addendum circle, left involute with pressure angle α_{01} , right involute with pressure angle α_{02} , trochoidal profile fillet, in the case of rounded edge tools, for one side and the other side of the tooth, etc.) we then proceed, based on the multitude of parameters obtained through the calculations, to the automatic discretization of the structure and to the subsequent structural analysis.

The result obtained is shown in the figures above; as you can see, we obtain an appropriate density of elements which gradually move from the outside inwards, compatibly with the stresses arising in the structure when engaging with the tooth of the companion gear. Noteworthy is the orderly increase in the density of the elements in the fillet areas and their obvious branching off at the base of the tooth structure under analysis (non significant stresses).

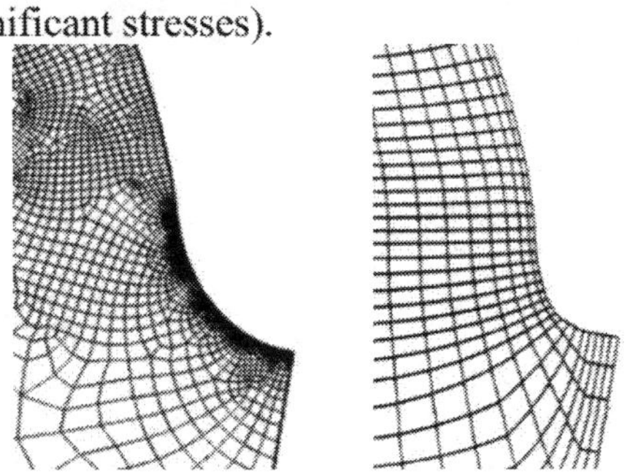

Figure 5. Increase in the density of the elements obtained with Commercial Software (left) and with the software under analysis

The increase of density of the elements obtained with the software under analysis produces elements having similar geometry of the other elements but with smaller dimension. This effect produces regular elements in the whole mesh. The commercial software produces a mesh with irregular elements characterized by different form and geometry. Figure 5 shows the described difference.

In this particular case, as is evident, the "parallels" cannot simply be arcs of a circle, but rather complex curves which are identified, defined and described automatically from time to time.

Similarly to the case of traditional gear discretization, also in this case we identify a certain number of points gradually separated from each other; a meridian is drawn through each point.

Each meridian consists of a curve passing through all points of the same general index placed on each of the parallels examined above.

In reality, it is no easy task to provide a generalized modelling system for this type of toothing.

Thanks to the system adopted, it is therefore possible to select the areas in which to automatically increase the density of discretization as a function of the stress fields during engagement; said fields may vary depending on the application of the gears.

In the case of gears for gear pumps, it is necessary to know the position of the force transferred between the two engaging teeth and to bear in mind that below or above the contact point between the teeth the stress field is represented by a hydrostatic pressure which is high on one side and equal to the suction pressure on the other side; hence, in such case the stress field is more complex and must be handled with great precision

6 Conclusions

These innovative wheels (different pressure angles for the two sides of the tooth: α_{01} and α_{02}) have some considerable advantages. The tooth is more robust: with other factors being equal, they can transmit a higher torque; on equal torque to be transmitted, their overall size is reduced.

The geometrical difference varies as the angles α_{01} and α_{02} of the two sides of the tooth, the left side and the right side, vary [10]. The variation in the second pressure angle, permits to vary the robustness of the tooth and hence to increase transmission performance [11]. However, the quantification of this increase in performance cannot be estimated other than through a prior FEM analysis of the tooth [12]. This analysis is in any case limited by the absence of an automatic and efficient mesh generator for the tooth structure.

The procedure adopted allows to conduct the structural analysis of traditional teeth and of innovative teeth accurately and rapidly: the use of advanced analysis software is made easier by starting with an accurate subdivision into elements.

The procedure under analysis is of a general nature and allows to obtain reliable results with short calculation times [13].

The mathematical model is complex and that depends in particular on the need to have a general tool that can be used for the various applications for which the gears of a given assembly are intended: transmission gears, gears for gear pumps, etc.

References

1. A. Sanders, D. R. Houser, A. Kahraman, J. Harianto, S. Shon, "An experimental investigation of the effect of tooth asymmetry and tooth root shape on root stresses and single tooth bending fatigue life of gear teeth." Proceedings of the ASME Design Engineering Technical Conference, 8, pp. 297-305. (2011).
2. R. Thirumurugan, C.C.C. Deepak, "Design and analysis of asymmetrical spur gear drive for automobile gear box application", Applied Mechanics and Materials, 592-594, pp. 2277-2281, (2014).
3. A. Kapelevich, "Direct gear design for asymmetric tooth gears" Mechanisms and Machine Science, 34, pp. 117-143, (2016)
4. S. Ekwaro-Osire, I. Durukan, F.M. Alemayehu, "Experimental and probabilistic analysis of asymmetric gear tooth. Conference Proceedings of the Society for Experimental Mechanics Series, 5, pp. 207-212, (2011)
5. G. Di Francesco, S. Marini, "Structural Analysis of Asymmetrical Teeth: Reduction of Size and Weight" *Gear Technology*, Randall Publishing inc. Sept- Oct. 1997, Elk Grove Village IL. U.S.A.(1997)
6. G. Di Francesco, S. Marini, "Asymmetric teeth: bending stress calculation", Gear Technology (Mar-Apr 2007), Randall Publishing inc., Elk Grove Village, IL USA (2007)
7. G. Di Francesco, S. Marini, "Tensioni di flessione di dentature asimmetriche e dimensionamento", Organi di Trasmissione (April 2005), Tecniche Nuove, Milan (2005)
8. G. Di Francesco, S. Marini, "Asymmetrical gear wheels: automatized procedure for the design", International Conference on Gears, Munich 2005
9. G. Di Francesco, S. Marini, "Ruote a denti asimmetrici: procedura automatizzata per il dimensionamento", Organi di Trasmissione (February 2008), Tecniche Nuove, Milan (2008)
10. G. Di Francesco, S. Marini, G. Grasso, C. Clienti, "Asymmetric gear wheel thermographic analysis", Colloque Photomecanique, Étude du comportement des matèriaux et des structures, Albi (2004)
11. G. Di Francesco, S. Marini, "Asymmetrical gear wheel design: determining the bending stress in relation to the degree of asymmetry", TMT 2004, Trends in the development of machinery and associated technology, Neum (2004)
12. G. Di Francesco, S. Marini, "Dentature asimmetriche: evoluzione e procedura analitica per il dimensionamento" *Organi di Trasmissione*, Tecniche Nuove, Milano, (1998)
13. G. Di Francesco, S. Marini, "Adattamento della procedura ISO C alle dentature asimmetriche", XXXIV Convegno AIAS, Milan (2005)

MATEC Web of Conferences **51**, 01007 (2016)
DOI: 10.1051/matecconf/20165101007

Calculation of the Stiffness in the Roll Tensioning of the Circular Saw Blade

Vo Tung Linh [a], and Zang Yong

School of Mechanical Engineering, University of Science and Technology Beijing, 100083, China

Abstract: The circular blade has been widely used in some projects such as cutting stone, wood and other projects. Owing to its particularity and wide use, it has an important position in cutting industry. Roll tensioning is considered as an effective method which can be used to improve the stiffness and performance of the circular saw blade. The effect of rolling position and width in the roll tensioning is obvious. In this paper the calculation of the maximum stiffness at different rolling position and width in the rolling were carried out through the finite -element. The results show that three ideal points are found. And when rolling position is ∅950mm and rolling width is 20mm, the maximum stiffness of the circular saw blade whose minimum deformation is 0.028mm is found. The roll tensioning can increase the stiffness of the saw blade. It will provide a theoretical basis and guidance for the actual production.

1 Introduction

The circular saw blade has been widely used in some projects such as cutting stone, wood and other projects. Owing to its particularity and wide use, it has an important position in cutting industry. The circular saw blade is a typical rotating sheet. Usually the thickness diameter ratio is very small. So the stiffness and stability of the saw blade are poor. Higher dynamical stiffness and stability are required to ensure the safety of processing, cutting precision and outturn rate. It has become a major topic concerned and researched by some scholars at home and abroad[1-3] .

At present, some researches show that a suitable distribution of residual stresses in the prestressed saw blade is better to improve the stability of the blade. It is considered as the most economical and effective method. This method is called as roll tensioning. The roll tensioning is an important technology in the processing of the saw blade. The local plastic deformation which changes the distribution of residual stresses in the saw blade occurs in rolling and hammering. Among these methods, the roll tensioning is recognized as an effective method for tensioning. It has a wide application in the industry. The blade tensioning process was studied by many scholars at home and abroad. R.G.Parker and C.D.Mote.Jr[4] proposed a tensioning method. The tensioning load was perpendicularly put on the edge of the hole. Under the different size of the pressure, number of holes and shape of the hole, the natural frequency of the circular saw blade was studied after rolling. The results showed that it can increase the natural frequency and avoid the excessive adaptation. G.S.Schajer and K.J.Kishimoto[5,6] proposed a heating tensioning method. With the nonuniform heating on the circular saw blade, the temporary tensioning is generated. It can eliminate the dish

[a] Corresponding author : tunglinh77@gmail.com

phenomenon caused by excessive tensioning. N. Nicotetti[7] comprehensively studied roll tensioning by experimental and finite element method. Tangential and normal gages were placed on the saw blade. Distribution of strain is obtained. Y.M.Stakhiev[8] studied the effect of rolling diameter, roller shape and rolling pressure on the tensioning. Stakhiev[9] believed that increasing the tensioning load, tensioning decreases at the initial stage and increases gradually. Y.M.Stakhiev[10] introduced a tensioning machine invented by V. Dunaeva, V. Dunaev, Y. Stakhiev. Its principle is that the pressure is put on the circular saw blade to obtain tensioning through a vibrator. At the same time the vibration sensor is used to detect vibration. G. S. Schajer[11] made a deep research on the roll tensioning, and proposed some methods to improve the rolling defects. And the rolling position and width has a very large impact effect on tensioning processing. But there is little research on rolling position and width. In this paper the calculation of the maximum stiffness at different rolling position and width in the rolling were carried out through the finite - element. It will provide a theoretical basis and guidance for the actual production.

2 Material model of circular saw blade

In this paper, the material of the saw blade is 65Mn. According to the relevant information, the mechanical properties of 65Mn are obtained as seen in Table 1.

Table 1. Mechanical properties of 65Mn.

Material		65Mn
Density（kg/m^3）	ρ	7810
Elastic modulus (N/m^2)	E	2.0E+11
Poisson's ratio	v	0.3
Shear modulus (GPa)	G	82
yield limit (MPa)	σ	800

Owing to the plastic deformation generated in the rolling zone of the saw blade, the stress intensification of the material should be considered in this research. But the strain produced by rolling is very small. Therefore, in this paper the bilinear kinematic hardening criterion (Bkin) can be used to simulate rolling. The stress-strain curve of the material is shown in Figure 1. The yield limit of 65Mn was 800Mpa, and the shear modulus of 65Mn was 82Gpa.

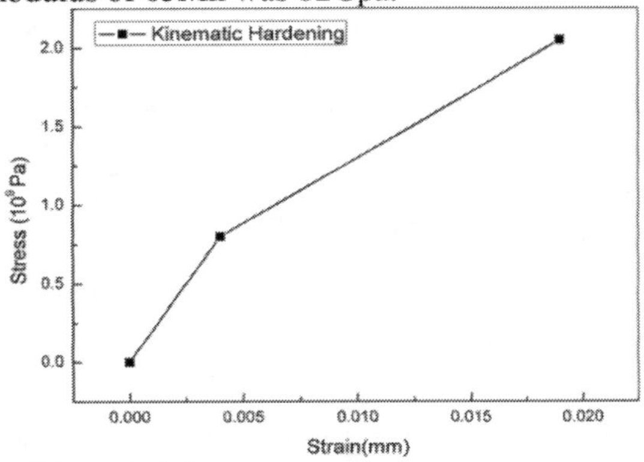

Figure 1. The stress-strain curve of the material.

3 Calculation of the stiffness of the circular saw blade in roll tensioning

3.1 Mesh model

Mesh model of the saw blade is shown in Fig.2. In the mesh model the hexahedral element is adopted. First, planar element is used to divide flat mesh of the saw blade. Then the flat mesh is extruded to generate body mesh. The mesh sensitivity is fine. From Fig.2 it can be seen that the mesh in the edge and centre of the saw blade is dense. There are 4125 units and 8929 nodes.

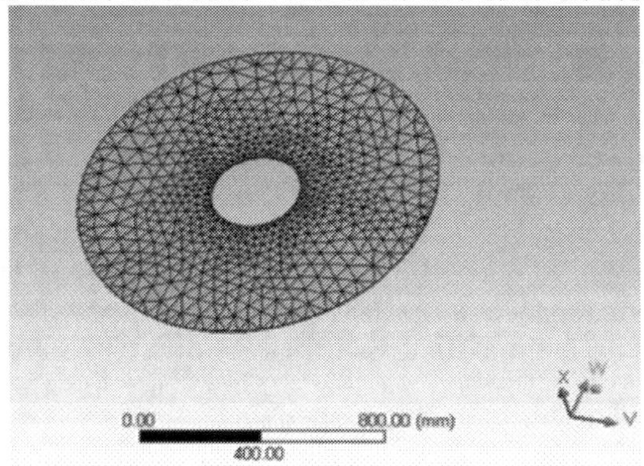

Figure 2. Mesh model of Circular saw blade

3.2 Results and discussion

In order to investigate the changes of stiffness after rolling, the explicit dynamic FEM is adopted to analyze rolling model. Through this method the distribution of residual stress after rolling can be accurately obtained. As the initial stress load, the residual stress is introduced in subsequent analysis. Because many various conditions including different rolling position and width are took into account in subsequent analysis. It is time consuming. So the algorithm is not appropriate for this analysis. In order to simplify the calculation, when the effect of rolling position and width on the stiffness of the saw blade are analyzed, rolling pressure is directly load on the saw blade to generate the plastic deformation and the residual stress is introduced. Then radial and tangential loads are put on the circular saw blade to investigate the radial deformation of the circular saw blade. The stiffness can be obtained through calculation. In order to get the changes of the stiffness, three load steps can be done in ANSYS. In the first load step the rolling pressure is directly load on the saw blade to generate the plastic deformation. In the second load step, rolling pressure is removed to solve the distribution of residual stress after rolling. The radial and tangential loads are put on the circular saw blade to investigate the radial deformation of the circular saw blade and calculate the changes of the stiffness in the third load step.

Different plant uses different rolling width. According to the investigation, the rolling width varies from 5mm to 20mm. Rolling position varies from ∅600mm to ∅1100mm. Using the finite element ANSYS software, the optimization of rolling position and width of circular saw blade is carried out under 900MPa pressure. The maximum radial deformation of the saw blade which is the optimization objective is set at the minimum. Fig.3 shows the result of the optimization. From Fig.3 it can be seen that three optimization points are found. And when rolling position is ∅950mm and rolling width is 20mm, the maximum stiffness of the circular saw blade whose minimum deformation is 0.028mm is found.

Table of Schematic D4: Optimization

	A	B	C	D
1		P2 - XYPlane.D6	P3 - XYPlane.L5	P1 - Total Deformation Maximum (mm)
2	Optimization Study			
3	Objective	No Objective	No Objective	Minimize
4	Target Value		5	
5	Importance	Default	Default	Default
6	GDO Sample Set 1			
7	Candidate A	950	20	0.028436
8	Candidate B	910	20	0.028464
9	Candidate C	870	20	0.028531

Figure 3. Result of the optimization.

Fig.4 shows the distribution of feasible point. In Fig.4 the horizontal coordinate shows the radial deformation of the saw blade. And the longitudinal coordinate show the rolling position. From Fig.4 it can be seen that when rolling position is ∅950mm, the radial deformation of the saw blade is minimal. The minimum deformation is 0.028mm. When rolling position is ∅700mm, the radial deformation of the saw blade is maximal. The maximum deformation is 0.03mm.

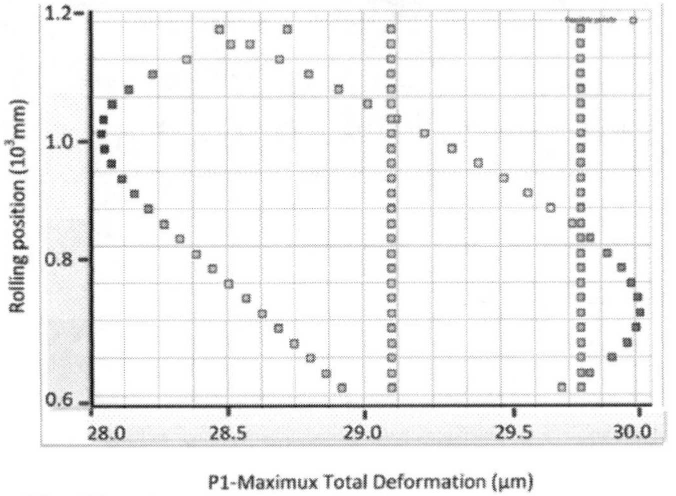

Figure 4. The distribution of feasible point.

Fig.5 shows 3D response of the deformation. In Fig.5 the X-axis shows the rolling position, the Y-axis shows the rolling width and the Z-axis shows the radial deformation of the saw blade. From Fig.5 it can be seen that with the increasing rolling position, the deformation of the circular saw blade increases and then decreases. With increasing of the rolling width, the deformation of the circular saw blade decreases. The results of the calculation will be used for subsequent experiments.

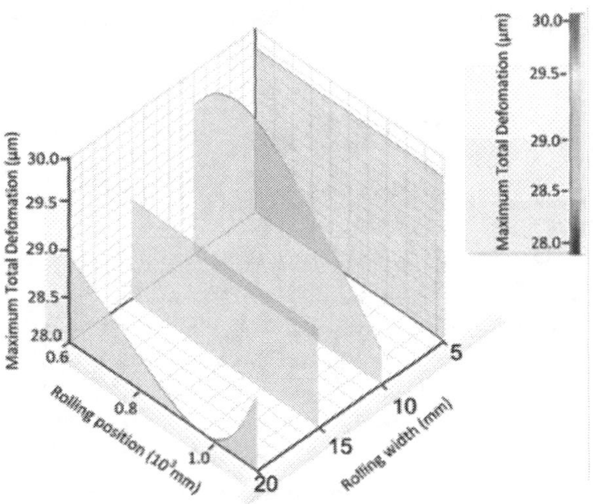

Figure 5.. 3D response of the deformation.

Generally rolling is carried out to increase the stiffness of the circular saw blade. In order to testify the correctness of the above optimization, the radial deformation of the circular saw blade without rolling is analyzed. Fig.6 shows the deformation of the saw blade without rolling. From Fig.6 it can be seen that the maximum deformation of the saw blade is 0.03mm. The maximum deformation of the saw blade without rolling is larger than the maximum deformation of the saw blade with rolling. It indicates the stiffness of the saw blade with rolling is larger than the stiffness of the saw blade without rolling. The roll tensioning can increase the stiffness of the saw blade.

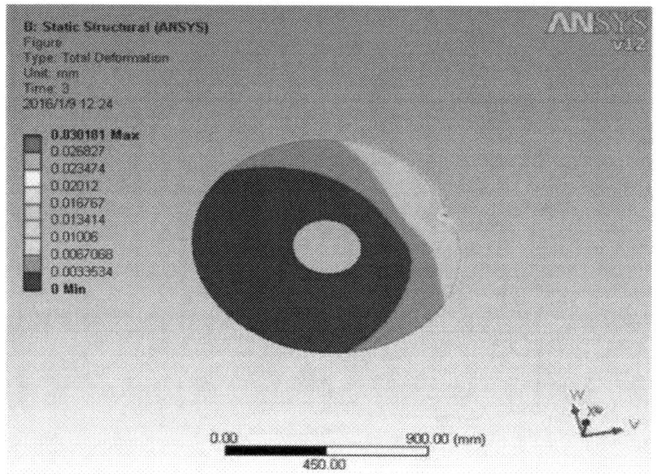

Figure 6. The deformation of the saw blade without rolling.

4 Conclusions

In this paper the calculation of the stiffness of the circular saw blade in the rolling were carried out through the finite-element. The results show that three appropriate points are found. And when rolling position is ∅950mm and rolling width is 20mm, the maximum deformation of the saw blade is 0.028mm. The maximum deformation of the saw blade without rolling is 0.03mm. The maximum deformation of the saw blade without rolling is larger than the maximum deformation of the saw blade with rolling. The appropriate roll tensioning can increase the stiffness of the saw blade.

References

1. B. Gospodaricˇ • B. Bucˇar • G. Fajdiga. Active vibration control of circular saw blades. Eur. J. Wood Prod, **73**:151~158 (2015)
2. Wen-Tung Chang & Chih-Hsien Su & Dong-Xie Guo & Geo-Ry Tang & Fang-Jung Shiou. Automated optical inspection for the runout tolerance of circular saw blades. Int J Adv Manuf Technol, **66**:565~582 (2013)
3. Mats Ekevad , Luís Cristóvão & Birger Marklund. Wear of teeth of circular saw blades. Wood Material Science and Engineering, **7**: 150~153 (2012)
4. R. G. Parker and C. D. Mote. Jr. Asymmetric tensioning of circular saws.Holz als Roh-und Werkstoff, **47**:143~151 (1989)
5. G. S. Schajer. Understanding saw tensioning. Holz als Roh- and Werkstoff, **42**:425~430 (1984)
6. G. S. Schajer .North American techniques for circular saw tensioning and leveling: Practical adjustment methods. Holz als Roh- and Werkstoff, **50**:153~157 (1992)
7. N. Nicoletti, D. Fendeleur, L. Nilly, M. Renner. Using Finite Elements to model circular saw roll tensioning. Holz als Roh- and Werkstoff, **54**:99~104 (1996)
8. Y.M.Stakhiev. Research on circular saws roll tensioning in Russia: Practical adjustment methods. Holz als Roh- and Werkstoff, **57**:57~62 (1999)
9. Stakhiev .Y . Cavity oscillation and equilibrium stability loss of flat thin disk irregular heating[J]. Woodworking machine and tool, **2**:27~30 (1993)
10. Y.M.Stakhiev. Y.M.Stakhiev. Coordination of saw blade tensioning with rotation speed: myth or reality? Holz Roh Werkst, **62**:313~315 (2004)
11. Schajer. G. S, KISHIMOTO.K. J . High -speed circular sawing using temporary tensioning[J]. Holz als Rol-und werkstoff, **54**(4) :361~367 (1996)

MATEC Web of Conferences **51**, 01008 (2016)
DOI: 10.1051/matecconf/20165101008

A Multi-Leveled Coordinate for Multibody Mechanics

Fan Shen[1], Siyuan Rong[1,a], Jiafu Liu[2], Naigang Cui[1]

[1]*Department of Space Engineering, Harbin Institute of Technology, Harbin, Heilongjiang, China*
[2]*College of Aerospace Engineering, Shenyang Aerospace University, Shenyang, Liaoning China*

Abstract. For a proper way of modelling and simulation, as well as the precise and completeness, and a new multi-level method named homogeneous tensor coordinate(HTC) was developed by improving the absolute nodal coordinate formulation(ANCF) and the homogeneous coordinates(HC). In HTC, a new form of 2×2 matrix was employed instead of angles, when the capacity of expression and calculation was increased at the same time. HTC has good performance in traditional rigid body mechanics, which especially suitable for multibody modelling. Therefore, the robotic engineering may benefit from this finding, and a new formulation for scientific research may also affect the physics although some mathematical problems are raised in matrix theory.

1 Introduction

As a cornerstone of modern science, classical mechanics was a mature and reliable theory[1]. However, the accuracy of classical mechanics is not able to meet the increasing demands of modern research or engineering. As a representative of the analytical modelling, floating frame method does not have a sufficiently general form and accuracy for questions as highly-flexible problem for the angle components[2]. And for some particularly difficult problems there is even not a unified theory, for that there is various limited models for turbulence. The accuracy and completeness can be concluded to a basic question in science: how to describe an object in proper approximation. Multi-order method could be one of the answers, which means, use a specific value in a specific scale.

The inaccuracy may be caused by the incompleteness. The problem will be confusion while using a basic tool, but may be simple if we use a complicated modelling method. It can be a selectable way to solve a question precisely by changing our basic theories.

In most differential equation, multi-orders are used to describe the characteristics of different objects in various scale. And in the history of maths, physics and engineering, there are many expressions composed with multiple components, such as plural and quaternion[3], which use linear superposition of the base band to represent, when vector[4] and tensor[5] are represented by a matrix form. The control equation[2] is also a common form of multi-leveled applications in particular, which is not composed with multiple dimension, but with different orders of a single dimension. Modern control theory constituted in the equations above, and reached a high maturity not only in forms but also in theory. Taylor polynomial expansion[6] was always used in computation with nonlinear or uncertain values, which has different substrates and will represents approximated value more accurately.

[a] Corresponding author : rage2004@163.com

More and more approximated linear formulations were derived and used in multi-orders, for that there is enough theoretical basis and application environments already. Therefore, an advanced maths method for describing multiple levels was proposed. Meanwhile, a new coordinate in matrix form was raised for the practice of this method by referring the Homogeneous coordinates (HC)[7] and Absolute Coordinate Nodal Formulation (ANCF)[5].

ANCF is a method for FEM raised by Prof A.A.Shabana[2,8,9], which uses all the first and second order component values of a node in 3-D coordinate. It abandoned the classical angle component θ, because the angles are actually functions resulted by x, y, z in nature. ANCF is an accurate and complete method which uses 12 components to describe a node, and has solved many practical problems[10,11,6]. However, the length of array and the unusual rotational description made it difficult to describe the coordinate, there should be a more simple form instead.

HC is an excellent tool with completeness August Ferdinand Möbius created[12], which could use matrix multiplication instead of vector. Due to the special definition and operations, it is not widely used. However, its idea was inherited by many other methods. Such as discrete time transfer matrix method, Xiaoting RUI uses linearized transfer matrix for multibody system dynamics, although the result still have the problem of premature linearized and was not satisfied enough.

Thus, to merge the advantages in HC and ANCF, an improved coordinate is represented in this paper. The new method uses the absolute precise tensor components in ANCF and abandoned the angle description, when a two-dimensional matrix is constructed according to the idea of HC, for easier derivation and operation. The method was successfully applied in rigid body mechanics, which has a more complete and concise physical description and a more accurate mathematical form.

2 A Multi-Levelled Expression

To give out a multi-leveled expression, we need an array could be written as

$$S=\left\{s,\frac{\partial s}{\partial x},\frac{\partial s}{\partial y},\cdots,\frac{\partial^2 s}{\partial x^2},\frac{\partial^2 s}{\partial x\partial y},\frac{\partial^2 s}{\partial y^2},\cdots\right\} \tag{1}$$

There is a gradient operator

$$\nabla=\begin{pmatrix}\frac{\partial}{\partial x} & \frac{\partial}{\partial y} & \frac{\partial}{\partial z}\end{pmatrix} \tag{2}$$

Then we define a potential function ϕ , and $\nabla\phi=(\underline{x},\underline{y},\underline{z})$, according to the homogeneous coordinates , set the zero-order value to be 1, we obtain

$$S=\{1,\nabla\phi,\nabla\nabla\phi,\nabla\nabla\nabla\phi,\cdots\} \tag{3}$$

The complete multi-levelled equation should be

$$f(S_1,S_2,\cdots)=0 \tag{4}$$

For not all orders could be considered, we need to determine a appropriate number of orders to use. Particular problems and mathematical algorithms should be examined to give out the answer.

Although the operation of this method is unclear , for simple operation, its significance is obvious. In basis i , we have $s_i^j=\nabla_i\phi_j$, $s_{ii}^j=\nabla_i\nabla_i\phi_j$, then the complete form should be

$$S_i^j=\{1,s_i^j,s_{ii}^j,\cdots\} \tag{5}$$

$S(n)$ is the n th-order component in S, such as $S(2)=s_{iij}$, according to the algorithm of tensor, easily we have

$$S_i^j(n)S_j^k(0)=S_i^j(n),(n=1,2,3,\cdots) \tag{6}$$

$$S_i^j(n+1)S_j^k(1)=S_i^j(n),(n=1,2,3,\cdots) \tag{7}$$

$$S_i^j(n+2)S_j^k(2)=S_{ii}^k(n),(n=1,2,3,\cdots) \tag{8}$$

Transfer rules for any order could be derived from the algorithm of tensor, but for higher order there is no clear definition and value to use out of equation(6), (7) and (8), deeper theory will be discussed in further study.

3 The Homogeneous Tensor Coordinate

A brief form is needed as the ideal form had been given, so we take the order 0, 1 and 2 for construction.

A space Cartesian coordinate system was defined by x_i y_i z_i with the subscribe i , consider the potential function and gradient operator, with the tensor-like expression, the matrix could be defined as

$$\mathbf{T}^j=\phi_j \tag{9}$$

$$\mathbf{T}_i^j=\nabla_i\phi_j=\begin{pmatrix}\dfrac{\partial\phi_j}{\partial x_i}\\ \dfrac{\partial\phi_j}{\partial y_i}\\ \dfrac{\partial\phi_j}{\partial z_i}\end{pmatrix},\mathbf{T}_{ii}^j=\nabla_i\nabla_i^T\phi_j=\begin{pmatrix}\dfrac{\partial^2\phi_j}{\partial x_i\partial x_i} & \dfrac{\partial^2\phi_j}{\partial x_i\partial y_i} & \dfrac{\partial^2\phi_j}{\partial x_i\partial z_i}\\ \dfrac{\partial^2\phi_j}{\partial y_i\partial x_i} & \dfrac{\partial^2\phi_j}{\partial y_i\partial y_i} & \dfrac{\partial^2\phi_j}{\partial y_i\partial z_i}\\ \dfrac{\partial^2\phi_j}{\partial z_i\partial x_i} & \dfrac{\partial^2\phi_j}{\partial z_i\partial y_i} & \dfrac{\partial^2\phi_j}{\partial z_i\partial z_i}\end{pmatrix} \tag{10}$$

Combine 3 matrix above, appropriately, then we have a 4 × 4 matrix. With the gradient symbol, we defined operator L_i as

$$\mathrm{L}_i=\begin{pmatrix}1 & 0\\ \nabla_i & \nabla_i\nabla_i^T\end{pmatrix} \tag{11}$$

This is an extension of homogeneous coordinates. Let the potential function to be 1, which places at its upper left corner, then the addition of matrix is changed to multiplication like HC, with a series of similar expression and computational advantages. There are 12 remaining non-zero value, which correspond to the components of ANCF, with a second-order accuracy in one node.

With respect to the authors of ANCF and HC, in this paper we call it homogeneous tensor coordinate(HTC), whose operations and meanings will be given later in the paper.

HTC follow all the matrix algorithms with a series of operational advantages. Defining $\mathbf{R}_i^j=\mathrm{L}_i\phi_j$, the algorithms is presented as below.

1. The addition and subtraction of HTC. When the matrix have the same subscript it is easy to known by block matrix algorithms that

$$\mathbf{R}_i^j\pm\mathbf{R}_i^k=\begin{pmatrix}\mathbf{T}^j & 0\\ \mathbf{T}_i^j & \mathbf{T}_{ii}^j\end{pmatrix}+\begin{pmatrix}\mathbf{T}^k & 0\\ \mathbf{T}_i^k & \mathbf{T}_{ii}^k\end{pmatrix}=\begin{pmatrix}\mathbf{T}^j\pm\mathbf{T}^k & 0\\ \mathbf{T}_i^j\pm\mathbf{T}_i^k & \mathbf{T}_{ii}^j\pm\mathbf{T}_{ii}^k\end{pmatrix} \tag{12}$$

Obviously, these two operations are allowed, although the significance of addition remains uncertain when T^i and T^k are all non-zero.

2.The multiplication and division of HTC. Since the HTC is the matrix form, it will be directionality. Consulting the vector multiplication can be divided into ordinary multiplication, dot product and cross product. Multiplication operate could be

$$\mathbf{R}_i^j\mathbf{R}_j^k=\begin{pmatrix}\mathbf{T}^j & 0\\ \mathbf{T}_i^j & \mathbf{T}_{ii}^j\end{pmatrix}\begin{pmatrix}\mathbf{T}^k & 0\\ \mathbf{T}_i^k & \mathbf{T}_{ii}^k\end{pmatrix}=\begin{pmatrix}\mathbf{T}^j\mathbf{T}^k & 0\\ \mathbf{T}_i^j\mathbf{T}^k+\mathbf{T}_{ii}^j\mathbf{T}_j^k & \mathbf{T}_{ii}^j\mathbf{T}_{jj}^k\end{pmatrix} \tag{13}$$

$$\mathbf{R}_i^j\cdot\mathbf{R}_i^k=\mathbf{R}_i^{jT}\mathbf{R}_i^k=\begin{pmatrix}\mathbf{T}^{jT} & \mathbf{T}_i^{jT}\\ 0 & \mathbf{T}_{ii}^{jT}\end{pmatrix}\begin{pmatrix}\mathbf{T}^k & 0\\ \mathbf{T}_i^k & \mathbf{T}_{ii}^k\end{pmatrix}=\begin{pmatrix}\mathbf{T}^{jT}\mathbf{T}^k+\mathbf{T}_i^j\cdot\mathbf{T}_i^k & \mathbf{T}_i^{jT}\mathbf{T}_{ii}^k\\ \mathbf{T}_{ii}^{jT}\mathbf{T}_j^k & \mathbf{T}_{ii}^{jT}\mathbf{T}_{jj}^k\end{pmatrix} \tag{14}$$

Cross product is only for an array, so it can be

$$\mathbf{R}_i^j\times\mathbf{R}_i^k=\begin{pmatrix}1 & 0\\ \mathbf{T}_i^j & 0\end{pmatrix}\times\begin{pmatrix}1 & 0\\ \mathbf{T}_i^j & 0\end{pmatrix}=\begin{pmatrix}1 & 0\\ 0 & \tilde{\mathbf{T}}_i^j\end{pmatrix}\begin{pmatrix}1 & 0\\ \mathbf{T}_i^j & 0\end{pmatrix} \tag{15}$$

where

$$\tilde{\mathbf{T}}_i^j T_i^k=\mathbf{T}_i^j\times\mathbf{T}_i^k \tag{16}$$

Division is the inverse operation of multiplication, which can consult the block matrix.

3.The differential and Integral of HTC. According to the differential law of matrix and tensor, we could differentiate each element in the HTC when the matrix needs to be differentiated.

$$\frac{\partial\mathbf{R}_i^j}{\partial t}=\begin{pmatrix}\dfrac{\partial\mathbf{T}^j}{\partial t} & 0\\ \dfrac{\partial\mathbf{T}_i^j}{\partial t} & \dfrac{\partial\mathbf{T}_{ii}^j}{\partial t}\end{pmatrix} \tag{17}$$

We can see from the paragraph that HTC can be used to represent a multi-levelled information, whose low-level can provide an easy form for the theoretical equation, when the high-level provides a sufficient high-precision. This indicated would give rise to two problems: 1. Commonly used first-order vector become a second-order matrix, third-order operation may also appear in using HTC, when there is still no higher-dimensional matrix theory yet. a new requirements of matrix analysis theory is put forward. 2. It is predictable that a coupling effect will occur in the state variables which will be 2-D matrix instead of 1-D array. It is also a problem had never be dealt with.

However, we have to face these problems rather than to avoid. Compared to the classic expression, this multi-levelled expression is more in line with the objective world. Since the invention of calculus, it is the first time a clear representation defined for the hierarchical data. It is worth using such a matrix for the need of more accurate research. And in nature, there is no variable non-independent, the additional equation to keep the balance between translational and rotational movement may be pass off with the proposal of HTC.

4 HTC for Rigid Body Mechanics

4.1 Expression in mathematics

4.1.1 Basic form

Let $x_i=\dfrac{\partial\phi}{\partial x}$, the coordinate for a node could be written as

$$\mathbf{R}_o^i=\begin{pmatrix}1 & 0\\ \mathbf{x}_o^i & \boldsymbol{\theta}_o^i\end{pmatrix} \tag{18}$$

where

$$\mathbf{x}_o^i = \begin{pmatrix} x_i \\ y_i \\ z_i \end{pmatrix}, \quad \boldsymbol{\theta}_o^i = \begin{pmatrix} \frac{\partial x_i}{\partial x_o} & \frac{\partial x_i}{\partial y_o} & \frac{\partial x_i}{\partial z_o} \\ \frac{\partial y_i}{\partial x_o} & \frac{\partial y_i}{\partial y_o} & \frac{\partial y_i}{\partial z_o} \\ \frac{\partial z_i}{\partial x_o} & \frac{\partial z_i}{\partial y_o} & \frac{\partial z_i}{\partial z_o} \end{pmatrix} \tag{19}$$

According to the definition and algorithms of HC, the translational matrix can be written as $\mathbf{X}_o^i$. Similarly, consulting the definition of ANCF, the angle definition can be given as $\boldsymbol{\Theta}_o^i$

$$\mathbf{X}_o^i = \begin{pmatrix} 1 & 0 \\ \mathbf{x}_o^i & \mathbf{I} \end{pmatrix}, \quad \boldsymbol{\Theta}_o^i = \begin{pmatrix} 1 & 0 \\ \mathbf{o} & \boldsymbol{\theta}_o^i \end{pmatrix} \tag{20}$$

4.1.2 Equivalence for small deformation

When only small spatial displacement occur, We can decompose the matrix into $\mathbf{R}_o^i = \hat{\mathbf{R}}_o^i + \tilde{\mathbf{R}}_o^i$ as

$$\hat{\mathbf{R}}_o^i = \begin{pmatrix} 1 & 0 \\ \mathbf{o} & \hat{\boldsymbol{\theta}}_o^i \end{pmatrix}, \quad \tilde{\mathbf{R}}_o^i = \begin{pmatrix} 1 & 0 \\ \mathbf{x}_i & \tilde{\boldsymbol{\theta}}_o^i \end{pmatrix} \tag{21}$$

where

$$\hat{\boldsymbol{\theta}}_o^i = \begin{pmatrix} \frac{\partial x_i}{\partial x_o} & \frac{1}{2}\left(\frac{\partial x_i}{\partial y_o} + \frac{\partial y_i}{\partial x_o}\right) & \frac{1}{2}\left(\frac{\partial x_i}{\partial z_o} + \frac{\partial z_i}{\partial x_o}\right) \\ \frac{1}{2}\left(\frac{\partial y_i}{\partial x_o} + \frac{\partial x_i}{\partial y_o}\right) & \frac{\partial y_i}{\partial y_o} & \frac{1}{2}\left(\frac{\partial y_i}{\partial z_o} + \frac{\partial z_i}{\partial y_o}\right) \\ \frac{1}{2}\left(\frac{\partial x_i}{\partial z_o} + \frac{\partial z_i}{\partial x_o}\right) & \frac{1}{2}\left(\frac{\partial y_i}{\partial z_o} + \frac{\partial z_i}{\partial y_o}\right) & \frac{\partial z_i}{\partial z_o} \end{pmatrix} \tag{22}$$

$$\tilde{\boldsymbol{\theta}}_o^i = \begin{pmatrix} 0 & \frac{1}{2}\left(\frac{\partial x_i}{\partial y_o} - \frac{\partial y_i}{\partial x_o}\right) & \frac{1}{2}\left(\frac{\partial x_i}{\partial z_o} - \frac{\partial z_i}{\partial x_o}\right) \\ \frac{1}{2}\left(\frac{\partial y_i}{\partial x_o} - \frac{\partial x_i}{\partial y_o}\right) & 0 & \frac{1}{2}\left(\frac{\partial y_i}{\partial z_o} - \frac{\partial z_i}{\partial y_o}\right) \\ \frac{1}{2}\left(\frac{\partial x_i}{\partial z_o} - \frac{\partial z_i}{\partial x_o}\right) & \frac{1}{2}\left(\frac{\partial y_i}{\partial z_o} - \frac{\partial z_i}{\partial y_o}\right) & 0 \end{pmatrix} \tag{23}$$

$\hat{\mathbf{R}}_o^i$ is a matrix for deformation in elastic mechanics, which does not exist in rigid body movement, therefore it is 0. $\tilde{\mathbf{R}}_o^i$ can be written in Euler angles form approximatively as

$$\tilde{\mathbf{R}}_o^i = \begin{pmatrix} 1 & 0 & 0 & 0 \\ x_i & 0 & -\psi_i & \vartheta_i \\ y_i & \psi_i & 0 & -\theta_i \\ z_i & -\vartheta_i & \theta_i & 0 \end{pmatrix} \tag{24}$$

Equation (24) is composed with translational matrix and angle matrix. The node make the translational movement first, then finish the rotational movement. Two movement constitute a coordinate expression, so we have $\tilde{\mathbf{R}}_o^j = \tilde{\mathbf{X}}_i^j \tilde{\mathbf{\Theta}}_o^i$,where $\tilde{\mathbf{\Theta}}_o^i$ is a translational movement, when $\tilde{\mathbf{\Theta}}_o^i$ is for rotating. Note that $\tilde{\mathbf{R}}_o^j \neq \tilde{\mathbf{\Theta}}_o^j \tilde{\mathbf{X}}_i^j$.

4.1.3 Equivalence for large deformation

Considering the normal condition, we use Euler angle for example, according to the order $x(\alpha) \to y(\beta) \to x(\gamma)$, and write $\mathrm{s}\beta = \sin\beta$ and $\mathrm{c}\theta = \cos\theta$, a successful transformation in HTC could be

$$\bar{\mathbf{R}}_o^i = \mathbf{C}_x(\gamma)\mathbf{C}_z(\beta)\mathbf{C}_x(\alpha)\mathbf{R}_o^i \tag{25}$$

where

$$\mathbf{C}_x(\gamma) = \begin{pmatrix} 1 & 0 & 0 & 0 \\ 0 & 1 & 0 & 0 \\ 0 & 0 & \mathrm{c}\gamma & -\mathrm{s}\gamma \\ 0 & 0 & \mathrm{s}\gamma & \mathrm{c}\gamma \end{pmatrix}, \quad \mathbf{C}_z(\beta) = \begin{pmatrix} 1 & 0 & 0 & 0 \\ 0 & \mathrm{c}\beta & 0 & \mathrm{s}\beta \\ 0 & 0 & 1 & 0 \\ 0 & -\mathrm{s}\beta & 0 & \mathrm{c}\beta \end{pmatrix} \tag{26}$$

The translation matrix (25) kept the translational and rotational displacement in one matrix, which has an operational result consistent with classical mechanics theory.

4.2 The operation and meaning in mechanics

As the coordinate defined, we will then derive its algorithm.

4.2.1 Minus and multiplication

We can use the subtraction between $\mathbf{R}_o^i$ and $\mathbf{R}_o^j$ to represent the displacement of nodes. In HTC form it can be written as

$$\bar{\mathbf{R}}_o^{i,j} = \mathbf{R}_o^j - \mathbf{R}_o^i \tag{27}$$

The transfer of coordinates can be achieved with the HTC matrix multiplication. Using subscript as the basis of coordinate, superscript as the symbol, according to the algorithm of the matrix and tensor, it is easy to prove

$$\mathbf{R}_o^j = \mathbf{R}_o^i \mathbf{R}_i^j \tag{28}$$

4.2.2 Intergrate and difference for order changing

For the differential form of HTC, differentiate equation will be

$$\partial \dot{\mathbf{R}}_o^i = \frac{\partial \mathbf{R}_o^i}{\partial t} \tag{29}$$

According to the matrix algorithms, the homogeneous item become 0 after differentiation, which means it is no longer a displacement in physical meaning, but a velocity coordinate instead. If we differentiate the equation with the time variable t , then we have

$$\partial \dot{\mathbf{R}}_o^k = \partial \dot{\mathbf{R}}_o^i \mathbf{R}_i^j \mathbf{R}_j^k + \partial \mathbf{R}_o^i \dot{\mathbf{R}}_i^j \mathbf{R}_j^k + \partial \mathbf{R}_o^i \mathbf{R}_i^j \dot{\mathbf{R}}_j^k \tag{30}$$

We could learn from equation that the homogeneous item will be 0 in the sum of three matrix, that is, the HTC matrix is a description for speed now. Meanwhile, according to the physical meaning of rigid body motion, if the R_o^i, R_i^j, R_j^k were kinematic pairs, the 3 items in addition corresponds to convected speed caused by 3 pairs, while the result is its real speed of body.

However, this represents have some shortcomings: 1. The homogeneous item of acceleration differentiated by velocity is the same as velocity in HTC, which will be 0 and may bring some confusion in description. 2. Having the value 0, homogeneous item in velocity and acceleration can not be integrated to displacement which has value 1 there. Strict mathematical definition needs further study.

4.2.3 Dot-multiply for kinetic and potential energy

For classical mechanics , kinetic and potential energy is needed. According to kinetic equation in vector coordinates and the dot product operation in HTC defined, the kinetic energy for HTC can be written as

$$\mathbf{E}_V = \frac{1}{2} m \dot{\mathbf{R}}^T \dot{\mathbf{R}} \tag{31}$$

For rigid coordinates, we define $\dot{\mathbf{R}}_a = \dot{\mathbf{\Theta}}_a \mathbf{X}_a + \mathbf{\Theta}_a \dot{\mathbf{X}}_a$, then we have

$$\mathbf{E}_V = \frac{1}{2} m \left(\mathbf{X}_a^T \dot{\mathbf{\Theta}}_a^T + \dot{\mathbf{X}}_a^T \Theta_a^T \right) \left(\dot{\Theta}_a X + \mathbf{\Theta}_a \dot{\mathbf{X}}_a \right) = \mathbf{E}_{X_a X_a} + \mathbf{E}_{\Theta_a \Theta_a} + \mathbf{E}_{X_a \Theta_a} + \mathbf{E}_{\Theta_a X_a} \tag{32}$$

where

$$\begin{cases} \mathbf{E}_{\mathbf{X}_a \mathbf{X}_a} = \frac{1}{2} m \dot{\mathbf{X}}_a^T \mathbf{\Theta}_a^T \mathbf{\Theta}_a \dot{\mathbf{X}}_a \\ \mathbf{E}_{\mathbf{X}_a \mathbf{\Theta}_a} = \frac{1}{2} m \dot{\mathbf{X}}_a^T \mathbf{\Theta}_a^T \dot{\mathbf{\Theta}}_a \mathbf{X}_a \end{cases}, \begin{cases} \mathbf{E}_{\mathbf{\Theta}_a \mathbf{\Theta}_a} = \frac{1}{2} m \mathbf{X}_a^T \dot{\mathbf{\Theta}}_a^T \dot{\mathbf{\Theta}}_a \mathbf{X}_a \\ \mathbf{E}_{\mathbf{\Theta}_a \mathbf{X}_a} = \frac{1}{2} m \mathbf{X}_a^T \dot{\mathbf{\Theta}}_a^T \mathbf{\Theta}_a \dot{\mathbf{X}}_a \end{cases} \tag{33}$$

The $\mathbf{E}_{\mathbf{X}_a \mathbf{X}_a}$ is the translational kinetic energy, when $\mathbf{E}_{\mathbf{\Theta}_a \mathbf{\Theta}_a}$ is the pure rotational kinetic energy. The remaining two item $\mathbf{E}_{\mathbf{X}_a \mathbf{\Theta}_a}$ $\mathbf{E}_{\mathbf{\Theta}_a \mathbf{X}_a}$ is the coupled kinetic energy. Especially, for a rotational energy, if there is $\mathbf{X}_b$, and

$$\mathbf{E}_V = \frac{1}{2} m \mathbf{X}_a^T \dot{\mathbf{\Theta}}_a^T \mathbf{X}_b^T \mathbf{X}_b \dot{\mathbf{\Theta}}_a \mathbf{X}_a \tag{34}$$

The $\mathbf{X}_b^T \mathbf{X}_b$ will consist the Parallel Axis Theorem, when m, $\mathbf{X}_a^T$, $\mathbf{X}_a$ consist the moment of inertia. The results are consistent with the classical equations.

4.3 Summary

From this section we can conclude that HTC has a perfect description for solid body mechanics. The multi-levelled and tensor-formed values can be described in one HTC. In case that the node have a physical meaning, the HTC follow the general operating rules of matrix.

Therefore we can learn from this paper that: the angle may be the root of many problems such as nonlinearity when the rotational description are used in the same level together with translational description. If we improve this to the HTC description, more accurate answers may come true in the future research. That may be a help for highly-flexible mechanics and so on.

However, there is also some problem raised as well as new directions provided: 1. Multi-levelled description needs strict definition and demonstrate in mathematics. 2. The matrix may have to be extended to more than 2 dimensions, in order to accommodate the higher operational needs, which need new matrix analysis theory to accomplish.

5 Conclusion

A level-based mathematical expression and algorithm was raised, a new tensor-formed matrix called HTC was defined and the operating rule was also given by referring the ANCF and HC, which agrees well with the current theory in rigid body mechanics. This method provides some innovation for related subjects, although there is still many theory need to develop.

The article made some arrangements in classical theory of Newtonian mechanics. Based on the multi-level thought, HTC can be gradually improved in follow-up works, modelling and simulation may become more comfortable especially in multibody mechanics. The usage of HTC in other object is still unclear and more development is needed.

Acknowledgement

This work was supported partially by the National Natural Science Foundation of China (Project Number: 11272101, 11302134), Research Fund for Doctoral Program of Higher Education of China (Project Number: 20102302110051), the 12th Five-Year Plan Projects for civil aerospace of China, the innovation fund of China aerospace science and technology group.

References

1. I. Newton, *Mathematical principles of natural philosophy*. A. Strahan (1802)
2. A. A. Shabana and T. Note, Definition of the slopes and the finite element absolute nodal coordinate formulation, *Multibody System Dynamics*, **1**(3), pp. 339–348 (1997)
3. P. G. Tait, *An elementary treatise on quaternions*. University Press (1890)
4. A. A. Shabana, Flexible Multibody Dynamics: Review of Past and Recent Developments, *Multibody System Dynamics*, **1**, 189–222, June (1997)
5. A. A. Shabana, H. A. Hussien, and J. L. Escalona, Application of the absolute nodal coordinate formulation to large rotation and large deformation problems, *Journal of Mechanical Design*, **120**(2), 188–195 (1998)
6. A. A. Shabana and R. Y. Yakoub, Three Dimensional Absolute Nodal Coordinate Formulation for Beam Elements: Theory, *Journal of Mechanical Design*, **123**(4), 606–613, 2000.
7. L. E. Garner, *An outline of projective geometry*. North Holland New York and Oxford (1981)
8. G. G. Sanborn and A. A. Shabana, A rational finite element method based on the absolute nodal coordinate formulation, *Nonlinear Dynamics*, **58**(3), 565–572 (2009)
9. G. G. Sanborn and A. A. Shabana, On the integration of computer aided design and analysis using the finite element absolute nodal coordinate formulation, *Multibody System Dynamics*, **22**(2), 181–197 (2009)
10. H. Sugiyama and A. A. Shabana, Application of plasticity theory and absolute nodal coordinate formulation to flexible multibody system dynamics, *Journal of Mechanical Design*, **126**(3), pp. 478–487 (2004)
11. R. Y. Yakoub and A. A. Shabana, Three Dimensional Absolute Nodal Coordinate Formulation for Beam Elements: Implementation and Applications, *Journal of Mechanical Design*, **123**(4), 614–621, (1999)
12. F. Berchtold and J. Hausen, Homogeneous coordinates for algebraic varieties, *Journal of Algebra*, **266**(2), 636–670 (2003)

MATEC Web of Conferences **51**, 01009 (2016)
DOI: 10.1051/matecconf/20165101009

Work Hours And Sorting Balancing Study of One Auto Mixed Line

Cao Shouqi , Chen Ying [a] and Chen Chengming

College of Engineering Science and Technology, Shanghai Ocean University, Shanghai, China

Abstract. To deal with the former sub-frame assembling mixed line balance of One Automotive Industry Corporation, the study has used the methods from Industrial Engineering such as ECRS and motion analysis principles to find the problem factors and has improved the stations and their work hours and decreased the number of workers to reduce the personnel cost. Meanwhile it has used FLEXSIM to simulate and got the best production scheduling plan to improve the production efficiency.

1 Introduction

Mixed line production has a strong development since 1980s and it is mostly used in aircraft and automobile manufacturing. And the automobile production mode has changed to lean production of small amount and multispecies from mass amount and single species, then the study of mixed line balance consists of the work hour balance study and multispecies sorting balance study.[1]

This paper is based on Industrial Engineering theories and FLEXSIM simulating to analyze the balance problems of sub-frame mixed assembling production line. It aims at to reduce the work hours of every worker and get a best sorting plan to improve the whole production line.

2 Mixed Line Production and Balance Study Overview

2.1 Front Subframe Mixed Line Production Overview

Front sub-frame is one part of underpan and is the constituent part of front back-axle. There are two types productions named as A and B and 13 stations on this assemble line. Through the analysis of relevant videos and spot observation, there are some main problems that influenced the production efficiency such as follows:

(1) Work hours load imbalance

Some workers finish their work quickly when the other workers still work at the same time.

(2)Motorcycle type production ratio is not the best

Type A is assembled on station 1 for a long time, then the next type must wait on station 2 with not being assembled, so should choose one sorting plan which is the best and meeting the real situation to decrease waiting time and avoid overload to increase efficiency.

[a] Corresponding author : cy07120810@163.com

2.2 Balance study methods overview

2.2.1 Industrial Engineering theories

For the work hours balance study, there are some methods as follows to solve the production balance in Industrial Engineering:

(1) ECRS analysis theory

This theory mainly aims at work allocation for each worker during the operating. Eliminate the not necessary motions and combine two or more easy motions and rearrange the motions order and simplify the complicated motion.

(2)Ten principles of motion analysis

The details are as follows:

1) The motions on both hands should be the same time and symmetrical.
2) Body movements should be the lowest level and achieve satisfaction.
3) Use objects' kinetic energy;
4) Tools and materials should be put close to workers and sorted in the best work order.
5) Use elements' and materials' self-weight to workers (Gravity feed).
6) Should have proper lighting equipments and the height of workbench and chairs should be good to workers.
7) Delete hands motions in possible and replaced by fixtures or feet tools, then workers can be relaxed.
8) Should combine two or more tools be one if possible.
9) The work load should be balanced; Let workers do less if possible.
10) Tools and materials should be put in the work stations in advance. [2-4]

2.2.2 Balance calculation formula

The balance rate computational formula as follow:

$$\text{Balance rate } \eta = 1 - \frac{N*C - t_{总}}{N*C} * 100\% \tag{1}$$

In the formula, $t_{总}$ is the total assembled time, N is the number of stations, C is operating beat.

2.2.3 FLEXSIM simulating

FLEXSIM is a software developed by a American company. Some researchers used it to simulate the manufacturing process and count the whole assemble time to analyze through model the production line and set the corresponding parameters.[5]

3 Mixed Line Balance Study of Front Subframe

For the work hours balance study, this paper set the product beat as A=66s/stage and B=60s/stage and each station has one worker and every two stations cannot start at the same time and if the last station's work is not finished, then the next station will not begin.

3.1 Work hours balance study

3.1.1 Initial work hours data

Table1 is the operating time of A and B: because of every worker's different proficiency and some others elements, so if the station work hours has ±10s different with the beat, it is acceptable. If the work hour is over or under the beat for 10s or more, it should be considered to be improved.

Table1. A and B work hours

Station	A's work hour(s)	B's work hour(s)
1	54.20	63.20
2	61.06	62.83
3	80.81	50.92
4	63.07	45.91
5	70.47	73.90
6	62.97	59.01
7	67.59	74.31
8	74.16	54.51
9	64.92	67.11
10	68.07	62.45
11	53.25	58.08
12	78.56	64.69
13	64.04	48.16
Total	863.17	785.08

According to the beat, compare the station work hour with their type's beat and calculate the balance rate as follows:

(1) During A's processing, the time of station 3 and station 12 are over the beat like 13s and the time of station 1 and station 11 are under the beat like 12s. Station 1 and 13 are adjacent and so station 11 and 12 do, so need adjust these stations' work and facility layout.

(2) During B's processing, operating time of station 5 and 7 and 9 are all over the operating beat and they are choke point stations that need be improved, meanwhile, the operating time of station 4 and 13 are smaller and also need get reassignment.

(3) Initial balance rate

Put the work hour data of A and B, then get their balance rate as follows:

Balance rate of A: $\eta_{A1}=1-\frac{13*66-863.17}{13*66}*100\%=100.6\%>100\%$。

Balance rate of B: $\eta_{B1}=1-\frac{13*60-785.08}{13*60}*100\%=100.65\%>100\%$。

From the result, both A and B's work hours balance rate is not normal. Then analyze each station's process and find the problem and optimize based on the Industrial Engineering principles.

3.1.2 The analysis of productions' work hours

Because every station has a lot process and the methods for improvement is the same, so take the improvement of A's station 11 and 12 for example.

Table 2. Process of A's station 12

Work content	Time(s)
1. Get material a	37.41
2. Tie up the power steering machine oil filling pipe to power steering pump	3.60
3. Get material b	6.11
4. Install inlet oil pipe to storage tank assembly	2.44
5. Install power steering oil pot to power steering pump	4.79
6.Connect air conditioning evaporation tube and radiator	6.60
7. Tidy engine lines	3.80
8. Install quick release buckle to sub-frame	5.16
9. Self-check	8.65

From Table2, this station's time of getting materials is 37.41s and close to 1/2 of all time. So the distance of materials and station should be adjusted and decrease the time. The details are as follows:

Table 3. Some operation details of station 12

Get materials	Objects	Time(s)	Left hand operating state(0 is operating/1 is waiting)	Right hand operating state(0 is operating/1 is waiting)
Get material 1	Pneumatic gun	1.29	1	0
Get material 3	Steering liquid storage pot to binning pump fuel sucking pipe	1.80	0	1
Put down material 4	Cock	1.00	0	0
Get material 8	Clip	5.03	0	1
Get material 9	nipper	1.11	0	0
Put down material 10	nipper	0.49	1	0
Get material 11	Component	0.84	0	1
Get material 16	Nut	1.69	1	0
Get material 17	Line assembly	2.16	0	1
Get material 18	Lube	0.92	0	0
Put down material 22	Electric gun	0.84	1	0
Get material 23	White paint pen	1.85	0	1

A. Analyze Table3's data and get the conclusion: the motions about getting material 3 and 4, because of the left hand waiting of motion 3 and both hands can operate together, so can integrate their time and the final time is 1.80s; Same to above, combine the motion 8 and 9 and the final time is 5.03s. And then use the same method and combine the motion 10 and 11 and the final time is 0.84s. As the same, combine the motion 16 and 17 and the final time is 2.16s and combine the motions of putting down material 22 and 23 and the final time is 1.85s. After all the improvement, the all operating time of station 12 is 73.43s.

B. Because the operating time of station 11 is short and the residual quantity with the beat is 12s, so transfer second process of station 12 to the worker of station 11,then the operating time of station 12 changes to 66.51s and time of station 11 changes to 60.17s.

As well, use the same methods to improve the other stations of type A and B that need improved.

3.1.3 Stations' work hours balance analysis after improved

Calculate and adjust the process and facility layout towards different motorcycle types again and again, then get the total assembled work hours after improvement as A is 781.64s and B is 708.98s.

Through the formula, the after work hour balance rate of type A and B are as follows:

A's after balance rate: η_{A2}=98.69%

B's after balance rate: η_{B2}=98.47%

3.1.4 Station work hour balance study conclusion

The conclusion is that has used the Industrial Engineering methods to adjust and improve station process and decreased the number of stations and decrease the personnel cost and the station work hours are all close to the beat.

3.2 Productions process sorting balance study

This paper assumes that front sub-frame assembled line is stable and process continuously, the date need for type A and B of the market are 270 stages and 180 stages and the day work hour is 8h.

3.2.1 Production beat

In the mixed line process, the mixed beat

$$r = \text{Work hours} \ / \ \text{Sum of all type needs} \tag{2}$$

According to the above assumption, this mixed beat $r = \frac{8*60*60s}{270+180} = 64$ s/Stage .

3.2.2 Sorting combinatory analysis

All the sum needs of type A and B is 450 stages and use the cyclic sorting method to find the minimum cell of two types' needs ratio, according to the number in the minimum cell to sort and then recycle it. Under this assumption, the number of sorting combinatory is $n = \frac{5*4*3*2*1}{(3*2*1)*(2*1)} = 10$.

3.2.3 Sorting simulation analysis

There is 10 different sorting plans and if calculate one by one will be a huge work. So it uses FLEXSIM to simulate. Set the objective function $f = \sum_{i=1}^{N} IDLE_i + \sum_{i=1}^{N} OVER_i$ and $IDLE_i$ is the waiting time， $OVER_i$ is overload time. The sorting plan with the minimum f is the best.

(1) Build simulation model

According to entities in FLEXSIM, build the model for simulation as the follow Figure1:

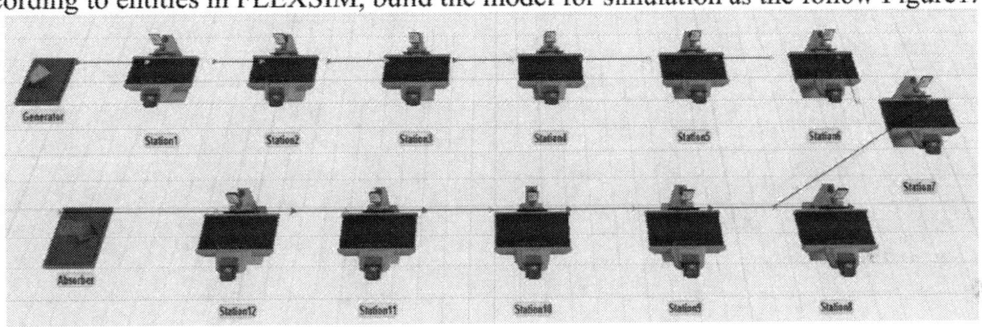

Figure1. Model building

(2) Sorting combinatory parameter determination

Set the relevant parameter in FLEXSIM according to each type's beat and every sorting combinatory. During the model building, generator 3 is the output pot of all the processed products. And input every arrival's time when the arrival method is "Arrival time schedule"

(3) Operating time parameter determination

Set "Deal depends on the type of entities" and set every generator's case and face to the real.

(4) Set the simulation time and get the final conclusion.

3.2.4 Choose result of sorting plans

According to simulation analysis through FLEXSIM, get every objective function value as follows:

Table 4. Every sorting plan's function value

Sorting plan	Objective function f
12211	25721.31
21112	25655.72
21121	26305.08
21211	26313.52
22111	25664.18
11122	25686.35
11212	26313.18

11221	25708.47
12112	26342.21
12121	26348.44

From Table 4, the plan which has the minimum objective function value is 2-1-1-1-2 and choose this plan can get rid of long waiting time or high overload, then improve the work self-use ratio and decrease the time cost and then improve the production efficiency.

4 Conclusion

After the Industrial Engineering methods use and simulating the production sorting with FLEXSIM. The results are:

(1) A's balance rate had changed to 98.69% from the first 100.6%, B's balance rate had changed to 98.47% from the first 100.65%.

(2) The number of stations had changed to 12 from 13, the workers had been decreased and then the cost of the workers and time were both decreased and improved the worker use ratio.

(3) Through the simulation with FLEXSIM, set the parameter function and got a best plan that had fewer waiting and overload time.

References

1. X.Y.WU. Study of assembling production balance [D].Shanghai: Shanghai Jiaotong University(2007).
2. S.P YI, F.GUO. Basic Industrial Engineering [M].Beijing: China Machine Press(2011).
3. F.GUO, K.M.ZHANG. Industrial Engineering and Management,**2**,120 (2005).
4. R.Q.CHEN, S.H.MA. Production management [M]. Beijing:Higher Education Press(2007).
5. J.R.ZHANG. Simulation Study Based on FLEXSIM mixed flow assembly line simulation of a car [D].Shanxi: Changan University(2009).

MATEC Web of Conferences **51**, 01010 (2016)
DOI: 10.1051/matecconf/20165101010

A Scenario Tree based Stochastic Programming Approach for Multi-Stage Weapon Equipment Mix Production Planning in Defense Manufacturing

Xuan Li[1], Yu Zhou[2,a], Tianjun Liao[3], Yajun Hu[2]

[1]*Science and Technology on Information Systems Engineering Laboratory, National University of Defense Technology, Changsha, 410073, P. R. China*
[2]*School of Materiel Management and Safety Engineering, Air Force Engineering University, Xi'an 710051, P. R. China*
[3]*State Key Laboratory of Complex System Simulation, Beijing Institute of System Engineering, Beijing, 100101, P. R. China*

Abstract. The evolving military capability requirements (CRs) must be meted continuously by the multi-stage weapon equipment mix production planning (MWEMPP). Meanwhile, the CRs possess complex uncertainties with the variant military tasks in the whole planning horizon. The mean-value deterministic programming technique is difficult to deal with the multi-period and multi-level uncertain decision-making problem in MWEMPP. Therefore, a multi-stage stochastic programming approach is proposed to solve this problem. This approach first uses the scenario tree to quantitatively describe the bi-level uncertainty of the time and quantity of the CRs, and then build the whole off-line planning alternatives assembles for each possible scenario, at last the optimal planning alternative is selected on-line to flexibly encounter the real scenario in each period. A case is studied to validate the proposed approach. The results confirm that the proposed approach can better hedge against each scenario of the CRs than the traditional mean-value deterministic technique.

1 Introduction

The multi-stage weapon equipment mix production planning (MWEMPP) is crucial to maintain the military power of one country and has the properties of high-risk, multi-period and tremendous investment [1]. Therefore, the MWEMPP would be particularly important in the national defence industry. However, the combat capability requirements evolve with the ever more demanding mission tasks. Meanwhile, the exact values of the evolving CRs are unknown. That is, each CR has two uncertain parameters, i.e. time and quantity, and they are evolving with the period. The possible value ranges of the time and quantity can be predicted, but the corresponding distributions are unknown in each period. Thus, MWEMPP contains a multistage programming problem under the uncertain CRs.

There are some methods for the uncertain multistage programming problem, such as dynamic programming [2] and markov decision process [3]. But they need strict conditions on the probability of the event and the relations of the variables in each period. With the continuous increase of

[a] Corresponding author : zhouyu_gfkd@126.com

computing power, the stochastic programing with multilayer scenario tree is more and more popular to model the multistage uncertain problem [4-6]. The multilayer scenario tree can model the arbitrary stochastic evolutions of the uncertain parameters as a group of discrete scenarios without the need to know the detailed probability distribution. Moreover, there is no limitation for the number of the uncertain parameters. Hence, the uncertain parameters of the CRs can be easily interpreted as a scenario tree. Furthermore, the stochastic programming method can adjust the development time and the Development budget to hedge against the various scenarios in each period.

Based on these surveys, we seek to propose a scenario tree based stochastic programming approach to model and solve MWEMPP problem. First, the multilayer scenario tree is built to model the evolution of the uncertain CRs. Hence, a specific computational process is proposed for the scenario tree generation. Finally, a multi-stage stochastic programming model is developed to solve effectively solve MWEMPP problem. A case is provided to illustrate the advantages of the proposed stochastic programming approach. To the best of our knowledge, this approach has not been proposed in other fields, and it is first proposed here.

2 Problem statement

Suppose that there are C CRs and M weapon categories, in which each CR is supported by more than one weapon category and vice versa. Each weapon category contains a number of weapon models. For each weapon model, the capability values depend on the corresponding development time and Development budget. In each weapon category, the weapon models have different capability values when they have equivalent development time and Development budget. The weapons development processes are shown in Fig. 1.

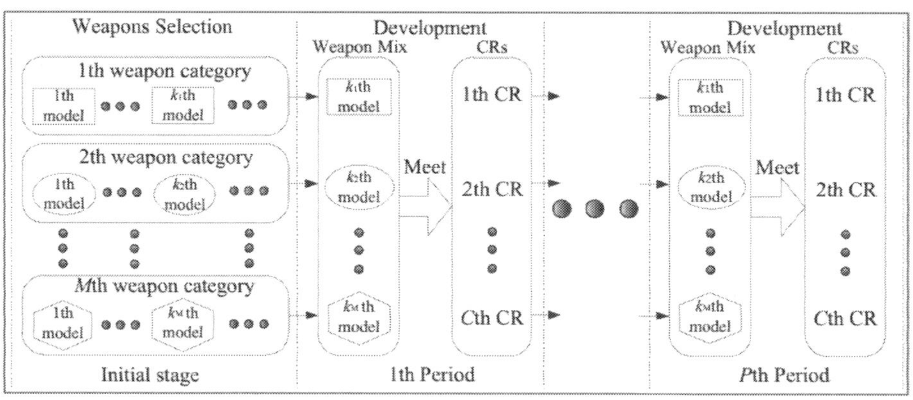

Figure 1. The processes of MWEMPP

As shown in Fig. 1, the M weapon models must be selected from each category at the initial stage to form a weapon mix. Then the weapon mix is produced to meet the evolving CRs continuously in each period. Therefore, the corresponding weapon equipment mix contains two kinds of optimizations, one of which is to determine the weapon mix and the other one is to determine the development time and Development budget in each period. The objectives are to minimize the capability gaps between the weapon mix and the CRs in each period. In this context, the capability deficiencies and surpluses between weapon mix and the CRs are all recognized as the capability gaps.

3 The multistage stochastic programming approach

3.1 Scenario tree modeling for the uncertain CRs

We begin to use the scenario tree to model the uncertain CRs based on the formulations of them. All the possible evolutionary processes of CRs can be modeled by a scenario tree, as shown in Fig. 2.

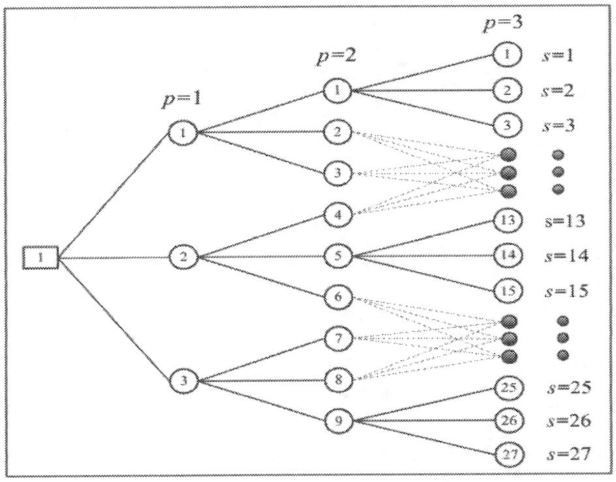

Figure 2. An example of a 3-periods scenario tree

Each circular node represents one possible state of the CRs corresponding to previous period. The path from the root node (rectangular node) to a leaf node (circular node) is called a scenario, i.e., a possible evolutionary process of the CRs over the entire planning horizon. For instance, the fifteenth scenario consists of the second node in the first period, the fifth node in the second period and the fifteenth node in the third period. Thus, the uncertain CRs can be converted into a number of certain scenarios in a scenario tree. Furthermore, the amount of CRs is assumed to be up to 5 kinds in the context of defense manufacturing, so that the scenario tree is generated in a manageable size.

3.2 Scenario tree generation

It is possible to obtain the time requirement interval $[Rtl_{p,c}, Rtu_{p,c}]$ and the quantity requirement interval $[Rql_{p,c}, Rqup_{e,c}]$ of the CRs in each period, but the possible values of them are not available due to the confidential causes. Therefore, we propose a practical approach combining Monte Carlo simulation to build the scenario tree. The process of this approach is presented as follows:

Step 1: Determine the branch Zp, the time requirement interval $[Rtl_{p,c}, Rtu_{p,c}]$ and the quantity requirement interval $[Rql_{p,c}, Rqu_{p,c}]$ in each period.

Step 2: Divide the intervals of time requirement and quantity requirement into Zp parts, respectively.

Step 3: Initialize p=0 and z_p=0. Select one part randomly out from the Z_p parts of the time requirement without replacement, then within which generate an integer number randomly for $Rt_{p,z,c}$ in this part.

Step 4: Select one part randomly out from the Z_p parts of the quantity requirement without replacement, then within which generate a real number randomly for $Rq_{p,z,c}$ in this part.

Step 5: Set $z_p = z_p + 1$, repeat steps 3~4 until $z_p = Z_p$.

Step 6: Set $p = p+1$, repeat steps 2~5 until $p=P$.

Step 7: Use the $\sum_{p=1}^{P} Z_p$ group values of the CRs to generate the scenario tree.

3.3 Multistage stochastic programming model

For the multi-period uncertain production planning problem, the traditional method is the mean-value deterministic programming [7]. This technique optimizes a group of fixed variables to cope with each possibility of the CRs. Contrary to the MVDP model, the multistage stochastic programming (MSP) model generates different variables in each node to hedge against the variations of the CRs. To build the weapons multistage stochastic programming model, the constraints of "non-anticipativity" have to be considered firstly [8]. Hence, the MSP model is formulated as follows:

$$\min \sum_{p=1}^{P}\sum_{n_p=1}^{N_p}\pi(n_p)\times\left(\sum_{c=1}^{C}(w^{+}_{p,n,c}d^{+}_{p,n,c}+w^{-}_{p,n,c}d^{-}_{p,n,c})\right) \tag{1}$$

$$s.t.\quad WC_{p,n,c}-d^{+}_{p,n,c}+d^{-}_{p,n,c}=Rq_{p,z,c} \tag{2}$$

$$d^{+}_{p,n,c},\ d^{-}_{p,n,c}\geq 0 \tag{3}$$

$$z_p=\begin{cases}Z_p; & \mathrm{mod}(n_p/Z_p)=0\\ \mathrm{mod}(n_p/Z_p); & \mathrm{mod}(n_p/Z_p)\neq 0\end{cases} \tag{4}$$

$$N_p=\prod_{p'=1}^{p}Z_{p'}\qquad P,N_p,Z_p\in ¥^{+} \tag{5}$$

where $\pi(n_p)$ is considered as the unconditional probability to realize the possibility of the CRs in the n_pth node, or the relative importance among the nodes in the same period. Constraint (2-4) means that the weapon mix generates distinctive capability value $WC_{p,n,c}$ to meet each possible value $Rq_{p,z,c}$. Constraint (5) describes the relationship between n_p and z_p. The mod(*) represents the remainder operation.

4 Case study

4.1 Problem description

The C4ISR weapon equipment mix production planning [9] which is one important kind of MWEMPP problem is studied to validate the proposed hybrid approach. Thus, the C4ISR weapons need to be planned to satisfy multiple CRs of joint operations. Suppose that there are five weapon categories to be planned to meet seven CRs in the future. The data on the C4ISR weapons are show in Table 1.

Table 1: The data on the C4ISR weapons

	NCWM	OPB	CR1	CR2	CR3	CR4	CR5	CR6	CR7
RSate	5	0~300	√	√	√	√			
AWACS	4	0~200	√	√		√			√
MHAUAS	6	0~100	√	√	√		√		
BLOSDL	3	0~80				√	√	√	
C2S	3	0~100					√	√	√

All of the notations in Table 1 are listed in the following: (1) CR represents one capability requirement. (2) CR1 is the target identification. (3) CR2 is the location tracking. (4) CR3 is the constantly monitoring. (5) CR4 is the information transition and distribution. (6) CR5 is the interoperability. (7) CR6 is the self-adaption. (8) CR7 is the command and control. (9) NCWM is the numbers of the candidate weapon models in each C4ISR weapon category. (10) OPB is the optional development budget for each C4ISR weapon category. (11) RSate is the reconnaissance satellite. (12) AWACS is the airborne warning and control system. (13) MHAUAS is the medium to high altitude unmanned aerial system. (14) BLOSDL is the beyond line of sight data link. (15) C2S is the command and control system.

4.2 Validation of the MSP model

We use the MVDP, MSP, ideal multi-stage programming (IMP) model to solve the addressed case, respectively, in order to find which one has more advantages to hedge against the uncertainties of the CRs, in order to investigate the degree that the MSP model diminishes the uncertainties of CRs. The respective global capability gap of the three models is shown in Fig. 3.

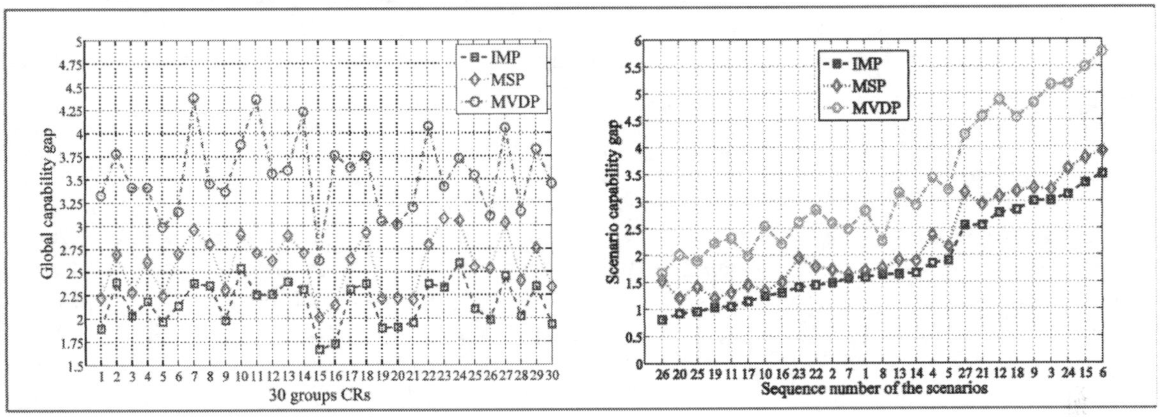

Figure 3. The capability gap of the three models

As shown in Fig. 3, for each group of CRs, IMP model gets the smallest global capability gap, and it represents the ideal values without considering the uncertainties of the CRs. The MSP model is closer to the ideal values than the MVDP model under each group CRs. This is because MSP takes different recourse actions to encounter each possible CRs, and MVDP model generates the fixed solutions to encounter all scenarios in each period. However, because of the constraints of the 'non-anticipativity', the MSP model cannot offset the negative effects of uncertainties completely. Thus, it never achieves the level of the IMP model.

As shown of the scenario capability gap, the scenarios are sequenced by the values of the capability gaps generated by the IMP model in ascending order. Compared with the MVDP model, the MSP model is closer to the ideal capability gap curve, except individual scenario such as the 26th and the 23th. Furthermore, the extreme scenarios which are from the 27th to the 6th bring little influences to the close degree of the curves between the MSP model and the IMP model. In contrast, the distance of the curves between the MVDP model and the IMP model enlarges. It means that the MSP model have the better robustness than the MVDP model.

According to the above analysis, the MSP model outperforms the MVDP model, but there are no free lunches. Next, we change the quantity of the budget to test the variation of the global capability gap corresponding to the three models. The results are shown in Fig. 5.

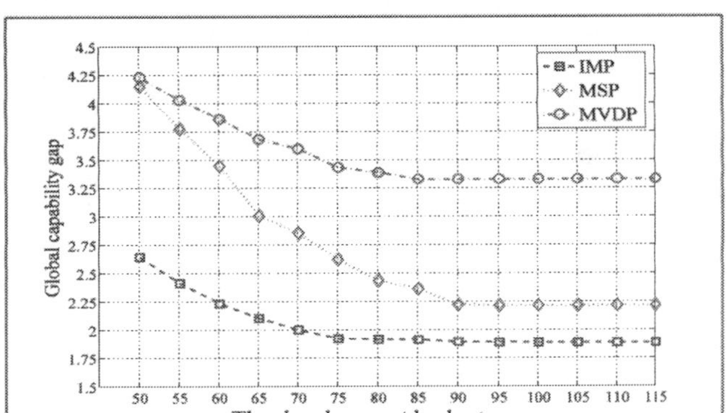

Figure 4. The variations of the global capability gap with the budget

All of the three models obtain a smaller global capability gap with a larger budget. When the budget increases to a certain value, the gaps do not reduce any more. This is because the more budget are, the larger optional ranges of variables are. Thus, the alternative space contains better solutions. Furthermore, the gap of the MSP reduces more than MVDP model and IMP model when the budget increases. Obviously, only if the budget is more than 90 normalized amounts, the MSP model can obtain the optimal global capability gap which is close to the ideal value. This is because the MSP model can adjust the development time and the Development budget to avoid capability deficiency and surplus in different scenarios. When the capability requirements are large, the MSP model needs more budgets to adjust the development time and the Development budget. Thus, it naturally needs more budget.

5 Conclusions

We comparatively studied the MSP model with the traditional MVDP model and the ideal IMP model. The primary conclusions are as follows: (1) Both of the global capability gap and the scenario capability gap generated by the MSP model are smaller than those generated by the MVDP model. (2) The budget imposes a large influence to the MSP model. When the budget decreases, the MSP model obtains a larger global capability gap and closes to the MVDP model. When the budget increases, it derives a smaller global capability gap and closes to the IMP model. (3) The interval sizes of the CRs differs from each other in some extreme cases, thus, it means that we need to study the scenario tree with flexible branches to model the more widely evolution of the uncertain CRs.

Funding

This research is supported in part by the National Natural Science Foundation of China under contract Nos 71401167, 71571185.

References

1. Y. Zhou, J. Jiang, Z. Yang, Y. Tan. Proc IMchE Part B: J Engineering Manufacture, 2014.
2. Lew, A., & Mauch, H. Berlin, Germany: Springer-Verlag, 2006.
3. Ferreira, K. D., & Wu, D. D. Int. J. Development Economics, 134, 95-107, 2011.
4. Ahmed, S., & Sahinidis, N. V. Operations Research, 51(3), 461-371, 2003.
5. Alfieri, A., & Brandimarte, P. Germany: Springer-Verlag, 2005.
6. Heitsch, H., & Römisch, W. Mathematical Programming, 118, 371-406, 2009.
7. Birge, J. R. New York, USA: Springer-Verlag Inc, 1997.
8. Ahmed, S., King, A. J., & Parija, G. Journal of Global Optimization, 26(1), 3-24, 2003.
9. Y. Zhou, K. Yang, Y. Tan, Z. Yang. Proc. of the 7th SOSE, Genoa, Italy, 2012.

MATEC Web of Conferences **51**, 01011 (2016)
DOI: 10.1051/matecconf/20165101011

Axial Compression Properties Nonlinear Analysis on Square Double Skin Steel Stub Short Columns Filled with Recycled Concrete

Song Bing, Li Baishou[a]

Department of Structural Engineering, Yanbian University, Yanji, 133002, China

Abstract. Taking the mixing amount of diatomite calcined and vitrified micro bubbles(VMB) as the main changing parameters, experiment studies the properties of the vitrified micro bubbles recycled concrete blocks; then this paper adopts the finite element software ANSYS to analyze the square double skin steel stub short columns filled with recycled concrete under axial compression. According to the vertical stress distribution, strain and bearing capacity of the steel tube and core concrete, we make a contrastive axial compression properties analysis on the different hollow ratio χ(0,0.35)and the VMB content(0%,100%,130%) of square double skin steel stub short columns filled with recycled concrete. The result shows that: Compressive strength of VMB recycled concrete increases with the increase of diatomite calcined content, when mixing amount of diatomite calcined is 3%,the compressive strength of 130% VMB content test specimen can reach 32.45 MPa;Because of the inner circular steel tube is setted which strengthening component buckling capacity and improving the ductility of the component, stress distribution of hollow components is more balance than solid components, and their axial displacements decrease by 5.6% compared with the solid components when they reach ultimate bearing capacity; When the hollow ratio is same, ultimate bearing capacity of 130% VMB content test specimen compared with the content is 0% only reduces by about 3.5%; When the VMB content is same, ultimate bearing capacity of hollow components compared with solid components increases by about 2.5%, which reducing weight as well as improving the anti-seismic performance.

1 Introduction

Currently, vitrified micro bubbles(VMB) is widely used as a new sort of environment-friendly and inorganic lightweight insulation materials, which can be used to make VMB recycled concrete when it is added into recycled concrete. Its strength and thermal insulation property can reach the ideal performance that is tested by the experiment[1].Recycled concrete filled square steel tubes (CFST) is the structure style casting recycled concrete into the steel pipe, core concrete bears three directions force due to the restraint of the steel tube, improving the bearing performance and the recycled concrete engineering application value; Besides the same advantages with CFST, concrete filled double-skin steel tube(CFDST)could also expand cross section, with a higher bending rigidity lighter weight and better performance in earthquake resistant and fire resistant. Compared with those

[a] Corresponding author : bsli3399@163.com

structural parts in style of SHS outer and SHS inner, SHS outer and CHS inner has better malleability [2-3].

Therefore, by means of experiment this paper studies the properties of VMB recycled concrete blocks,and adopts ANSYS to make an analysis on vertical stress distribution, strain, displacement of steel tube and core concrete and bearing capacity of square double skin steel stub short columns filled with recycled concrete to compare the influence of different hollow ratio and VMB content to the axial compression properties of square double skin steel stub short columns filled with recycled concrete.

2 General situations of the experiment

2.1 The experiment of vitrified micro bubbles recycled concrete

Experiment uses 42.5 ordinary Portland cement; recycled coarse aggregate with a broken jaw crusher whose screening particle size is 5~20mm,bulk density is 1280 kg/m^3 and bibulous rate is 4.8%; sand fineness modulus is 2.9; fly ash uses the original ash from Yanji heating plant, basically reaching the level II for fly ash fineness requirements;Jilin linjiang City Tianyuan catalyst co.LTD., production of 325 mesh diatomite calcined, its performance parameters are shown in table 1;Vitrified Micro Bubbles produced by Linghai City Longyan Building Materials Factory,stacking density is 128kg/m^3 whose performance parameters are shown in table 2;poly carboxylic acid high efficiency water reducing agent produced by Yanji Fangsheng Building Materials Company, water reducing rate is 25% or higher, which contains 0.8% of gas composition;mixing water is ordinary tap water.

After trial mixing,eventually water-binder ratio of VMB recycled concrete mixture ratio is 0.45, sand ratio is 30%. Results of each test block of experiment parameters and maintenance of indoor after curing 28d are shown in table 3.

Table 1. Performance of diatomite calcined

Fe_2O_3 content/%	S_iO_2 content/%	Specific surface area /(cm^2/g)	Ignition loss /%	Bulk density /(g/cm^3)
≤1.5	≥92	600 thousand	≤0.5	≤0.42

Table 2. Performance of vitrified micro bubbles

Granularity /mm	Density /(kg/m^3)	Heat conductivity coefficient /[W/(m·k)]	Surface vitrified /%	Water absorption /%
0.5~1.5	50~200	0.0284~0.054	≥95	20~50

Conclusions according to table 3:when the diatomite calcined content is same, increasing the VMB content, or when the dosage of VMB is same, increasing dosage of diatomite calcined, the compressive strength of concrete has a trend of increase; when diatomite calcined is 3% and the dosage of VMB is 130%, compressive strength of block reaches 32.45 MPa and its density is just 1942.45 kg/m^3, compared with the ordinary concrete dry apparent density of 2000~2500 kg/m^3, it can achieve reducing building self quality and improving anti-seismic performance.

Table 3. Experimental parameter and test result of VMB recycled concrete

Test specimen number	Material utilization amount/(kg/m^3)								Compressive strength /MPa	Dry density /(kg/m^3)
	Cement	Fly ash	Diatomite calcined	Recycled coarse aggregate	Sand	VMB	water reducer	water		
BRC-0-3	349.11	93.5	14.2	1171.35	504	0	10.38	206	34.5	2118.05
BRC-100-3	458.18	122.6	18.4	1114.95	479.5	127.8	13.51	270	30.8	1969.35
BRC-100-2	464.3	122.6	12.31	1114.95	479.5	127.8	13.51	270	26.49	1965.43

BRC-100-0	476.49	122.6	0	1114.95	479.5	127.8	13.51	270	22.36	1909.12
BRC-130-3	458.19	122.6	18.4	1114.95	479.5	154	13.51	270	32.45	1942.45
BRC-130-2	464.25	122.6	12.31	1114.95	479.5	154	13.51	270	27.56	1910.28
BRC-130-0	476.49	122.6	0	1114.95	479.5	154	13.51	270	25.95	1923.42

Note: BRC-130-3,BRC represents VMB recycled concrete,130 represents the VMB volume contains in the total volume is 130%,3represens that the diatomite calcined content is 3%,others can get by the parity of reasoning.

2.2 Design of test specimens

In order to study CFDST, length of the test specimen must be proper, the specimen's length-width ratio is 3[4].Steel type is Q235,main parameters of test specimen are shown in the table 4, $\chi = D_i/(B-2t)$, $\xi = A_{so}f_y/(A_{co}f_{ck})$, in the formula, A_{so} is the outer steel pipe cross-sectional area, f_y is outer steel pipe yield strength,A_{co} is the area contained by the outer steel tube,f_{ck} is standard value of recycled concrete compressive strength.

Table 4. Test specimen number and main parameters

Test specimen number	B×t×L /(mm³)	D_i×t_i /(mm²)	Hollow ratio χ	Confining factor ξ	F_y /(MPa)	F_{cu} /(MPa)	N_{ANSYS} /(kN)
S-K1-0	180×2.5×540	——		0.5891		34.5	1110
S-K1-100	180×2.5×540	——	0	0.6599		30.8	1008
S-K1-130	180×2.5×540	——		0.6265		32.45	1061
S-K2-0	180×2.5×540	62×2.5		0.5891	235	34.5	1126
S-K2-100	180×2.5×540	62×2.5	0.35	0.6599		30.8	1041
S-K2-130	180×2.5×540	62×2.5		0.6265		32.45	1090

Note: All the specimens'diatomite calcined content of recycled concrete are 3%;K1and K2 represent hollow ratios of test specimens are 0 and 0.35; 0,100,130 mean VMB content accounted for 0%,100% and 130% of the total concrete volume.

3 Establishment of the finite element model

3.1 The selection of element types

Steel uses the Solid45 element to simulate, E_s=2.06×10^{5}MPa, V_s=0.3;Concrete uses the Solid65 element to simulate, elastic modulus is calculated by recycled concrete elastic modulus formulas that is put forward by the reference [5], V_t=0.2.

$$E_C = \frac{10^5}{2.2+34.7/\mathrm{fcu}}/(0.2811\delta+1.065) \tag{1}$$

In the folumar: δrepresents the replacement rate of recycled coarse aggregates, f_{cu} is the cube compressive strength of recycled concrete.

3.2 Constitutive relations of materials

(1)Constitutive relation of steel

The steel uses dual linear kinematic hardening model (BKIN), which is shown in the figure 1 (a).

(2)Constitutive relation of concrete

As CFST, the interaction between steel tube and concrete makes the working performance of the core concrete complicate.Therefore, the paper uses theory[6] that core recycled concrete stress-strain relationship model, as shown in figure 1 (b).

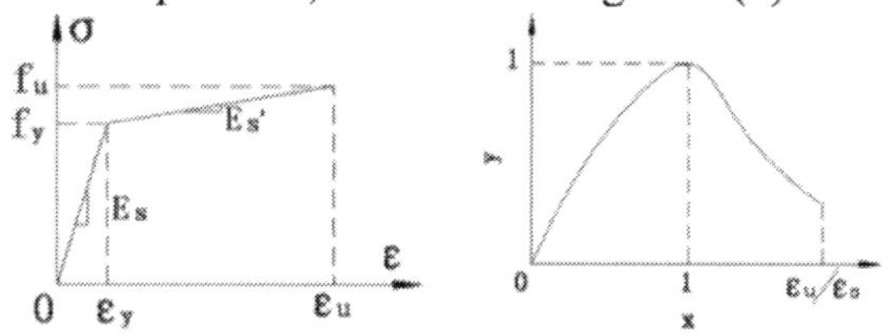

(a) Constitutive relation of steel (b)Constitutive relation of concrete

Figure 1. Material stress-strain relations

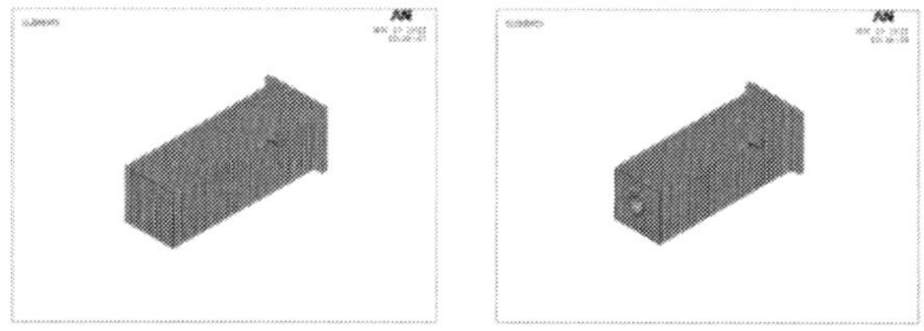

(a) Solid concrete filled steel tube (b) Concrete-filled double skin steel tube

Figure 2. Finite element model

3.3 Modeling and Meshing

Regardless of slip between core concrete and steel tube, we add 15mm rigid plate in order to prevent the phenomenon of stress concentration on both ends of pillars.Quality of meshing directly affects the precision and speed of calculation, so hexahedron unit division side length of solid45 and solid65 are 15 mm. As shown in figure 2.

3.4 Control of Convergence

(1) Loading and boundary conditions

We add axial load on the upper plate,which generated by the rigid plate into uniformly distributed load on the model, at the same time we add x, y, z three directions of displacement constraints on the column bottom plate.

(2) Concrete crush Settings

The open crack of concrete shear transfer coefficient and closed joint shear transfer coefficient are respectively set as 0.35 and 0.9 , axial tensile strength is 3 MPa, close the concrete crush option.

(3) Nonlinear analysis options

Open the large deformation of static analysis and the automatic time step length, number of substeps is set as 200 , maximum equilibrium iteration times is set as 40, the Newton - Laposen (N - R method) is used to solve and convergence tolerance is set as 0.05.

4 Calculated result analysis of finite element

Post-processing function of ANSYS is used to study stress, strain and displacement fringe of specimen S-K1-0, S-K2-0 and specimen S-K1-130, S-K2-130 .

4.1 Stress analysis of test specimens

As shown in figure 3(a),(b)and(c),von Mises stress nephogram of the solid components in the square steel tube corner distributes complexly and the value of stress is the maximum;von Mises stress of the core concrete stress is approximately elliptic constraint zone, the middle is larger than the side of constraint and load; The axial stress of steel from the load and constraint transits to the middle and distributes unevenly;In figure 4(a),(b)and(c),outer steel tube axial stress distribution of hollow components transits smoothly,the setting of inner steel stub makes the stress distribution more evenly; The end of load core concrete axial stress is bigger;Axial stress distribution of inner steel tube is larger near the middle of the component.

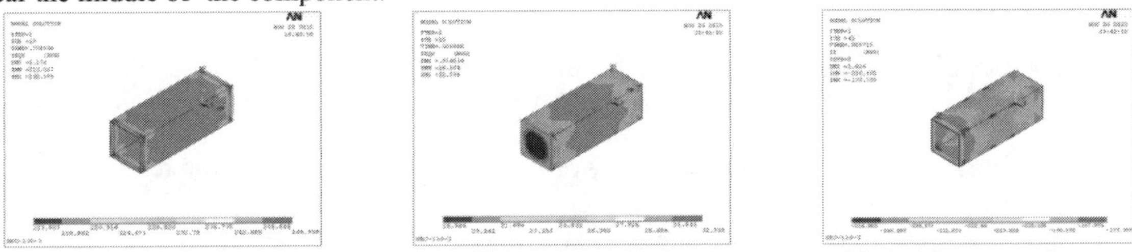

(a) Von Mises stress nephogramof steel tube

(b) Von Mises stress nephogram of concrete

(c) Axial stress nephogram of steel tube

Figure 3. S-K1-0 CFST stress nephogram

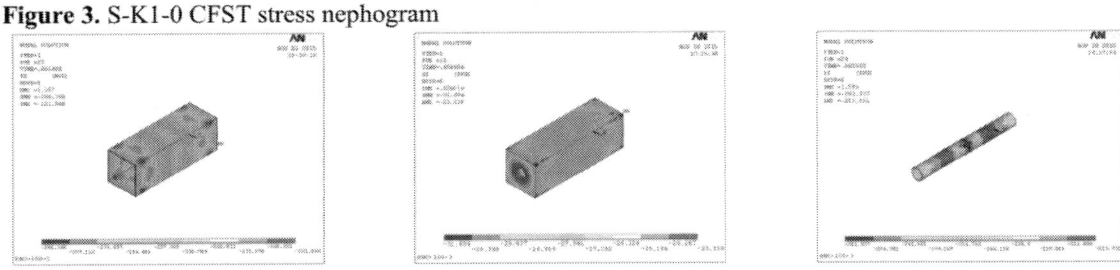

(a) Axial stress nephogram of outer steel tube

(b) Axial stress nephogram of concrete

(c)Axial stress nephogram of inner steel tube

Figure 4. S-K2-0 CFDST stress nephogram

4.2 Strain and displacement analysis of test specimens

Figure 5 is the strain and displacement nephogram of specimen S - K1-130 and S - K2-130 that have arrived the ultimate bearing capacity. As shown in figure 5 (a) and (b), axial strain of the components corner in the load and constraint is apparently greater than other regions, especially the middle is apparently greater than its adjacent areas, local buckling occurres, axial displacement is 1.7mm; as shown in figure 5 (c) and (d), hollow components appear the local buckling,but do not change apparently in the middle, which indicats the inner steel tube can prevent or delay local buckling of the components,axial displacement is 1.605 mm that decreases by 5.6% compared with the solid components,setting of inner steel tube enhances the deformation capacity and improves the ductility.

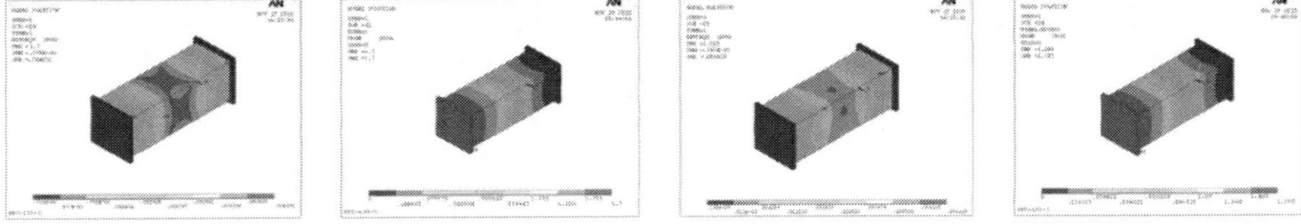

(a)S-K1-130 axial strain contours

(b) S-K1-130 Displace vector sum

(c) S-K2-130 axial strain contours

(d) S-K2-130 Displace vector sum

Figure 5. Component stress and displacement nephogram

4.3 Ultimate bearing capacity analysis of test specimens

The ultimate bearing capacity of 6 different hollow rates and VMB content of CFDST are shown in table 3, results of ANSYS analysis are following: When the hollow ratio is same, the ultimate bearing capacity of 130% VMB component compared with content is 0% only reduces by 3.5%; When VMB content is same,though CFDST have reduced the amount of concrete,because of the inner steel tube,compared with solid components ,their ultimate bearing capacity increase by about 2.5%, which reducing weight and enhancing the anti-seismic performance of components.

5 Conclusions

(1)When the diatomite calcined content is same,with the increase of vitrified micro bubbles content, or when the vitrified micro bubbles content is same,with the increase of diatomite calcined content, compressive strength of recycled concrete shows a trend of increase; When diatomite calcined content is 3% and vitrified micro bubbles content is 130%,compressive strength of block can reach 32.45 MPa, and its density is just 1942.45 kg/m^3;
(2)Due to the inner steel tube setting of hollow components, the stress distribution is more evenly compared with solid components,their axial displacemen decreases by 5.6% when they reach ultimate bearing capacity,which improving the buckling performance of the component and the component defor-mation ability and increasing the ductility of components;
(3)When the hollow ratio is same, the ultimate bearing capacity of 130% VMB component compared with content is 0% only reduces by 3.5%; When VMB content is same,though CFDST have reduced the amount of concrete,because of the inner steel tube,compared with solid components ,their ultimate bearing capacity increase by about 2.5%, which reducing weight and enhancing the anti-seismic performance of components.

References

1. Chen Kaihua, Luo Surong, Zheng jianfeng. Cause vitrified microsphere, recycled concrete heat preservation and heat insulation performance testing research [C].The third session of the national academic seminar on recycled concrete. Qingdao, 2012:495~522
2. Han linhai, Yang Youfu. Modern concrete filled steel tube structure technology [M]. Second edition. Beijing: China building industry press, 2007
3. Wang Nannan. Hollow interlayer of square concrete-filled steel tube column experiment research and numerical simulation performance [D]. Hebei, hebei university of engineering, 2013
4. Zhong Shantong. Concrete filled steel tube structure [M]. Harbin: Heilongjiang science and technology press, 1994
5. Zhou Jinghai, He Haijin. Basic mechanical properties of recycled concrete test [J]. Journal of shenyang architectural university (natural science edition), 2010, 26 (3) : 464 ~ 468
6. Yang Youfu. Recycled concrete filled steel tube components load - deformation relations theory analysis [J]. Industrial construction, 2007, 37 (12) : 1 ~ 6

MATEC Web of Conferences **51**, 01012 (2016)
DOI: 10.1051/matecconf/20165101012

Design and Numerical Analysis of Spiral Pressure Recovery Blades in the Overflow Pipe of the Cyclone Separator

Qidong Chen[1,a], Yuegao Sun [2]
[1]Changshu institute of Technology, Changshu, 215500;
[2]Soochow University, Suzhou, 215000

Abstract. Spiral pressure recovery blades which can make the rotational flow convert into direct flow in overflow pipe of the cyclone separator are designed in the overflow pipe. The numerical calculation of the cyclone separator with different blades is carried out by using the Reynolds stress turbulence (RSM) model. Meanwhile, the trend of the total pressure of the inlet and outlet of the cyclone separator is discussed while the blades are inserted at different positions. Results show that the pressure of inlet and overflow outlet obviously reduced when the blade is inserted in the cyclone separator. It can make the inlet total pressure reduced more than 9% and the overflow outlet total pressure reduced more than 50% when four blades are inserted in the cyclone separator. The total pressure of the inlet and overflow outlet is the smallest when blades are inserted in the lowest of the overflow pipe.

1 Introduction

At present, the cyclone separator has been widely used in many fields, such as paper, coal, mineral processing, food, petroleum, chemical industry and so on. Many experts and scholars have done a lot of research on the performance of the cyclone separator. The influence of the pressure field and velocity field of the cyclone separator was studied by Jiang Minghu et al[1]. Wang Zunce et al [2] studied the influence of different inlet structure parameters of the cyclone separator on the pressure loss. The flow field of the cyclone separator was studied by Chen Qidong et al [3]. The influence of operation parameters on the separation performance of cyclone separator has also been studied and tested by many experts and scholars [4]. However, the study on the pressure recovery in the overflow pipe is relatively few. The fluid in the overflow pipe is a high speed rotational flow which has considerable rotational kinetic energy. If it is converted into direct flow, the mechanical friction loss can be reduced and dynamic pressure is converted to static pressure, and thus the pressure of the inlet and outlet of the cyclone separator is reduced.

In this paper, a new type of pressure recovery blade which can change the rotational flow into direct flow is designed. The influence of three different kinds of blade number on the inlet and outlet pressure is compared and the best position of the blade inserted into the overflow pipe is discussed.[a]

2 Design of pressure recovery blade

[a]Corresponding author: 18015565894@sina.cn

2.1 theoretical analysis

In order to convert rotational flow into direct flow, the attack angle of the front inlet edge of the blade must be maximally as the same as the attack angle for the fluid to enter the blade. Through the numerical calculation for the flow field of the cyclone without blades, Fig.1 shows 10 fluid tracks 0.045m away from the overflow outlet in the overflow pipe. According to the figure, the motion tracks of the fluid are not regular cylindrical spiral curves and the spiral distance thereof is constantly changed, so these tracks cannot be expressed by the standard spiral curve equation.

Figure 1. The track curves of the overflow pipe.

In the paper, the finite division theory is adopted to divide the whole fluid track into infinite small arcs which can be approximated into standard cylindrical spiral curves. The lower endpoint of an independent track is recorded as $f_0(x,y,z)$ and the upper endpoint is recorded as $f_n(x,y,z)$, and the endpoints of an optional arc taken from the above track are respectively recorded as $f_a(x,y,z)$ and $f_{a+1}(x,y,z)$. Meanwhile, the horizontal plane of point $f_a(x,y,z)$ is assumed as plane α, the projection of point $f_{a+1}(x,y,z)$ in plane α is assumed as $f_{a+1}(x,y,z)_\alpha$ and the projection of the above arc in plane α is assumed as $f_a f_{a+1\alpha}$. Since the arc is very small, the length thereof can be replaced by line $\overline{f_a f_{a+1\alpha}}$. Subsequently, the lead angle of the spiral curve can be calculated according to Formula (1) for the lead angle of the standard cylindrical spiral curve.

$$\tan\varphi_a = \frac{z_{a+1}-z_a}{\overline{f_a f_{a+1\alpha}}} = \frac{z_{a+1}-z_a}{\sqrt{(x_{a+1}-x_a)^2+(y_{a+1}-y_a)^2}} \tag{1}$$

Similarly, central angle can be approximately calculated according to Formula (2).

$$\theta_\alpha = \frac{2\overline{f_a f_{a+1\alpha}}}{O_\alpha f_a + O_\alpha f_{a+1\alpha}} = \frac{2\sqrt{(x_{a+1}-x_a)^2+(y_{a+1}-y_a)^2}}{\sqrt{x_a^2+y_a^2}+\sqrt{x_{a+1}^2+y_{a+1}^2}} \tag{2}$$

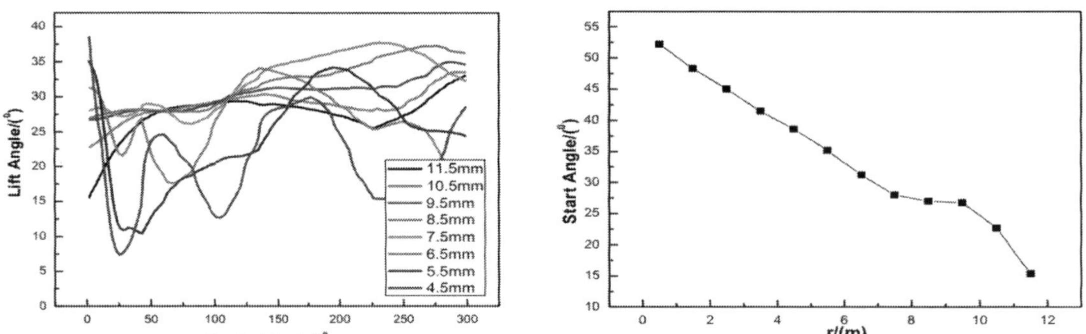

(a) Lift angle changes with the central (b)Initial lift angle change of 10track curves with r

Figure 2. Lift angle changes with the center angle and initial angle with the radial position change

After the coordinates of the division point of the track are exported, Formulae (1) and (2) can be adopted to calculate the lead angle and the central angle corresponding to each small arc. The variation curve of central angle along with lead angle is as shown in Fig.2(a). According to the figure, when θ is in the range of 0° ~30° , the lead angles of the tracks have a stable variation tendency, wherein the lead angles of the circumjacent tracks (for example r=11.5mm, 10.5mm, 9.5mm; r refers to the radial position) have linear increase tendency while the lead angles of the tracks near the central axis (for example r=4.5mm, 5.5mm) have reduction tendency. Additionally, the variation curves of the initial lead angles of 10 tracks at the same height and along the same radial direction along with radial position r are as shown in Fig.2(b). According to the figure, the initial lead angles have obvious reduction tendency. Therefore, when θ is in the range of 0° ~30° , the track curves are variable spiral surfaces. According to the above analysis, the spiral pressure recovery blade is reasonable.

2.2 Design of blade

In Solidworks software, 10 fluid tracks exported by the post-processing software are taken as the boundary, and then the boundary surface command is adopted to generate the front surface of the blade, and then the front surface is connected to the vertical plane through the transition surface, thus to establish the whole blade model. Specifically, the boundary dimension of the blade is as follows: the height of the whole blade is 30mm, the blade thickness is 0.5mm, the leading central angle of the bottom blade is 30°, the blade radius is 11.5mm, and the initial lead angle of the blade is gradually increased from the outside to the inside, with the variation range of 15°~90°. The pressure recovery blade models with different blade quantities are as shown in Fig.3.

(a)two blades (b)three blades (c)four blades

Figure 3. Different blade models.

3 Numerical calculation model

Without isotropy and eddy viscosity hypotheses, Reynolds stress model[5] can be used to simulate such complex three-dimensional motion as strong vortex flow. Liang Zheng, et al. [6] have comprehensively explained why Reynolds stress model is more applicable to the simulation of the internal flow field of the cyclone separator than other models (for example, standard $k-\varepsilon$ model[7] and RNG $k-\varepsilon$ model[8]). Specifically, ICEM software is adopted for hexagonal structure meshing, and the computational domain of each model is about 500,000 mesh points, and the hexagonal structure mesh diagrams of the five different models are as shown Fig.4.

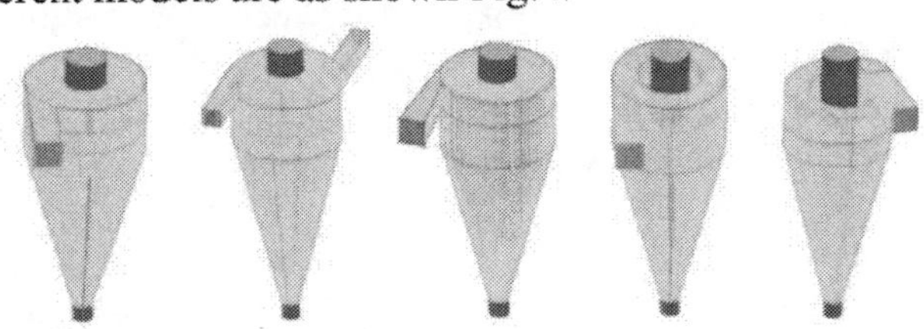

Figure 4. grid graph of different models.

Meanwhile, FLUENT software is adopted to calculate the internal flow field of the cyclone, and the velocity field and pressure field coupled SIMPLEC algorithm is adopted as the calculation method for transient calculation, wherein the time step is 0.001s and the residual error control value is set as 0.00001. Additionally, the pressure item is PRESTO algorithm and the second-order upwind scheme is adopted for other equations.

3.1 Geometrical physical model

Classic 75mm cyclone is adopted in the paper, wherein Model I is the single-inlet rectangular tangent feed pipe cyclone, Model II is the dual-inlet rectangular tangent feed pipe cyclone, Model III is the shrinkable rectangular tangent feed pipe cyclone, Model IV is the concentric annular feed pipe cyclone, Model V is the three-dimensional spiral feed pipe cyclone, and the structural parameters of the above five models[9] are as shown in Table 1. D is the diameter of the cylindrical barrel, a and b are respectively the length and the height of the rectangular inlet, h is the height of the cylindrical barrel, α is the taper angle, d_o is the diameter of the underflow port, d_u is the diameter of the overflow outlet, d is the diameter of the concentric annulus and z is the spiral distance.

Table 1. Structural parameters of five models.

	D (mm)	a (mm)	b (mm)	h (mm)	α (°)	d_o (mm)	d_u (mm)	d (mm)	z (mm)
Model I	75	17	27	75	20	12.5	25		
Model II	75	12	19	75	20	12.5	25		
Model III	75	17	27	75	20	12.5	25	40	
Model IV	75	17	27	75	20	12.5	25		
Model V	75	17	27	75	20	12.5	25		17

3.2 Boundary conditions

1) The velocity inlet is adopted as the inlet, and the inlet velocity is set as 2.28m/s.
2) The pressure outlets are adopted as the overflow outlet and the underflow outlet, the static pressure is set as 0 and is communicated with the atmosphere.
3) No-slip boundary is adopted for the wall surface, and the standard wall function is adopted for the wall treatment.
4) Physical property parameters of the mixture: the water density is 998 kg/m^3 and the viscosity is 0.001 $Pa\ s$.

4 Reliability analysis of results

American scholar Hsieh [10] adopted the laser Doppler tachymeter to measure and obtain the velocity distributions on different sections of a cyclones as early as 1988. Many experts and scholars[11-13] have adopted the above experimental data to verify the accuracy of the calculation results. In the paper, the post-processing software is adopted to output the radial position based distributions of the tangential velocity (Fig.5(a)) and the axial velocity (Fig.5(b)) at the position 60mm away from the highest point of the cylindrical barrel, and meanwhile such distribution data are analyzed and compared with Hsieh's experimental data.

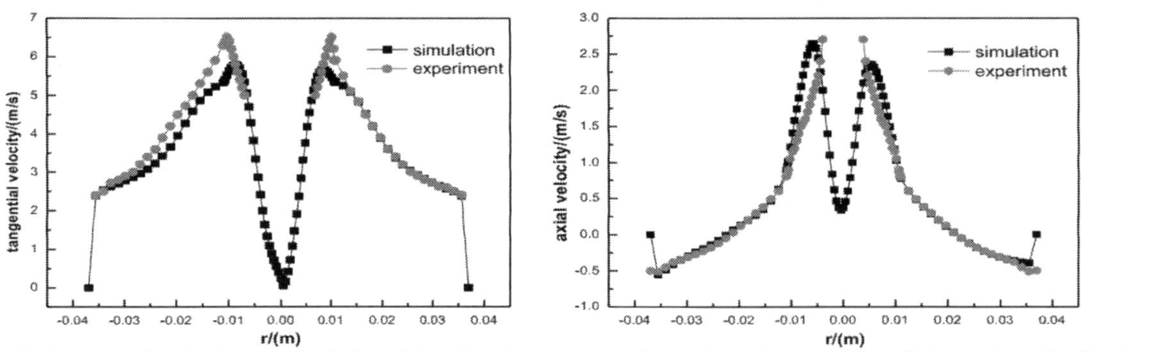

(a) Tangential velocity with radial position distribution (b) Axial velocity with radial position distribution

Figure 5. Comparison of tangential velocity and axial velocity between numerical simulation and experiment

According to the figure, the simulation result is basically consistent with the experimental data; the simulation experiment has wall error, because the wall boundary is not accurately processed in the numerical simulation and it is difficult to accurately measure the tangential velocity and the axial velocity at the position near the central position in the experiment.

5 Flow field analysis

5.1 The velocity field and pressure field of Model I under different blade number

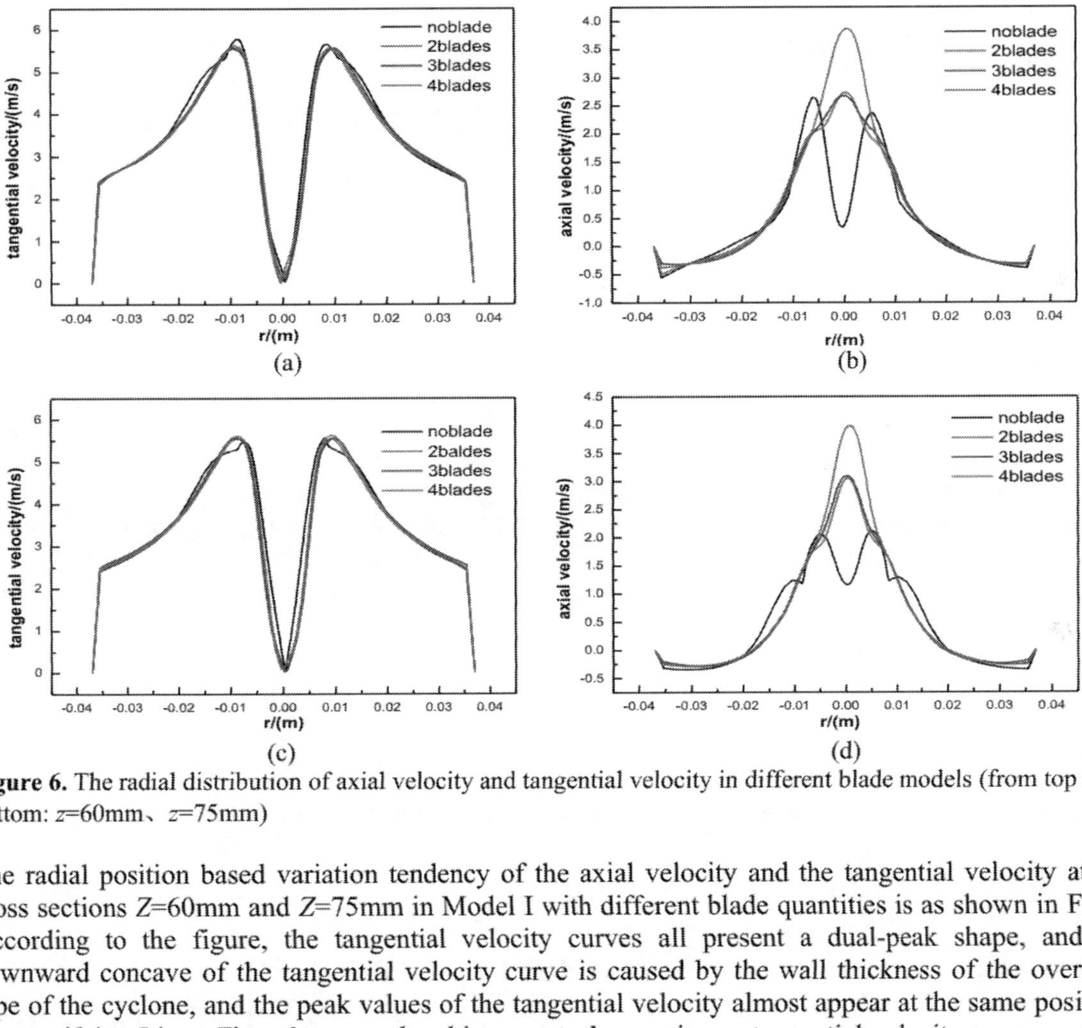

Figure 6. The radial distribution of axial velocity and tangential velocity in different blade models (from top to bottom: z=60mm、z=75mm)

The radial position based variation tendency of the axial velocity and the tangential velocity at the cross sections Z=60mm and Z=75mm in Model I with different blade quantities is as shown in Fig.6. According to the figure, the tangential velocity curves all present a dual-peak shape, and the downward concave of the tangential velocity curve is caused by the wall thickness of the overflow pipe of the cyclone, and the peak values of the tangential velocity almost appear at the same position, thus verifying Liang Zheng's research achievement: the maximum tangential velocity appears at the radial position slightly more than $2r_0/3$(r_0 is the diameter of the overflow pipe). According to the tangential velocity distribution diagram, in the forced vortex area, the tangential velocity of the cyclone inserted with blades is less than that of the cyclone not inserted with blades, thus indicating that the pressure recovery blades can weaken the rotating strength of the forced vortex and reduce the total energy for driving the equal fluid as well as reduce the total inlet pressure and the total outlet pressure; the tangential velocity curves of the cyclones respectively inserted with two, three and four blades are basically coincided with each other. According to the axial velocity distribution diagram, in the forced vortex area, the effect of the three-blade cyclone is similar to that of the two-blade cyclone;

the axial velocity of the four-blade cyclone is more than those of two-blade and three-blade cyclones, so the total outlet pressure of such cyclone can be effectively reduced.

5.1.1 Pressure fields of Model I with different blades

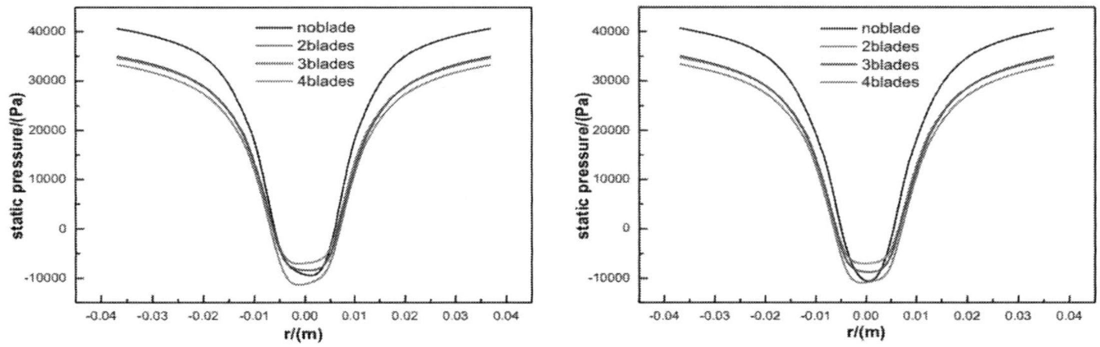

Figure 7. Static pressure distribution at different cross sections of the model 1 (from left to right: section z=60mm、z=75mm)

Under the conditions of respectively inserting two, three and four blades, the radial position based variation tendency of the static pressure of Model I at the cross sections *Z*=60mm and *Z*=75mm is as shown in Fig.7. According to the figure, the internal pressure of the cyclone separator is basically symmetrically distributed and the static pressure at the same cross section is gradually reduced from the wall surface to the central axis along the radial position. Since the central fluid is under high-speed revolution, thus the negative pressure generated in the central axis area is exactly one of the preconditions for generating air column inside the cyclone. The two curves for cross sections *Z*=60mm and *Z*=75mm are basically coincided with each other, thus indicating that the radial pressure gradient on different sections are almost the same. The static pressures at the positions with the same radius are the same, thus indicating that the axial pressure gradient is almost zero. After the pressure recovery blades are inserted, the static pressure in the standard free vortex area is obviously reduced. Additionally, the two-blade cyclone and the three-blade cyclone have equivalent static pressure reduction effect, and the four-blade cyclone can maximally reduce the static pressure, thus having the best effect.

5.2 Analysis of pressure fields of the blades at different heights in the overflow pipe

Inlet pressure and outlet pressure are important performance parameters of a cyclone. Under the condition of not influencing the separation efficiency, it is significant to reduce inlet pressure and outlet pressure. Three-blade pressure recovery blades are respectively inserted at five positions in the overflow pipes of the five models, wherein the five positions are respectively 5mm, 15mm, 25mm, 35mm and 45mm away from the overflow outlet surface. FLUENT software is adopted to calculate the total inlet pressure and the total overflow outlet pressure of the cyclone separator after the blades are inserted at the above five positions.

The total inlet pressure and the total overflow outlet pressure at each position in the five models are as shown in Fig.8. (Fig.8(a) shows the total inlet pressures of Models I, II, IV and V, Fig.8(b) shows the total inlet pressure of Model III, Fig.8(c) shows the total overflow outlet pressures of Models I, II, IV and V, and Fig.8(d) shows the total overflow outlet pressure of Model III). According to the figure, along with the increase of the insertion depth, the total inlet pressure and the total overflow outlet pressure of the cyclone separator are almost linearly reduced. When the bottoms of the inserted blades and the overflow pipe are at the same horizontal line, more rotational flows are converted into direct flows, thus minimizing the total inlet pressure and the total outlet pressure of the cyclone separator and realizing the best effect of the cyclone.

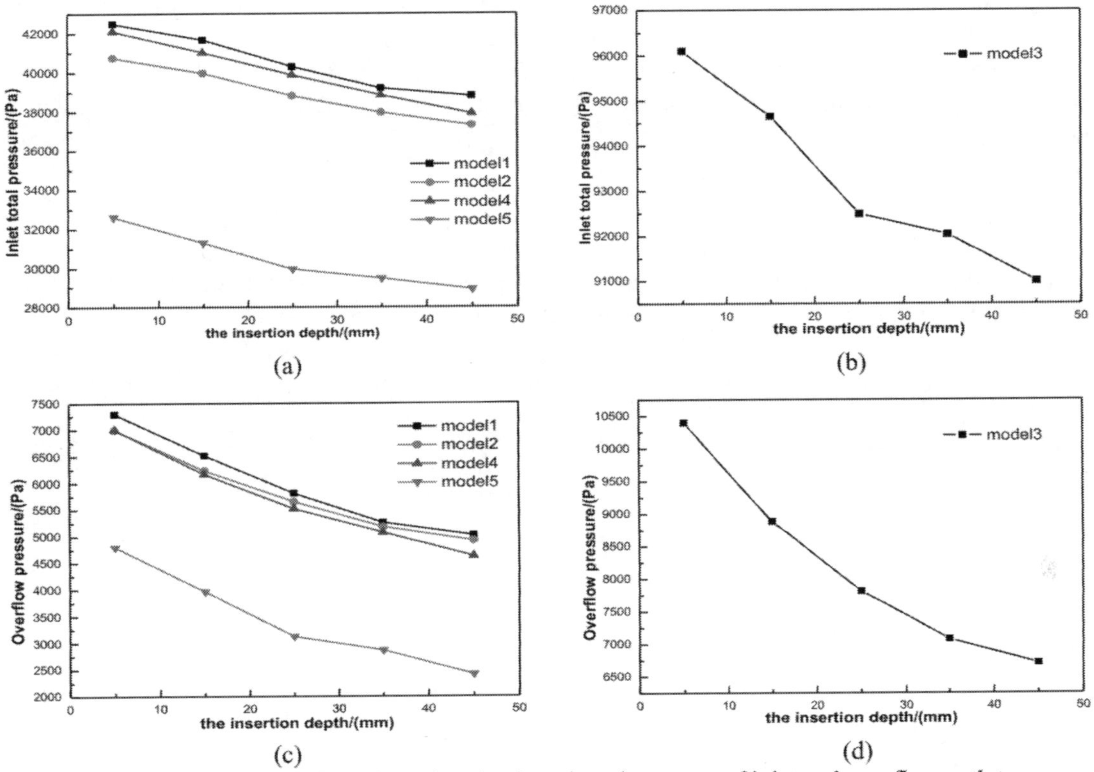

Figure 8. Relationship between blade insertion depth and total pressure of inlet and overflow outlet

5.3 Pressure fields of five different models with different blades

In order to realize the best effect of the cyclone, the best installation position of the pressure recovery blade shall be set at the nethermost end of the overflow pipe. The specific effects of the three blade structures installed at the nethermost ends of the overflow pipes in the five different models will be analyzed and discussed in this section.

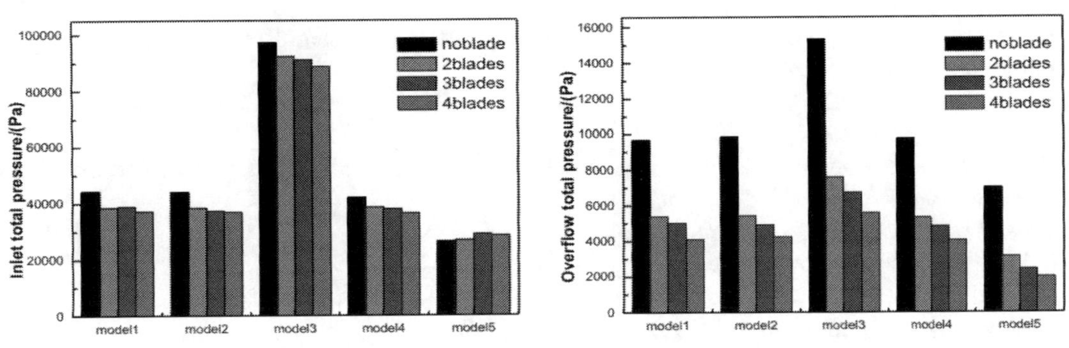

(a) Total inlet pressures (b) Total overflow outlet pressure

Figure 9. Inlet total pressure and overflow outlet pressure under five different models in different blade

The total inlet pressure and the total outlet pressure of the five different models with different blades are as shown in Fig.9. According to Fig.9(a), the pressure recovery blades can reduce the total inlet pressures of Models I, II, III and IV. For Models I, II and IV, the inserted blades can reduce the total inlet pressure at least by 13%; for Model III, the inserted blades can reduce the total inlet pressure by about 6%; for Model V, the inserted pressure recovery blade cannot reduce the total inlet pressure but increase it. According to Fig.9(b), the inserted pressure recovery blades can significantly reduce the total overflow outlet pressure, the two-blade cyclone can reduce the total outlet pressure by about 50%; the four-blade cyclone has the best effect for reducing the total inlet pressure and the total overflow outlet pressure, and the effect of the three-blade cyclone is only inferior to that of the four-blade cyclone.

6 Conclusion

1) The inserted pressure recovery blades can obviously reduce the total inlet pressure and the total overflow outlet pressure of the cyclone. The four-blade cyclone separator can reduce the total inlet pressure by about 9%~16% and reduce the total overflow pressure by about 50%~70%.
2) The cyclone separator inserted with the pressure recovery blades at the nethermost end of the overflow pipe has the best effect.
3) It is not suitable to insert the pressure recovery blades into the three-dimensional feed pipe cyclone separator.

References

1. JANG. Ming-hu, LIU. Dao-you, ZHAO. Li-xin, JIA. Gui-chun. The Effect of Cone Angle to Hydrocyclone Pressure Field and Velocity Field[J]. Chemical Machinery, **5**:572-576 (2011)
2. WANG. Zun-ce, GAO. Ye, Lu. Feng-xia, LIU. Xiao-ming, LI. Feng, The Effect of Liquid - liquid Hydrocyclone Inlet Structure Parameters to Pressure Characteristics[J]. Fluid Machinery, **2**:16-19 (2003)
3. CHEN. Qi-dong, ZUO. Zhi-quan, The Flow Field Numerical Analysis of Different Lateral Inlet Cyclone[J]. China Engineering Science, **2**:58-67+78 (2014)
4. YUAN. Hui-xin, YU. Jian-feng, LIU. Hong-bin, The Effect of Split Ratio to the Basic Performance of Oil-water Separation Hydrocyclone [J]. Petroleum Machinery, **9**: 17-19+2 (2000)
5. ZHANG Shi-shuai. Computational Fluid Dynamics and the Application [M].Wuhan: Huazhong University of Science and Technology Press (2011)
6. LIANG. Zheng, WANG. Jin-jin, REN. Lian-cheng, LI. Lian-ming, The Theoretical Research of Solid-liquid Separation Hydrocyclone Flow Field[M]. Beijing: Petroleum Industry Press, **2**:25 (2011)
7. Launder B E, spalding D B, Lectures in Mathematical Models of Turbulence [M]. Academic Press, London, England (1972)
8. Vahot V, Orszag S A. Development of turbulent models for shear flow by a double expansion technique[J].Phys Fluid, **4**(7):1510-1520 (1992)
9. ZHAO. Qing-guo, ZHANG. Ming-xian, Separation Technology of Hydroclone [M].Beijing: Chemical industry Press, **53** (2003)
10. K. T. Hsieh. Phenomenological Model of the Hydrocyclone [D].Salt Lake City: The University of Utah(1988)
11. K. W. Chu, B. Wang ,A. B .Yu, et, al. CFD -DEM Modelling of Multiphase Flow in Dense Medium Cyclones[J].Powder Technology, **193**(3):235-247 (2009)
12. J. A. Delgadillo,R. K. Rajamani. A comparative study of three turbulence-closure models for the hydrocyclone problem [J]. Int. J. Miner. Process, **77**:217-230 (2005)
13. XU. Yan-xia. The Numerical Simulation and Analysis of Hydrocyclone Separation Process[D]. East China University of Science and Technology(2012)

MATEC Web of Conferences **51**, 01013 (2016)
DOI: 10.1051/matecconf/20165101013

Application and Implementation of the Structured Assembly Process Planning in A Auto Enterprise Based on TS16949

Jiakun Sun [1,a], Ying Wang[1], Fupeng Yin[1] and Huihui Qiu [2]

1 School of Management Engineering, Shandong Jianzhu University, Jinan, 250101,China

2 Shandong Hoteam Software Co., Ltd. Jinan, 250101, China

Abstract. Process planning is the key link that connects the product design and manufacture. In order to continually improve the effectiveness and efficiency of process planning an auto enterprise in consistent with the process management requirements of TS16949, the requirements for assembly process planning is analysed and the main problems affecting the design and production efficiency are discussed. The infrastructure of the structured process planning and management platform is constructed combing the management technology and TS16949 quality system, then the structured PPPR (product, process, plant and resource)model is reintegrated to analyze the relationships among the objects with the object-oriented method. In the end, with the principles of maintaining conveniently and operating practically, the assembly process planning and management system is implemented according to the characteristics of the process in the enterprise, and the main function modules of the system are described.

1 Introduction

As one of the most important part in manufacturing enterprises, process planning is the key link that connects the product design and manufacture. As for the automotive industry, it has the characteristics of complex product structure, variety of components, polytrophic process routing and so on[1]. Consequently, the quality and the productivity is closely related to the level of process planning, and the application of CAPP (computer aided process planning)technology[2] is conducive to improve the standardization level and the efficiency of process planning. In recent years, structured process is becoming a tool to organize the product, process, plant, resource (PPPR)[3] and other data resources in a direct and structural manner, which makes it fast, effective and accurate to create and verify the process, and can ensure the standardization and reuse of the process data. It also gets rid of the previous application model where the process planning and management are mainly based on process cards or documents, which limits the support to the process management and be lack of relevance to process data management. Thus, it is becoming increasingly recognized and applied by manufacturing enterprises. The ISO/TS16949 is an ISO technical specification aimed at the development of a quality management system that provides for continual improvement, emphasizing defect prevention and the reduction of variation and waste in the supply chain. A supplier without a valid certificate has little chance of supplying a Tier 1 supplier and certainly no chance of supplying a car manufacturer with

[a] zpsunjk@sdjzu.edu.cn

standard parts[4]. Thus, the process planning system should be developed to integrate with TS16949 quality management system[5~6], and, in the application of the system, it must be in accordance with the TS16949 specification, establishing the basic process model of quality management system, combining with the enterprise standard and information technology, to support the standardization and normalization of the processes, so as to improve the quality management efficiency and effectively support the enterprise operation.

2. Requirement analysis

The automotive company, located in Wuhu City, Anhui Province, China, was established in 1997. After 18 years of development, it's products have covered the fields of passenger cars, commercial vehicles, etc. For the process planning and management, DELMIA[7] system is applied to support product design and complete the product process planning and verification through three-dimensional model. But there are still deficiencies in the integration of information with the downstream production department and in the combination of the organization and process management, which affect the design and production efficiency, the main problems are as follows:

1) Process planning and management is not standardized and process planning pattern cannot support process planning effectively.

For the process planning pattern, it mainly takes the process card as the core, which emphasis too much on the result of the process planning, so that it can't support the process effectively and only give limited backing to improve the process research, development and management. Meanwhile, the management of process data is chaotic, the data of the product design and process planning is isolated, the process of the process management is not standardized, hence the product design data can't be effectively utilized. Due to the lack of the structured model of management and the poor reuse level, it seriously affect the production schedule.

2) Process planning and management can't meet the fast and personalised market demand

As time goes by, the market requirements becomes fast-changing, and the automotive models are changing, also, the product configuration is becoming more and more complex. Assembly process is still the most time consuming stage among the four major stages of the automotive manufacturing.

3) TS16949 system file can't be managed quickly and effectively

Although the enterprise has passed the TS16949 certification and has a relatively complete management process as well, the TS16949 system management platform applied in the process planning department is not concordant with that of the production department and the functions is limited. As the core tool of TS16949 system，large number of process-related quality documents created by APQP[5] can't be arranged in accordance with the process specification, and the standard specification of the data organization and management can't be met. There are also many faults in the operation process of the TS16949 quality documents for different configurations of the same models, causing a lot of duplication of work.

To sum up, the problems summarized above increase the workload on the staffs, so the automotive enterprise is in urgent need of the support for a structured process planning and management system to improve the level of the standardization and structuralization for process planning and management and increase the efficiency of process planning.

3. System design

3.1 Infrastructure design

According to the requirements of the enterprise, with the principles of maintaining conveniently and operating practically, the infrastructure for the collaborative process planning and management platform is built combing the management technology and information technology. With the database and file server as the support, it could define different functions through the configuration of the object models and data models. The system also provides different interfaces for different users to achieve the process collaboration between the departments of process planning and product design. The system architecture is divided into four layers: basic support layer, maintenance layer, function layer and interface layer. The infrastructure is shown in Figure 1.

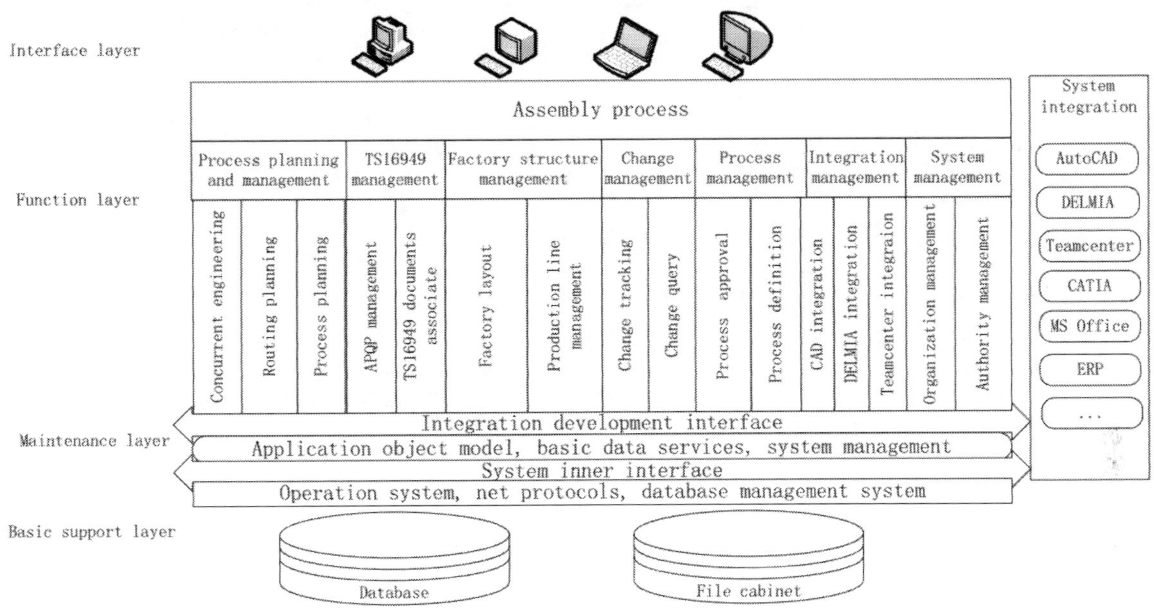

Figure 1. System infrastructure

1) Basic support layer, it is the basis of system development, including network protocols, operating system, a variety of file cabinets and databases. The database mainly used to store the basic data while the file cabinet is mainly used to store some unstructured documents, pictures and videos;

2) Maintenance layer, it is a prerequisite to ensuring the normal application of the system, including the application of the object models, basic data services and system management. Application object model maintenance is the main part of the layer, once the business is to be adjusted, the model will be modified or reconfigured. The system management is mainly to ensure the system security and maintain user's account information.

3) Function layer, it is to form various function modules: assembly process planning and management, TS16949 system management, factory structure management, change management, process management, integration management and system management with the support of the basic support layer and the maintenance layer according to business logic. New function modules can be customized through secondary development or application integration to meet the changing requirements of the enterprise.

4) Interface layer, it makes the interface more user-friendly, providing convenient operation through the application of the principle of man-machine engineering.

Based on the universal process information model, taking the basic system development idea of continual improvement and optimization, it establishes the process planning and management platform with the assembly process planning and management as the core function module that is relatively simple, easy to implement and maintainment.

3.2 PPPR model of assembly process

According to the characteristics of the enterprise, the PPPR model is reintegrated to analyze the relationship among the four objects: the product, the process, the plant and the resource, as shown in Figure 2.

ProductView contains all the attribute information and the structure information of an automotive model in the design phase, which mainly contains two entity objects: component and component structure. The component entity is used to show the attribute information of the components themselves, such as the basic information of name, type , unit, as well as the management information of the CAD diagram, quality requirements, etc. The component structure entity is used to reflect the features of the various models and the structural relations among the various components, which can provide complete information for every department, and it is the basis of PPPR model.

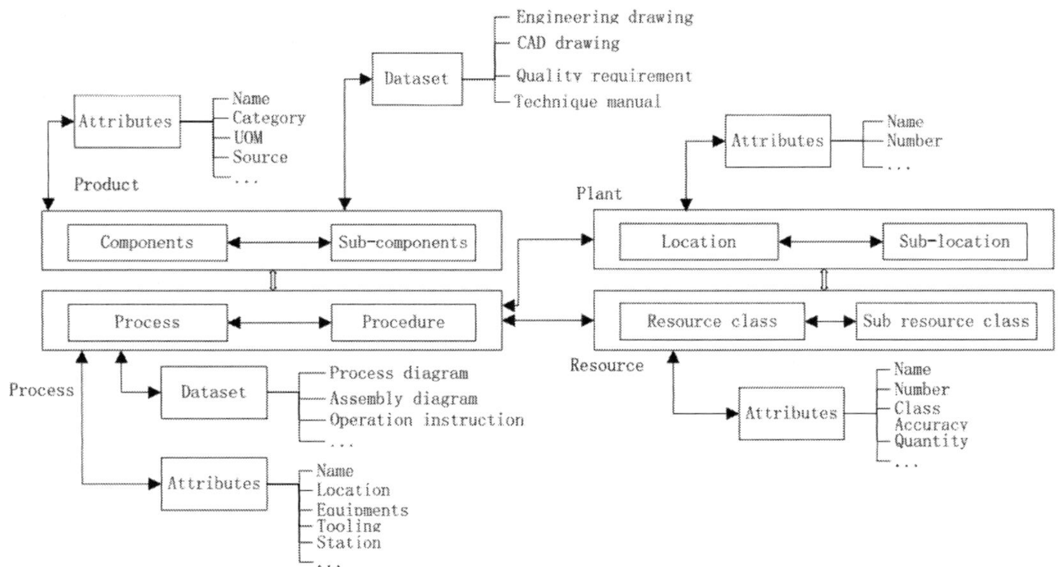

Figure 2. PPPR model

ProcessView is the view from assembly process planning, it contains the content of process planning for various components. The process of the components has independent process attribute information such as name, assembly location, equipment and tooling as well as corresponding technical document information. And, there exists the hierarchical relationships among the processes, the procedures and the operation units, for example, a process is composed of a set of procedures while one procedure consist of one or more operating units.

PlantView contains information about the location of the processed components, and each processing location has its own location attribute information such as location name, number, etc, and there are also the links among different position levels in the location model, such as the relationships among the factories, workshops, production lines and work stations.

ResourceView contains the information of the resources needed for component processing，such as various equipments, tooling, materials, etc，which could be managed through a classification system according to the processing capacity information (such as the processing accuracy, processing size, etc.), the resources' properties (such as name, tool size, etc.) and the relationship among the resources.

ProcessView is the center of the PPPR model，through which the information of man-hours of the components in productview, manufacturing resources information in resources view, the distribution information of the factories workshops, production lines in plantview and the process information in processview can be easily and quickly integrated. It can effectively avoid the information island by making full use of the independent and interrelated relationships among the objects, so as to form a structured process planning and management system which runs through the whole life cycle of a product.

4. System implementation and application

Process structure tree is the chain through which the process activities in manufacturing are connected according to the precedence relationship and the subordinate relationship between the process activities. The adjustment of the process structure tree is the foundation of the assembling process planning. Combining with PPPR model, the attributes information, location information, and resources information can be linked together to form the process object. Different hierarchy process objects are corresponding to the process structure tree, all resources are organized according to the purpose, and associated with the process objects, which makes the working process more clear.

1) To establish the assembly process structure tree

On the basis of DELMIA process structure according to the characteristic of enterprise, the process data can be gotten from DELMIA by system integration, then it will be reanalyzed and planned combining with the PPPR model, to establish the assembly process structure tree. The planning principles are as follows: A) the model of factory location should correspond with the location nodes in the process tree in DELMIA; B) location information of different level should correspond with the process objects of different levels. The assembly process structure tree is illustrated in Figure 3.

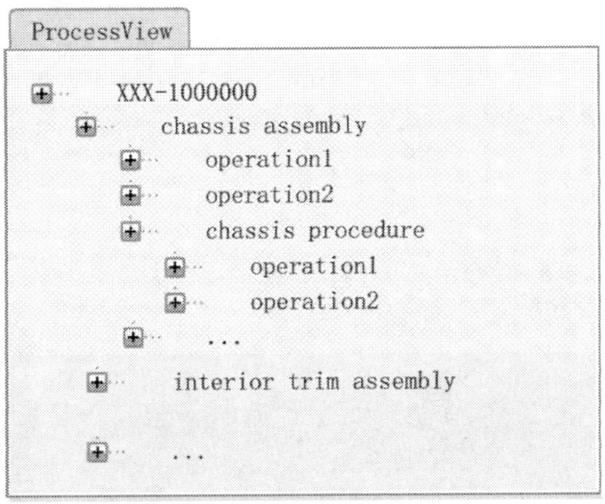

Figure 3. Assembly structure tree

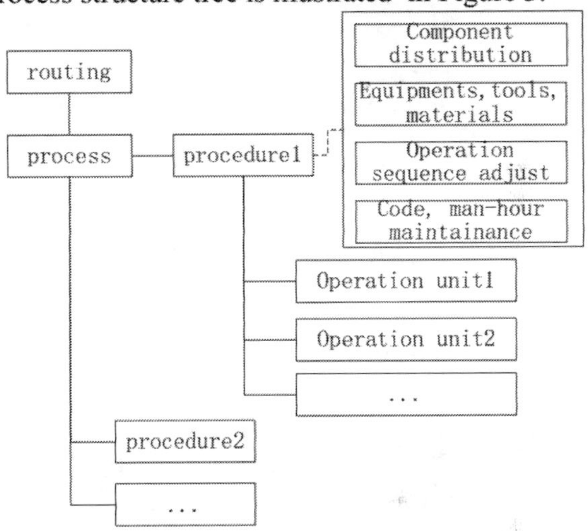

Figure 4. Consummate the process structure tree

2) To consummate the process structure tree

The process structure tree of the first edition needs to be consummated to satisfy the final assembly process planning. By the process object management model, the attribute information of the process objects in process structure tree is adjusted, if design change happens, new process object can be added according to the hierarchy structural relationship. In addition, according to the relationships among all the resources and process objects in PPPR model, the components, production materials and tools are distributed to the process objects of different levels. If the assembly sequence of the operation units under the procedure is changed, according to the relationship between the process object and the process structure tree, the order of the operating units can be adjusted, which can be reflected in the structure tree later. The completed process structure tree can meet the need of the final assembly process planning, as shown in Figure 4.

3) TS16949 documents management

The format of TS16949 documents is standardized, the professional terms and data commonly used in TS16949 quality system are stored in the repository through the application of CAPP knowledge repository, combing with the PPPR model, wherein each type of data resource has its attributes, it may also be classified and managed according to different resource properties. The data sets can be arranged in the form of a tree structure, which is easy to query and extract the relevant system resources. At the same time, the standard templates for each type of system documents, such as process flow diagrams, process FMEA[8] and control plans, are designed, thus, in every stage of APQP[8] implementation, the template of the corresponding class is invoked, the system will automatically pop up the properties box, the content can be selected to fill, then a complete system documents would be created after the execution; In addition, the associated linkage between the specific templates can be set, through which one template is filled, the other linked template will be filled with the similar content, when the data is modified or deleted, it also comply with this setup.

5. Summary

With assembly process planning and management as the object, the structured assembly process planning and management platform based on TS16949 quality system is developed. According to the requirements from an automotive enterprise assembly process, the main content and process of the assembly process planning and management is analyzed, and the overall structure of the CAPP system is constructed, the assembly process structure tree and process allocation model is designed based on PPPR management model. Also, the main functional modules of the system for process planning and management is analyzed and designed using the object-oriented method. In the end, the CAPP assembly process design and management system is implemented according to the characteristics of the process in the enterprise. Through the establishment and application of common process modules, final assembly process planning can be quickly completed. At the same time, the TS16949 specification is embedded in the structured process planning and applied in the automobile process planning and management. It can not only meet the requirement of the process management from TS16949 quality management system, but also improve the efficiency of the process planning and the use of R & D resources, reduce the burden of process workers.

Acknowledgements

This work was financially supported by Electronic and Information Industrial Fund (Contract No. 2012[407]), Ministry of industry and information technology, P.R.China

References

1. P.Wells, Sustainable business models and the automotive industry: A commentary. IIMB Manage. Rev., **25**,12(2013)
2. J.X.Xu, M.Z.Zou, X.G.Kong,etc. , Study on the development trend for modern CAPP technology. Manuf. Autom. **9**,5(2004)
3. J.Y.Lee, H.S.Kang, G.Y.Kim, S.D.Noh, Concurrent material flow analysis by P3R-driven modeling and simulation in PLM.Comput.Ind. **63**,15(2012)
4. C.Michael, ISO/TS 16949 QM in der Automobile industrie umsetzen. Carl Hanser Verlag, (2007)
5. C.P. Kartha. A comparison of ISO 9000:2000 quality system standards, QS9000, ISO/TS 16949 and Baldrige criteria. The TQM Magazine, **16**,10 (2004) ,
6. S.Kwon, K.K.Lee,Y.H.Park. A study on a methodology of integration lean DFSS and advanced product planning(APQP) in ISO/TS16949, Asian J.Qual, **8**,9(2013)
7. N.A.M.Salleh, S.Kasolang, A.Jaffar. Simulation of Integrated Total Quality Management (TQM) with lean manufacturing (LM) practices in forming process using Delmia quest. Procedia Eng, **41**, 6(2012)
8. A. P. Puvanasvaran, N. Jamibollah, N. Norazlin. Integration of POKA YOKE into process failure mode and effect analysis :a case study. Am.J.Appl.Sci., **11**, 11(2014)

MATEC Web of Conferences **51**, 02001 (2016)
DOI: 10.1051/matecconf/20165102001

A Study on Wind-Induced Smoke Extraction Systems for Building Fires

ShengBang LI[1], ChungHwei SU[1,a], Kuang-Chung TSAI[1]

[1]*Department of Safety, Health and Environmental Engineering, National Kaohsiung First University of Science and Technology, 82445 Kaohsiung City, Taiwan*

Abstract. Since smoke is one of the lethal factors at the scene of a fire, the effectiveness of a smoke control system is of great importance. Buildings are likely to be influenced by natural winds, so the effectiveness of a natural smoke extraction system is often questioned. This study discusses the feasibility of using the Venturi principle to improve a natural smoke extraction system for buildings under the effect of wind. The findings show that side smoke extraction is likely to be influenced by external air flow. Smoke extraction can clearly be improved if a negative pressure zone is formed outside the smoke vent.

1 Research motives and Purposes

1.1 Hazards of smoke at the scene of a fire

Populations are concentrated in cities in various countries where there is rapid growth in technology and the economy. In past years, as population densities increased, more accidental fires have caused casualties, increasing attention to the subject of building fire safety [1, 2].

Smoke at the scene of a fire can be lethal. The hazards of smoke include high temperature, toxicity and obscuration. The first two hazards can cause immediate death. The last hazard can negatively affect the evacuation of people and the work of rescue personnel [3]. "Smoke" is defined by the National Fire Protection Association (NFPA) as airborne solids and liquid particulates and gases evolved when a material undergoes pyrolysis or combustion, together with the quantity of air that is entrained or otherwise mixed into the mass" [4]. How to effectively control and remove the smoke at the scene of a fire is an important research subject.

1.2 Flow and control of smoke

At the scene of a fire, smoke moves with the air flow. The flow of the smoke can be influenced by a natural driving force and a forced driving force. The former one includes the stack effect, buoyancy, expansion and natural wind. The latter one includes an HVAC System and elevator piston effect [5].

A smoke control system can clearly enhance the safety of the residents in the buildings. When people are being evacuated, the smoke control system keeps the escape routes free from smoke that

[a] Corresponding author : such@nkfust.edu.tw, georgesu2000@gmail.com

obscures vision. These smokeless admission passages are created for the fire fighters to increase their rescue efficiency.

The control of smoke can be divided into "natural" and "forced" types. The natural smoke control system uses the elements inside the building to confine the flow of smoke to a region. This system is free of power consuming equipment. There is no need for emergency power when a fire happens [6].

1.3 Natural smoke extraction system under the effect of natural winds

Some scholars have proposed using the ventilation tower of a Venturi-shaped roof to increase the air rate and ventilation rate in the buildings. Some studies mention that this design could improve the indoor air quality and air change rate for cooling, especially in tropical regions [7]. However, this design has not been used for natural smoke extraction for buildings. This study will use Fire Dynamics Simulator (FDS) simulation software to analyze effective smoke removal from buildings on fire experiencing a strong ambient wind.

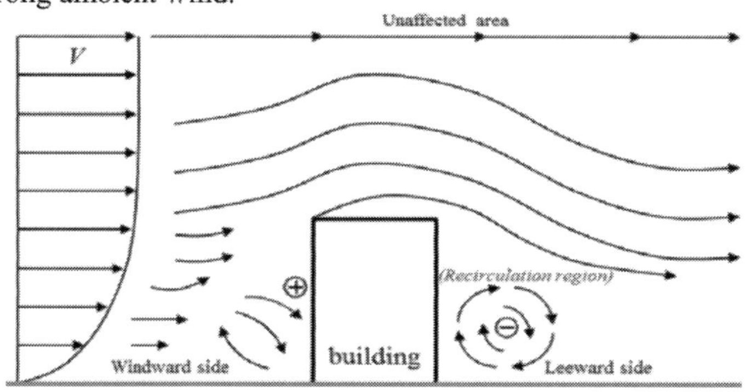

Figure 1. The impact of natural wind to buildings

2 Research method

2.1 Venturi principle

According to Bernoulli's law of fluid dynamics, static pressure decreases as the fluid velocity increases. Therefore, the velocity, increased through a reduced section of the fluid with any gain in kinetic energy, may be balanced by a pressure drop, as shown in Figure 2.

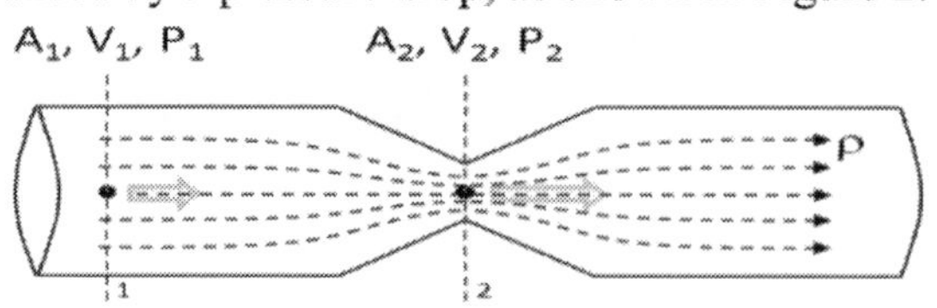

Figure 2. Venturi principle

Continuation equation $$Q = U_1 A_1 = U_2 A_1 \tag{1}$$

Bernoulli equation $$\frac{p_1}{\gamma} + \frac{U_1^2}{2g} = \frac{p_2}{\gamma} + \frac{U_2^2}{2g} + \Delta Z_2 \tag{2}$$

$$\frac{p_1 - p_2}{\gamma} - \Delta Z_2 = \frac{1}{2g}\left(U_2^2 - U_1^2\right) \tag{3}$$

The Venturi tube principle has been used in many experimental apparatuses for fluid. When the fluid flows through the flaring zone and throat zone in different sections of the Venturi tube, the different flow rates result in pressure differences [8].

The Venturi tube can be used to measure the fluid's flow rate. In engineering applications, a lot of equipment is set by using the Venturi principle, such as the foam mixing device for fire-fighting equipment. B. Blocken, T. van Hoff, et al. used the CFD and wind-tunnel experiment to research the effect of the different numbers of baffles on Venturi tube-shaped roofs of the same size [9].

2.2 Introduction to FDS software

This study used computer simulation software to build a fire simulation analysis, which is an extensively used method. The velocity, pressure, temperature and concentration represented by the grids of the fire location can be forecast accurately by the governing equation. The Field Model can simulate the actual conditions accurately, and this model is the one used most for simulation at present.

The FDS grid design must pay attention to both the efficiency and accuracy of the simulation. These two items are the main considerations in the simulation process. For the evaluation of an optimum grid, a fire source characteristic diameter computing mode is used to analyze the optimum grid size at the maximum heat release rate. McCaffery proposed the use of a minimum length scale for the fire plume as the characteristic fire diameter D * to determine the grid size [10]:

$$D^{*} = [\frac{\dot{Q}}{\rho_{\infty} C_{\infty} T_{\infty} \sqrt{g}}]^{2/5} \quad (4)$$

Herein:

$\dot{Q}$: total heat release rate, kW

g : gravitational acceleration, m/sec2

ρ_{∞}: air density, kg/m3

T_{∞}: space temperature, K

C_{∞}: air's specific heat, kJ/kg-K

When the model set grid size is about one-tenth that of the characteristic diameter (D*), the LES simulated time average axis velocity and temperature match McCaffery's experimental regression equation.

2.3 Wind scale

In 1926, the Beaufort wind scale was used to describe the effect of the natural wind speed on land and sea surfaces. The Taiwan Weather Bureau uses this standard at present, shown in Table 1 [11].

Table 1. Beaufort wind scale

Beaufort No.	Description	Wind speed (m/s)	Land conditions
0	Calm	< 0.3	Calm. Smoke rises vertically.
1	Light air	0.3–1.5	Smoke drift indicates wind direction. Leaves and wind vanes are stationary.
2	Light breeze	1.5–3.3	Wind felt on exposed skin. Leaves rustle. Wind vanes begin to move.
3	Gentle breeze	3.3–5.5	Leaves and small twigs constantly moving, light flags extended.

3 Fire scenario settings

In Taiwan, building fire safety equipment is installed according to the "Standard for Installation of Fire Safety Equipment Based on Use and Occupancy" [12]. The opening area of the smoke vent must be larger than 2% of the smoke compartment. The smoke must be discharged naturally.

This study simulates discharging the smoke by natural smoke extraction from a high-rise building on fire in a free flow field. The heat release rate of the fire source is 1 MW. The building size is 10 m(H) x 8 m(L) x 8 m(W), and the door size is 2 m(H) x 2 m(W), as shown in Figure 3. There are three vent window patterns, including:

Smoke vent on sidewall of building: According to regulations, the smoke vent is located 80 cm below the roof on the side of the building. The smoke vent area is 2% of the floor area.

Smoke vent at the top of building: The smoke vent is located at the top of the building. The opening area is 2% of the floor area.

Venturi-shaped roof on the building: The roof is located at the top. The area is 2% of the floor area. The smoke vent is located at the throat, and the throat height is 0.5 m [9].

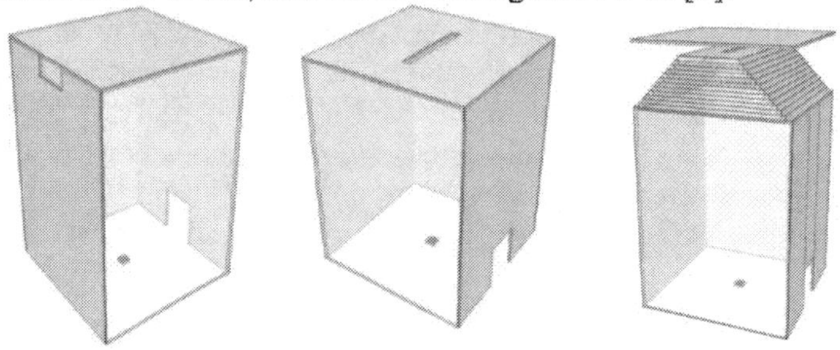

Figure 3. Exterior appearances of the simulation models

1. Space description: 48 m(L) x 40 m(W) x 45 m(H), 19 million cells.
2. Combustion regime: the fire source begins to burn when the air flow is stable after 30 seconds of simulation. The fire behavior is maximized immediately.
3. Fire source: the heat release rate is 1,000 kW in a fixed range. The maximum heat release rate is reached immediately. The chemical reaction of the fuel is the program's predetermined polyurethane.
4. Simulation time: 270 seconds.
5. Recorded items: three smoke vents are simulated under the natural ventilation effect. A flow rate variation of the smoke discharged from the smoke vent is observed. The pressure, temperature and flow rate at the monitoring points are recorded.

The fire scene simulation parameters are compiled in Table 2.

Table 2. The parameters of FDS simulation

Heat release rate(kW)	**D*(m)**	**0.1 D*(m)**	**Grid size(m)**
1,000	0.96	0.096	0.1
Zone	Model Size L(m)xW(m)xH(m)		Grid size (m)
Entire space	48×40×45		0.1×0.1×0.1
Zone #1	10×10×45		
Zone #2	11×10×45		
Zone #3	15×10×45		
Zone #4	9×10×45		
Zone #5	5×10×45		
Zone #6, #7	48×6×45		0.2×0.2×0.2
Zone #8, #9	48×9×45		0.6×0.6×0.6

4 Results and discussion

4.1 Preliminary test

The parameter setting conditions are imported into the executable file before simulation. The FDS is used for a 270 second simulation. The boundary condition of the natural wind velocity is Scale II of the Beaufort wind force scale, set as a mean of 2.4 m/s. In order to display the variation curve of natural wind from the ground to the upper space, the boundary is divided into 15 regions.

Equation $u = u_{ref}\left(z/z_{ref}\right)^{p}$ is used, where u is the average wind speed at height z; u_{ref} is the free flow field wind speed at height z_{ref} and p is the Hellman exponent. The wind speed decreases by degrees in each region, and the wind speed at the height of the building's roof is set at 2.4 m/s to simulate the pattern of the wind velocity gradient curve, from 4.24 to 1.10 m/s.

4.2 Smoke vent on side wall of building

A smoke vent is located on the wall within 80 cm below the building roof. The opening is 2% of the floor area. The fire source begins to burn when a steady state is reached after 30 seconds of simulation. Under the effect of buoyancy and makeup air inside the building, the fume flows upward. Under the effect of an external wind, the windward smoke vent has resistance; the smoke cannot be discharged through the smoke vent at the beginning, as shown in Figure 4 (A). The dense smoke accumulates downward under the ceiling.

The sustained combustion increases the smoke accumulated under the ceiling and the hot air driving force in the building. The smoke is discharged through the smoke vent. Figure 4 (B) shows the smoke extraction velocity profile at 240 sec. The outward air flow velocity at the smoke vent is 4.22 m/s. The outside smoke flow rate is increased by the natural wind. There is still resistance generated by the external natural wind at the smoke vent, extruding into the building. The smoke extraction effect is obviously inhibited.

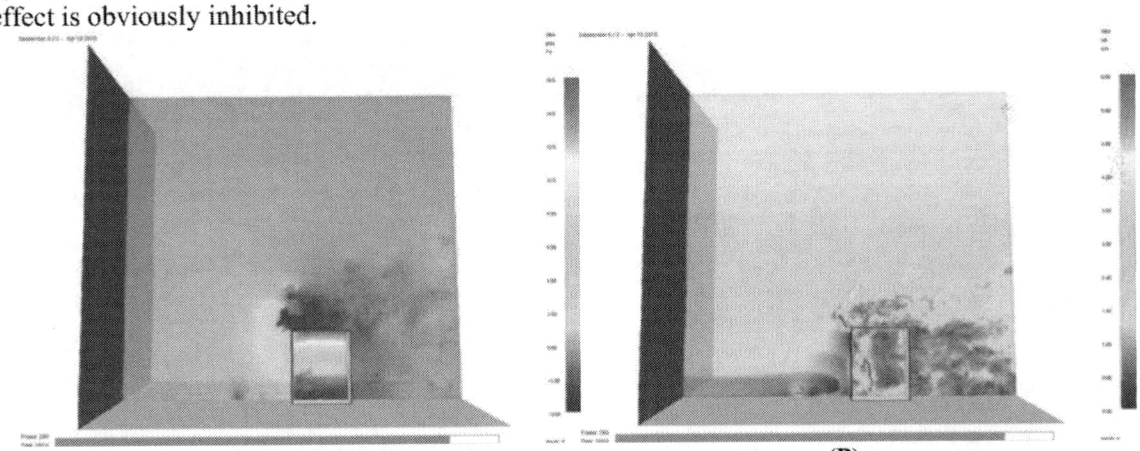

(A) (B)

Figure 4(A). Pressure distribution of smoke vent on sidewall(@240 second) **(B).** Velocity distribution of smoke vent on sidewall(@240 second)

4.3 Smoke vent on top of building

The air flow is in a steady state after 30 seconds of simulation. A negative pressure zone is formed externally above the building. The smoke vent is located in this region. The smoke flows up to the upper space rapidly under the thermal buoyancy. Figures 5 (A) and (B) show that at the initial stage, there is a negative pressure zone outside above the building. The high pressure in the building is released towards the smoke vent. There is a low pressure of 1.16 Pa outside above the building at 240 sec, forming a pressure difference from the internal high pressure, so that a 5.74 m/s high-velocity air blast is created, discharging the smoke and hot air from the building. As there is a negative pressure zone outside the smoke vent in this case, the smoke accumulates in the building slowly. The upper smoke vent has a good smoke extraction effect, but it needs improvement because when it rains, water or foreign materials enter the building.

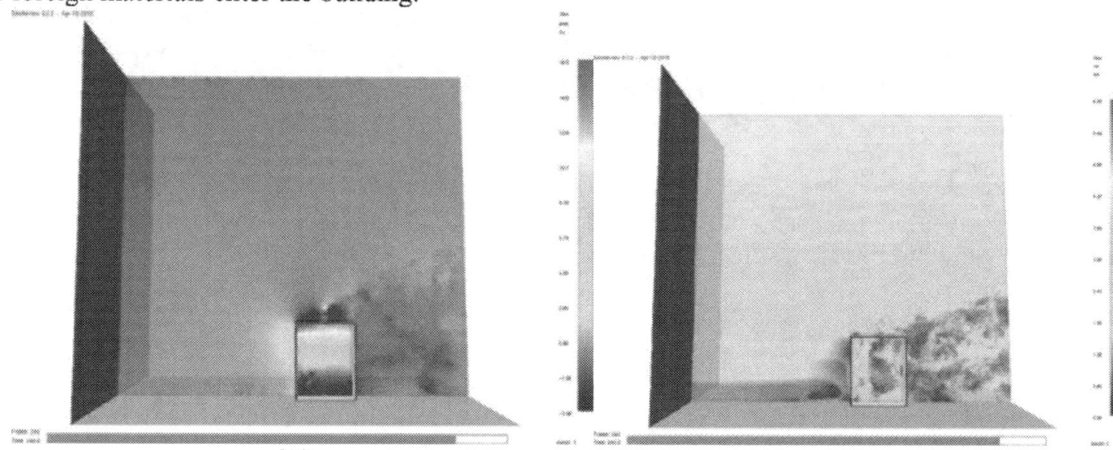

(A) (B)

Figure 5(A). Pressure distribution of upper smoke vent(@240 second) **(B).** Velocity distribution of upper smoke vent(@240 second)

4.4 Buildings with a Venturi-shaped roof

The Venturi-shaped roof is used for improvement in this study. The Venturi-shaped roof uses the Venturi principle to enhance smoke extraction. When the air flow is stable at 30 sec, a negative pressure zone is formed at the throat of the smoke vent. When the static pressure decreases, the dynamic pressure increases, resulting in a high flow velocity. At 66 sec of simulation, the maximum negative pressure 6.94 Pa is formed in the throat zone.

Figures 6 (A) and (B) show that the wind pressure resistance on the windward side generates a natural wind through the throat zone to form negative pressure due to its special design. The pressure increases gradually with the height inside the building to 240 sec. The high-pressure smoke flows towards the throat's negative pressure zone, discharging the smoke rapidly. The smoke is blown away from the building by the air flow at the throat.

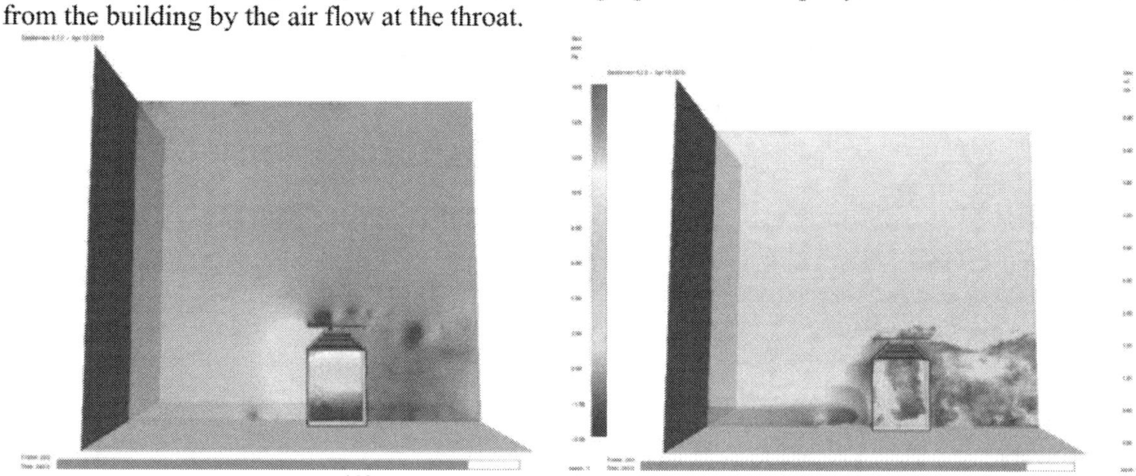

(A) (B)

Figure 6(A). Venturi-shaped roof pressure distribution(@240 second) **(B).** Venturi-shaped roof velocity distribution(@240 second)

4.5 Comparison of various cases

Figure 7 shows that some air is exhausted through the smoke vent at the steady stage of air flow, especially for the side smoke vent. When the fire begins to burn, the flow rate at the smoke vents at the top of the building and the Venturi-shaped roof increase rapidly; the 5.6 m/s extraction velocity is implemented immediately.

Figure 8 indicates the temperature of the smoke at the smoke vent. It is observed that the temperature rises gradually at the initial stage of smoke extraction. The smoke vents at the top of the building and the Venturi-shaped roof discharge smoke immediately when the smoke is formed. Smoke extraction through the side smoke vent is delayed. The curve jitter in the red circle of Figure 8 represents an eddy formed in the building, so that the discharged smoke volume and the temperature decrease. The temperature at the outlet increases with the exhaust's smoke level. The temperature at the side smoke vent is higher than the other two patterns as the air flow is relatively slow when the smoke extraction is stable.

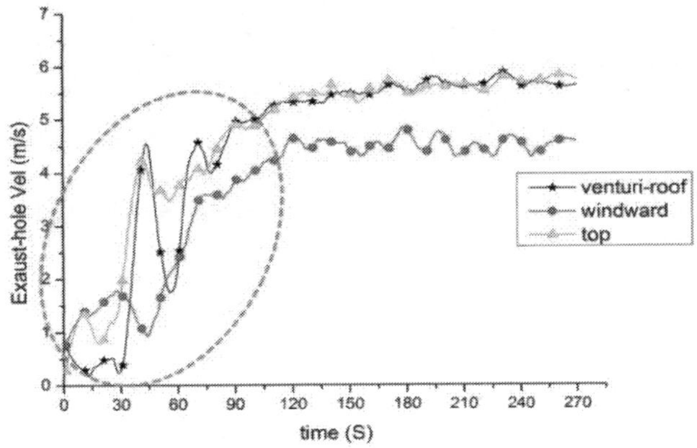

Figure 7. Velocity curves of three smoke vents

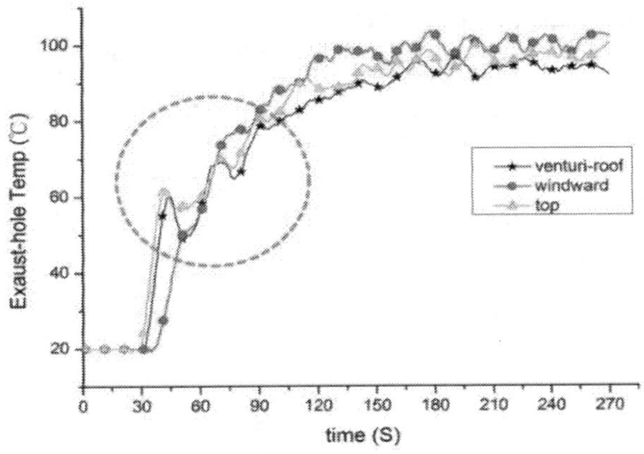

Figure 8. Temperature curves of three smoke vents

5 Conclusion

Fires cause more casualties in buildings with increased height. The effectiveness of building fire equipment has been studied a lot. Fire simulation software FDS is used in this paper to analyze the pressure, temperature, velocity and plume flow at the scene of a fire. The building is provided with three scenarios for smoke vents to evaluate the effectiveness of smoke extraction. The analysis's results show that when the smoke vent is located on the side, the smoke is slowly discharged through the smoke vent under the effect of the windward side; the smoke is likely to accumulate, resulting in high internal temperatures. If a negative pressure zone is formed outside the smoke vent, smoke extraction can obviously be improved.

Acknowledgement

The authors gratefully acknowledge the financial support from the Ministry of Science and Technology under Grant no. NSC 102-2221-E-274-001-MY3.

References

1. DiNenno, P. J. SFPE handbook of fire protection engineering (SFPE,2008)
2. Su, C. H., Tsai, K. C. and Xu, M. Y. Journal of Building Performance Simulation. Computational analysis on the performance of smoke exhaust systems in small vestibules of high-rise buildings, **8(4)**, 239-252 (2015)
3. Kobes, M., Helsloot, I., de Vries, B. and Post, J. G. Fire Safety Journal. *Building safety and human behaviour in fire: A literature review* , **45(1)**, 1-11 (2010)
4. National Fire Protection Association, NFPA 921: Guide for Fire and Explosion Investigations. *Chapter 3: Definitions*, pp.18 (2014)
5. W. Conshohocken. ASTM International, *ASTM E176-04 Standard Terminology of Fire Standards*, (2004)
6. Yamana, T. and Tanaka, T. Fire Science and Technology. *Smoke control in large scale spaces*, **5(1)**, 41-54 (1985)
7. van Hooff, T., Blocken, B., Aanen, L. and Bronsema, B. Building and Environment. *A venturi-shaped roof for wind-induced natural ventilation of buildings: wind tunnel and CFD evaluation of different design configurations* ,**46(9)**, 1797-1807 (2011)
8. Spoerel, W. E., Narayanan, P. S. and Singh, N. P. British journal of anaesthesia. *Transtracheal ventilation*, **43(10)**, 932-939 (1971)
9. Blocken, B., van Hooff, T., Aanen, L. and Bronsema, B. Computers and Fluids. *Computational analysis of the performance of a venturi-shaped roof for natural ventilation: venturi-effect versus wind-blocking effect*, **48(1)**, 202-213 (2011)
10. Baum, H. R. and McCaffrey, B. J. Fire Safety Science. *Fire induced flow field-theory and experiment*, **2**, 129-148 (1989)
11. Wheeler, D. and Wilkinson, C. The Mariner's Mirror. *From calm to storm: the origins of the Beaufort wind scale*, **90(2)**, 187-201 (2004)
12. National Fire Agency. *Standard for Installation of Fire Safety Equipments Based on Use and Occupancy. Ministry of the Interior, Taiwan* (2012)

MATEC Web of Conferences **51**, 02002 (2016)
DOI: 10.1051/matecconf/20165102002

Development of Compact Flexible Displacement Sensors Using Ultrasonic Sensor for Wearable Actuators

Tetsuya Akagi[1,a], Shujiro Dohta[1], Takafumi Morimoto[1], Yasuko Matsui[1] and So Shimooka[1]

[1] *Department of Intelligent Mechanical Engineering, Okayama University of Science, Japan*

Abstract. In position control of wearable actuator such as a rubber artificial muscle, a compact flexible displacement sensor is much attractive and required. In this paper, two types of flexible displacement sensor using the ultrasonic sensor were introduced. One is a built-in displacement sensor for rubber artificial muscle. Another is a sensor that can measure the sliding displacement on a flexible tube for flexible robot. Both sensors use ultrasonic displacement sensors. The construction, operating principle and measuring performance of two sensors were also described.

1 Introduction

Recently, pneumatic welfare systems to support a nursing care and a rehabilitation of the elderly and the disabled are actively researched and developed[1-4]. In the system, many wearable actuators such as a rubber artificial muscle have been used[5-7]. The systems using the wearable actuators had been sold in Japanese market. However, a position feedback control had not been embedded into the system because of lack of low-cost and compact flexible displacement sensor for these actuators. The target of this study is to develop a low-cost and compact flexible displacement sensor for a wearable actuator. In the previous study, the indirect displacement measurement using an inner diameter sensor of rubber artificial muscle had been proposed and tested[7-9]. To decrease the measuring error, the direct measurement is useful. In addition, the flexible robot using flexible pneumatic cylinders was proposed and tested in the previous study[10-11]. In the attitude control system of the robot arm, a compact flexible displacement sensor is also required. As a compact flexible displacement sensor, an ultrasonic sensor is selected because it has many advantages such as non-contacting measuring and relatively inexpensive.

2 Built-in displacement sensor for rubber artificial muscle

Figure 1 shows a schematic diagram of a tested measuring system of a rubber artificial muscle (Festo Co. Ltd., MXAM-20-AA, original length of 220 mm) with a built-in sensor and an embedded controller (Renesas Co. Ltd. H8/3664). For applying the typical ultrasonic sensor on the market (Parallax Inc. Ltd., 28015) into the pressurized environment, the smaller transmitter (T4008A1) and receiver (R4008A1) produced by Nippon Ceramic Co. Ltd were replaced and installed into the rubber artificial muscle. The measuring procedure of the ultrasonic sensor using the embedded controller is

[a] Corresponding author : akagi@are.ous.ac.jp

mentioned more detail in next section. The counting value corresponding to length of muscle is converted to the analogue signal in a D/A converter on the measuring board.

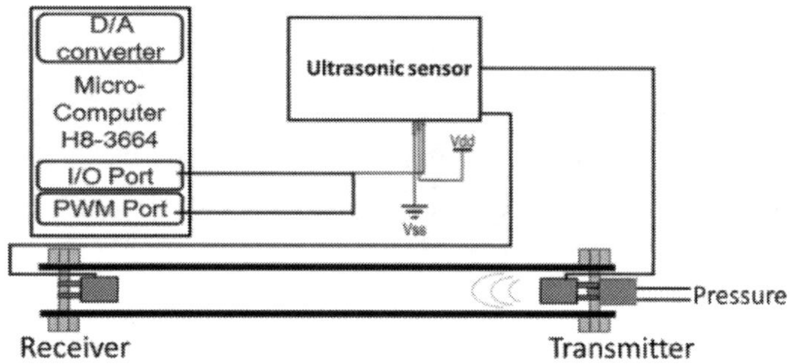

Figure 1. Tested measuring system of rubber artificial muscle.

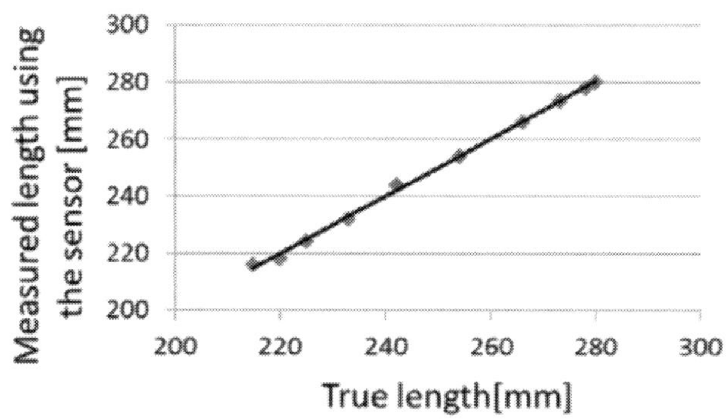

Figure 2. Comparison of measured and true length of the muscle.

Figure 2 shows the comparison between the measured length using tested sensor and the true length of the muscle measured by using a typical potentiometer on the market. In the experiment, the rubber artificial muscle was pressurized from 0 kPa (the length of the muscle: 280 mm) to 500 kPa (216mm). In Fig.2, the symbol shows the experimental result, the solid line shows the ideal case when the measured displacement is equal to true displacement. From Fig.2, it can be seen that the relation between the sensor output and the true length of the muscle is linear. This result means that the tested ultrasonic sensor is useful as a built-in displacement sensor of the muscle. The theoretical resolution calculated from the sampling period of counting and the speed of sound (343.26 m/s at 20 degree Celsius) of the sensor is about 0.2 mm.

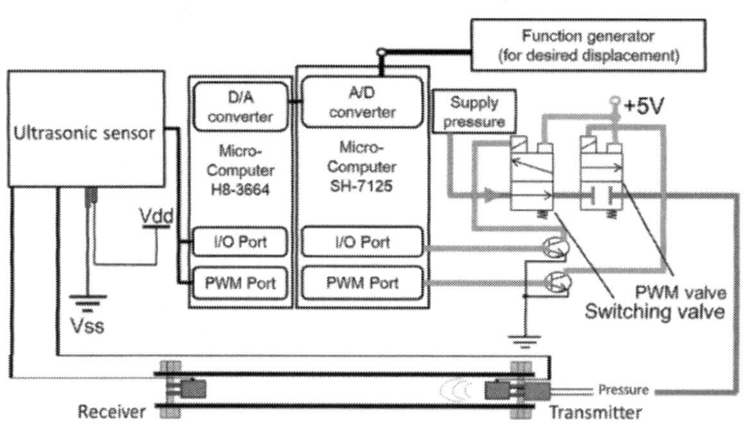

Figure 3. Positon control system using the rubber artificial muscle with a built-in displacement sensor.

Figure 3 shows the position control system using the tested muscle with built-in displacement sensor. The position control system consists of the muscle with built-in displacement sensor, the tested measuring system as shown in Fig.1, a position controller using embedded controller (Renesas Co. Ltd., SH7125) and a quasi-servo valve that consist of PWM controlled and switching on/off valves[12]. The position control is done as follows. First, the embedded controller for the ultrasonic sensor counts the time of the output signal from the tested sensor through I/O port. The embedded controller sends the measured displacement of the muscle length to another embedded controller as a voltage signal through the extra D/A converter. The displacement deviation from the desired is calculated by the controller for positioning and the control input for the quasi-servo valve is also calculated based on P control scheme. Finally, the control input is applied to the quasi-servo valve as PWM signal and on/off signal. The sampling period of control is 15 ms.

Figure 4 shows the transient response of displacement of the muscle controlled by using the tested system. In Fig.4, blue broken and red solid lines show the desired and controlled position, respectively. It can be seen that the position control can be realized by using the tested sensor.

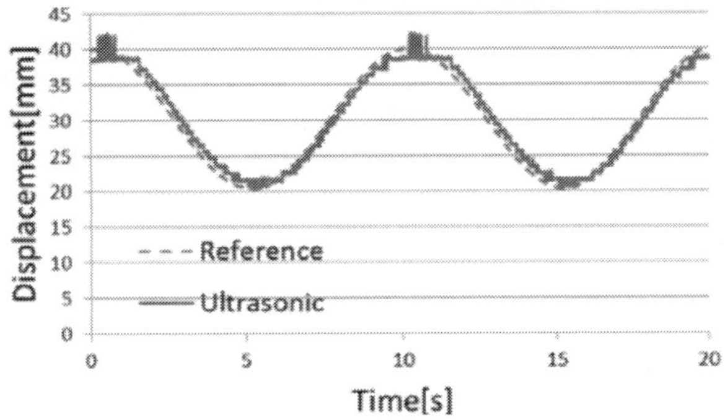

Figure 4. Transient response of displacement of the rubber artificial for positon tracking control.

3 Flexible displacement sensor using flexible tube

As an another sensor using ultrasonic sensor, the flexible sliding displacement sensor using flexible tube was proposed and tested. The sensor was developed to measure the displacement of the flexible robot arm for wrist rehabilitation device [11] as shown in Fig.5 on the left so as to construct the compact and low-cost measuring system with flexibility.

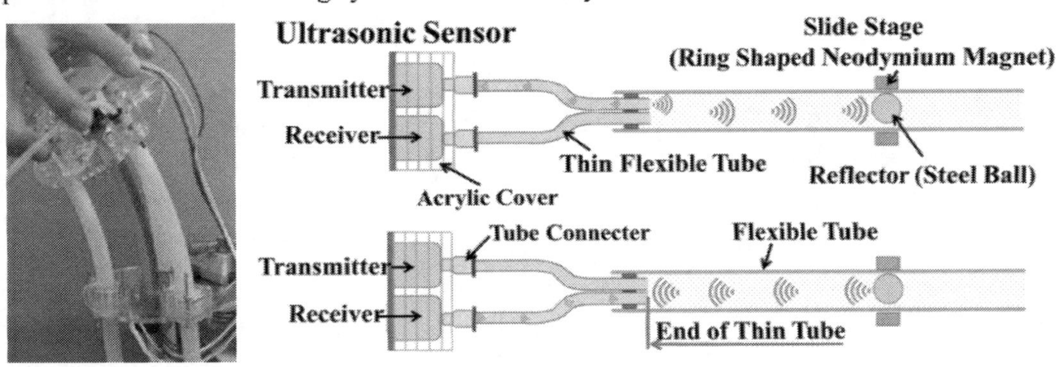

Figure 5. Flexible robot arm using flexible pneumatic cylinders and flexible sliding displacement sensor.

Figure 5 on the right shows the construction of the proposed sensor using same ultrasonic sensor. The tested sensor consists of the ultrasonic sensor mentioned above, an acrylic cover with two tube connecters and a thick flexible tube with an inner/outer diameter of 8/12 mm connected to two thin flexible tubes with an inner/outer diameter of 2.5/4 mm. As a sliding stage, the ring-shaped neodymium magnet with an inner/outer diameter of 13/25 mm, thickness of 5 mm and an acrylic cylindrical reflector with an outer diameter of 7.5 mm and length of 9 mm that a steel ball with an outer diameter of 5 mm were used. Figure 6 shows the construction and operating principle of the measuring system using the tested sensor. The measuring system consists of the tested sensor, the slide stage and an embedded controller (Renesas Co. Ltd., SH7125). I/O ports of the embedded controller are connected to the ultrasonic sensor to control the sensor. The cost of the tested sensor that includes the ultrasonic sensor and a magnet is low, that is about 35 U.S. dollars.

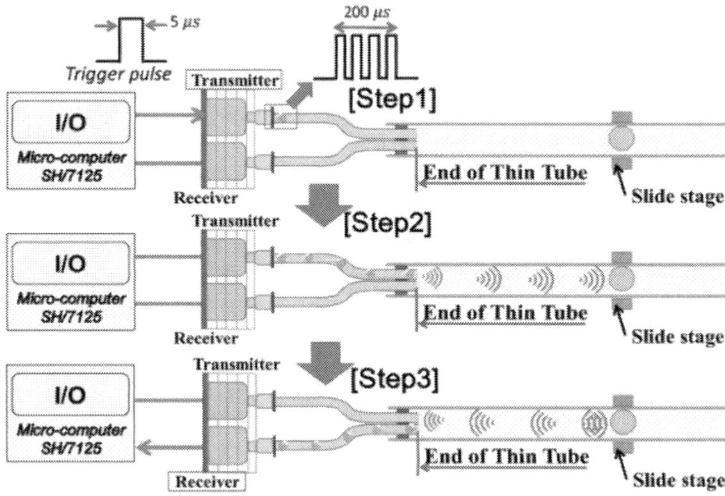

Figure 6. Construction and operating principle of measuring system using flexible sliding displacement sensor.

The measuring procedure is as follows. First, the embedded controller sends the trigger pulse for 5 μs. Then, the transmitter is driven for 0.2 ms with frequency of 40 kHz (Step 1 in Fig.6). The generated ultrasonic wave passes through the thin tube as a transmission tube. It reaches at the end of the tube, and is transmitted to the thick tube(Step 2 in Fig.6). After that, the ultrasonic wave reflects at the slide stage, and reaches at the ultrasonic receiver through another thin flexible tube (Step3 in Fig.6). The embedded controller can calculate the distance between the ultrasonic sensor and the slide stage by measuring the time (counting value) until the receiver detects the ultrasonic wave from the transmitter. By this method, the tested sensor can measure the displacement of slide stage even if the flexible tube bends.

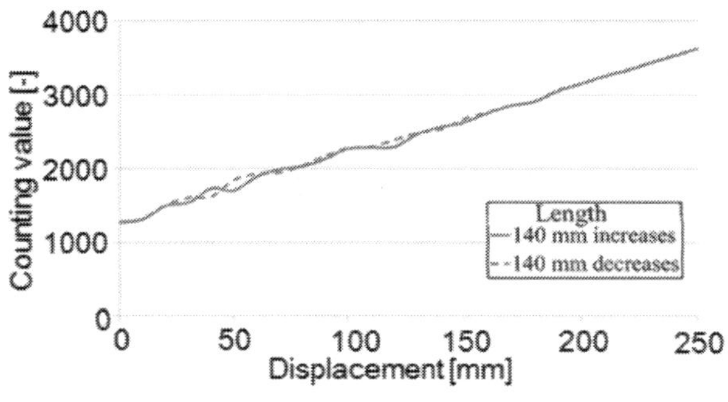

Figure 7. Relation between counting value and displacement.

Figure 7 shows the relation between the displacement of the slide stage and the counting value by using the thin tube with length of 140 mm. In Fig.7, the solid and broken lines show the case that displacement of slide stage increases and decreases, respectively. It can be found that it has relatively good linear relationship. The displacement resolution of the sensor can be calculated from the counting sampling period of counting (0.602 μs) and speed of sound (343.26 m/s), that is about 0.1 mm. The sampling period for measuring using the embedded controller is about 4 ms at in the case of measuring displacement of 0.2 m. It is also found that it could measure the displacement even if the thin and thick tubes bends.

4 Conclusions

Two types of flexible displacement sensor using the ultrasonic sensor and its measuring system using embedded controller were proposed and tested. The measuring performance of both measuring systems were investigated. As a result, it can be confirmed that the tested system could be applied to the positioning system of the rubber artificial muscle and the flexible robot arm using flexible pneumatic cylinder because of available resolution of 0.1 to 0.2 mm and relatively faster sampling period of 4 or 15 ms for rehabilitation device that requires slower motion.

Acknowledgement

This work was partially supported by the Ministry of Education, Culture, Sports, Science and Technology of Japan through a QOL Innovative Research Program (2012-) and Grant-in-Aid for Scientific Research (C) (Subject No. 24560315).

References

1. T. Noritsugu, M. Takaiwa, D. Sasaki, Development of Power Assist Wear Using Pneumatic Rubber Artificial Muscles, *Journal of Robotics and Mechatronics*, Vol. 21, 607-613(2009).
2. H. Kobayashi, T. Shiban, Y. Ishida, Realization of all 7 motions for the upper limb by a muscle suit, *Journal of Robotics and Mechatronics*, Vol. 16, 504-512(2004).
3. M. Ishii, K. Yamamoto, and K. Hyodo, Stand-Alone Wearable Power Suit -Development and Availability-, *Journal of Robotics and Mechatronics*, Vol. 17, No.5, 575-583(2005).
4. J. Piquion, A. Nayar, A. Ghazaryan, R. Papann, W. Klimek, and R. Laroia, Robot-assisted gynecological surgery in a community setting, *Journal of Robotics and Surgery*; Vol.3, Issue 2, 61–64(2009).
5. Y. Nagata, ed., *Soft Actuators -Forefront of Development-*, NTS Ltd., 291-335(2004).
6. T.Akagi, S.Dohta, Development of Wearable Pneumatic Actuator and Multiport Pressure Control Valve , *Journal of Robotics and Mechatronics*, Vol.17, No.5, 529-536(2005).
7. T. Akagi, S. Dohta, M.Ihara, Improvement of McKibben Artificial Muscle with Long Stroke Motion and Its Application, *Journal of System Design and Dynamics*, Vol.4, No.4, 538-551 (2010),.
8. T.Akagi, S.Dohta, Y.Kenmotsu, F.Zhao, M.Yoneda, Development of smart inner diameter sensor for position control of Mckibben artificial muscle, *Journal of Procedia Engineering*, Vol.41, 105 – 112(2012).
9. T.Akagi, S.Dohta, Y.Kenmotsu, S.Jinno, K.Taniguchi, Model Based Axial Displacement Measurement of Rubber Artificial Muscle Using Inner Diameter Sensor, *Lecture Notes in Electrical Engineering* 293, Vol 1, Springer, 349-357(2014).
10. T. Morimoto, Mohd Aliff, T.Akagi , S.Dohta, Development of Flexible Pneumatic Cylinder with Backdrivability and Its Application, *International Journal of Materials Science and Engineering*, Vol.3, No.1, 7-11(2015).
11. Mohd Aliff, S. Dohta, T. Akagi, T. Moimoto, Control of Flexible Pneumatic Robot Arm Using Master Device with Pneumatic Brake Mechanism, *JFPS International Journal of Fluid Power System*, Vol.8, No.1, 38-43(2015).
12. Y. Moriwake, T.Akagi, S.Dohta, F.Zhao, Development of low-cost pressure control type quasi-servo valve using embedded controller, *Journal of Procedia Engineering*, Vol.41, 143-149(2012).

MATEC Web of Conferences 51, 02003 (2016)
DOI: 10.1051/matecconf/20165102003

Development of Wearable Wrist Rehabilitation Device Using Flexible Pneumatic Cylinders

Naoki Kato, Shujiro Dohta, Tetsuya Akagi[a], Takafumi Morimoto, Yuma Soga

Department of Intelligent Mechanical Engineering, Okayama University of Science, 1-1, Ridai-cho, Kita-ku, Okayama, 700-0005,Japan

Abstract. Recently, the rehabilitation device using a soft pneumatic actuator for disabled are actively researched and developed because of the insufficient number of PT. In the previous study, the wrist rehabilitation device using flexible robot arm that consisted of three flexible pneumatic cylinders was proposed and tested. In this study, the wearable wrist rehabilitation device is proposed and tested. The attitude control based on the analytical model of the devise is also carried out. As a result, it is confirmed that the device can trace the desired position in the condition when a human wears the device.

1 Introduction

According to aging society and decreasing birth rate[1], lack of welfare workers and physical therapists ("PT" for short) becomes serious concern. The rehabilitation devices using soft pneumatic actuators for disabled are actively researched and developed[2-5]. In the previous study, the wrist rehabilitation device using flexible robot arm that consists of three flexible pneumatic cylinders[6] was proposed and tested. The attitude control of the robot arm while the hand is being put on the robot arm was carried out[7]. The robot arm with simple back drivability was also developed[8]. In both cases, the hands of patients must be put on the robot on the table. It is difficult to use it while lying down. In the next step, the wearable type rehabilitation device for wrist is required. In this study, a wearable wrist rehabilitation device using improved flexible robot arm that can attach with human hand and arm. The attitude control of the device while wearing on human arm is carried out.

2 Previous flexible robot arm for wrist rehabilitation

Figure 1 shows the flexible robot arm for wrist rehabilitation and construction of flexible pneumatic cylinder. The robot arm consists of two round stages and three flexible pneumatic cylinders. The cylinder consists of a flexible tube as a cylinder, the steel ball as a cylinder head, and a slide stage. The slide stage has two brass rollers set on the inner bore of the stage to press and deform the tube. The steel ball is held by two slide stages from both sides of the ball. When the pressure is applied to one side of the cylinder in the robot arm, the tube is moved up and down while holding the slide stage. Each flexible pneumatic cylinder is arranged so that the central angle of two adjacent slide stages becomes 120 deg. on the stage. An end of each flexible cylinder is fixed in the upper stage. The robot

[a] Tetsuya Akagi : akagi@are.ous.ac.jp

arm has the outer diameter of 100 mm and the length of 250 mm. The total mass of the robot arm is 380g. The operating principle of the arm is as follows. When the pressure is supplied on the three top ends of the cylinders, the robot arm extends or moves upward. On the other hand, the robot arm will contract or move downward when the pressure is supplied on the three bottom ends of the cylinders. The bending motion of the arm can be obtained when one cylinder is fixed and other two cylinders are pressurized from bottom ends. By this method, the arm is bent toward opposite side of the fixed cylinder. The robot arm can be also bent to every direction. In a rehabilitation, the arm is used while the upper stage of the robot is grasped by a human hand. However, the device can not be used while lying down. Therefore, it is necessary to develop a wearable type wrist rehabilitation device.

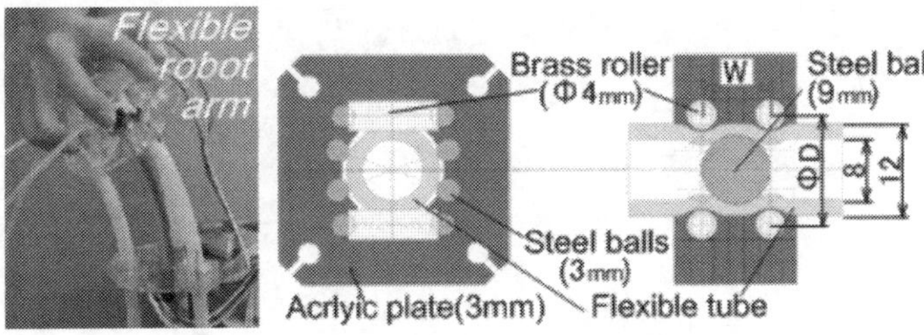

Figure 1. View of flexible robot arm and the inner construction of flexible pneumatic cylinder.

3 Wearable wrist rehabilitation device

Figure 2 shows the tested wearable type wrist rehabilitation device using flexible pneumatic cylinders. The fundamental construction of the device is similar to the flexible robot arm. Compared with the robot arm, the device is set so as to inverse the upper and lower stages. The end stage that is connected with the slide stages of the flexible pneumatic cylinders has a handle. The other stage that is connected with ends of the cylinders has a hole with inner diameter of 100 mm so as to insert a human arm. In other word, the patient arm passes through the holes of the base stage, and the patient holds the handle while its working. Each cylinder is also arranged with radius of 87.5 mm every 120 deg. from the center of the disk. The device also has the attitude control system that consists of an embedded controller (Renesas Co. Ltd., SH7125) and six quasi-servo valves[9] that each valve consists of two on/off control valves. The embedded controller and three quasi-servo valves are mounted on the base stage. On the end stage, there are three quasi-servo valves and an accelerometer for measuring the inclined angle of the stage. To supply the compressed air from base to end stage, a coil type tube that covers on a flexible pneumatic cylinder was used. By this method, the device requires only an electric power cable and an air supply pipe to drive it. The device has the outer diameter of 200 mm and the length of 420 mm. The total mass of the device including a controller and six quasi-servo valves is 1.14kg.

Figure 2. Construction of wearable wrist rehabilitation device using flexible pneumatic cylinders

4 Analytical model and attitude control of device

Figure 3 shows the definition of the cylinder length and an analytical model of the flexible robot arm, that is wearable rehabilitation device. In this model, the shape of the flexible pneumatic cylinder is assumed to be a circular arc when the device is bent. From the center of the device, the bending angle from X axis is called a bending direction angle α while the bending angle β is defined as the angle between the normal vectors from the center of the upper surface of the Z axis of robot arm. From the geometric relationship, the following equations about cylinder length L_1, L_2, L_3 and the radius of curvature R can be obtained.

$$L_1 = (R - r \cdot \cos\alpha) \cdot \beta \tag{1}$$

$$L_2 = \{R - r \cdot cos(\frac{2\pi}{3} - \alpha)\} \cdot \beta \tag{2}$$

$$L_3 = \{R - r \cdot cos(\frac{4\pi}{3} - \alpha)\} \cdot \beta \tag{3}$$

$$R = L/\beta \tag{4}$$

where, L means the length of human arm, and r = 87.5 mm is the distance from the center to the cylinder. Angles α and β can be obtained from the output voltage of the accelerometer.

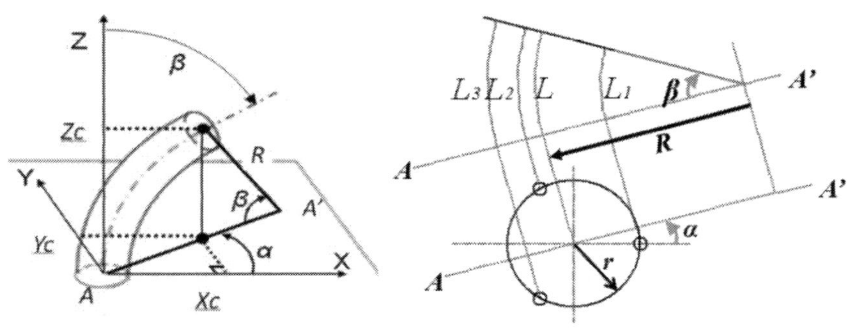

Figure 3. Analytical model of wearable device (flexible robot arm)

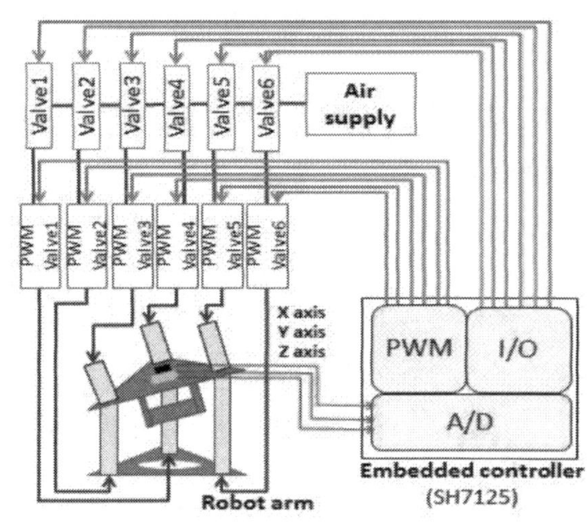

Figure 4. Schematic diagram of control system

Figure 4 shows the schematic diagram of control system of wrist rehabilitation device. The system consists of the flexible robot arm using three flexible pneumatic cylinders, an accelerometer, the embedded controller (Renesas Co. Ltd., SH7125) and six quasi-servo valves. The quasi-servo valve consists of two on/off type valves that one is a switching valve for supply or exhaust and another is PWM controlled valve for adjusting flow rate. The attitude control of the device is as follows. First, the embedded controller gets the output voltages from the accelerometer through A/D converter. Each length of the flexible pneumatic cylinder is calculated based on the model as shown in Fig.3. The embedded controller also calculates the deviation from the desired position for each cylinder. The quasi-servo valves are driven according to the control scheme. The desired position is set on the embedded controller previously.

Figure 5 shows the transient response of each cylinder length in attitude control. In the control, PID control scheme with sampling period of 7 ms was used. The length L was set to be 150 mm. In Fig.5, the broken and solid lines show the desired and controlled length of each cylinder, respectively. It can be seen that the cylinder can reach the desired length within about one second even if the force from the human wrist is applied to the device as a load. It can be confirmed that the tested device is useful to apply for self-rehabilitation device.

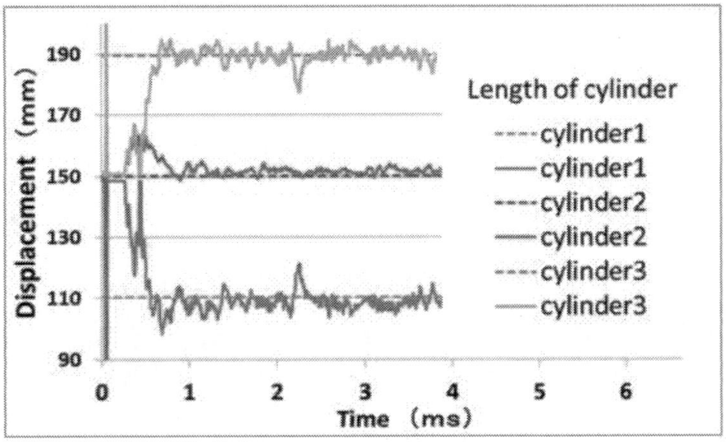

Figure 5. Transient response of cylinder length in attitude control

5 Conclusions

The wearable wrist rehabilitation device using three flexible pneumatic cylinders, an embedded controller, six quasi-servo valves and an accelerometer is proposed and tested. The attitude control based on the analytical model is also carried out. As a result, it is confirmed that the device can trace the desired position within about one second in the condition when a human wrist works as a load.

Acknowledgements

This research was supported by the Ministry of Education, Culture, Sports, Science and Technology of Japan through a Financial Assistance Program for QOL Innovative Research (2012-2016).

References

1. Ministry of Internal Affairs and Communications, Statistics Bureau, Statistics, Population Estimates, Result of the Population Estimates, Monthly Report. Result of the Population Estimates (online), Available: http://www.stat.go.jp/english/data/jinsui/tsuki/ (2015)

2. M. Ishii, K. Yamamoto, K. Hyodo. Stand-Alone Wearable Power Assist Suit -Development and Availability-. *Journal of Robotics and Mechatronics*, Vol.17, No.5, 575–583 (2005)
3. J. Piquion, A. Nayar, A. Ghazaryan, R. Papann, W. Klimek, and R. Laroia, Robot-assisted gynecological surgery in a community setting. *Journal of Robotics and Surgery*, Vol.3, Issue 2, 61–64 (2009)
4. T. Noritsugu, M. Takaiwa, D. Sasaki, Development of Power Assist Wear Using Pneumatic Rubber Artificial Muscles, Journal of Robotics and Mechatronics, Vol. 21, No. 5, 607-613(2009)
5. H. Kobayashi, T. Shiban, Y. Ishida, Realization of all 7 motions for the upper limb by a muscle suit, *Journal of Robotics and Mechatronics*, Vol. 16, 504-512(2004)
6. T. Akagi, S. Dohta. Development of a Rodless Type Flexible Pneumatic Cylinder and Its Application, *Transactions on Robotics and Automation of the JSME (C)*, Vol. 73, No. 731, 2108–2114(2007) (in Japanese)
7. T. Akagi, S. Dohta, Feifei Zhao, T. Fujikawa, Development and Attitude Control of Flexible Robot Arm Using Flexible Pneumatic Cylinder with Simple Structure, *International Journal of Automation Technology*, Vol.5, No.4, 523-530(2011).
8. T. Morimoto, T. Akagi, S. Dohta. Development of Flexible Haptic Robot Arm Using Flexible Pneumatic Cylinders with Backdrivability for Bilateral Control. *Lecture Notes in Electrical Engineering.345,Springer* (Applied)
9. F. Zhao, T. Akagi, S. Dohta, Development of Small-Sized Quasi-Servo Valve for Flexible Bending Actuator, *Proc. 2009 CACS Int. Automatic Control Conference*, 558-562 (2009)

MATEC Web of Conferences **51**, 02004 (2016)
DOI: 10.1051/matecconf/20165102004

Preliminary Design of Flexible Linear Stepping Actuator Driven by Pneumatic Balloons and Brakes

Yuya Eguchi, Tetsuya Akagi[a], and Shujiro Dohta

Department of Intelligent Mechanical Engineering, Okayama University of Science

1-1, Ridai-cho, Kita-ku, Okayama, 700-0005, Japan

Abstract. Recently, the requirement for development of soft actuator for the rehabilitation device and power assisting devices is increasing. However, development of a flexible actuator that has both larger force and longer stroke has not been carried out. In addition, it is also difficult to realize a flexible displacement sensor with long stroke while deforming its form according to the actuator's shape. In this paper, development of the flexible actuator with larger force and longer stroke that can adjust its stroke by giving stepping motion using pneumatic balloons and brakes is described. The fundamental performance of tested actuator is also investigated.

1 Introduction

According to aging society and decreasing birth rate in Japan[1], the requirement using soft actuator for the rehabilitation device and power assisting devices is rapidly increasing[2-5]. As a soft actuator, a rubber artificial muscle is well-known and actively used because of its larger generated force. However, the muscle cannot realize longer displacement, that is, the maximum stroke is limited until one fourth of its original length [6]. In addition, the generated force decreases according to the increase of the displacement of the muscle. In the previous study, as a long stroke flexible actuator, a flexible pneumatic cylinder that can be used even if the cylinder bends was developed [7]. Stroke of the cylinder is possible to be set longer. However, the generated force depends on sectional area of the cylinder and it is not so large. Therefore, a flexible actuator that can generate both larger force and longer stroke has been strongly desired. On the other hand, it is difficult to realize a flexible displacement sensor with long stroke while deforming its form according to the actuator's shape. In other words, the flexible actuator with larger force and longer stroke that can adjust its stroke without sensor is ideal. In this paper, to satisfy such a greedy demand, a flexible linear stepping actuator driven by pneumatic balloons and brakes is proposed and tested. The experiment for investigating the performance of the tested actuator is also carried out.

2 Previous flexible pneumatic cylinder

[a] Tetsuya Akagi : akagi@are.ous.ac.jp

Figure 1 shows the flexible pneumatic cylinder that was developed in our previous study[7]. The cylinder consists of a flexible tube (SMC Co. Ltd., TSU1208) as a cylinder, the steel ball with the outer diameter of 9 mm as a cylinder head, and a slide stage. The slide stage has two brass rollers with diamter of 4 mm set on the inner bore of the stage to press and deform the tube. The steel ball is held by two slide stages from both sides of the ball. When the supply pressure is applied to one side of the cylinder, the tube is moved forward and backward while holding the slide stage. The cylinder can move even if the cylinder bends. The cylinder also realize a longer stroke such as a more than 10 m. However, the frictional force of the cylinder is relatively large, that is about 10 N. The generated force of the cylinder is about 15 N when the supply pressure of 500 kPa is applied. The generated force of the cylinder depends on the sectional area of the cylinder. However, the frictional force increases according to increasing the diameter of the cylinder because of the sliding mechanism between the inner ball and rollers. In addition, it is difficult to realize a displacement sensor that can measure a longer stroke and deform the shape of the sensor according to deformation of the cylinder.

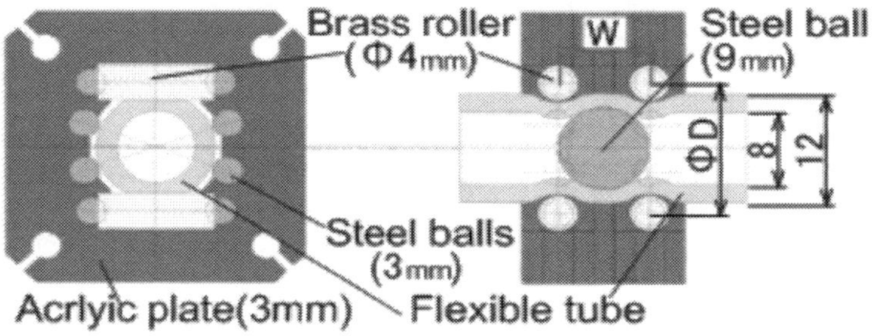

Figure 1. Construction of flexible pneumatic cylinder.

3 Construction of flexible linear stepping actuator

Therefore, in this study a novel pneumatic actuator with larger force and longer stroke that can adjust the displacement without sensor is proposed. Figure 2 shows the schematic diagram and view of the tested flexible linear stepping actuator driven by pneumatic balloons and brakes. The actuator consists of two pneumatic driven brakes set on both sides of actuator, a moving round stage with a pnematic driven brake, two doughnut-shaped balloons such as a diaphragm and a flexible tube as a rod. Figure 3 shows the pneumatic brake. The brake consists of silicone rubber tube with outer diameter of 4 mm surrounded by acrylic cover. When the supply pressure is applied to the brake, the tube in the rubber brake expands. At a same time, the rubber tube hold the flexible tube. The moving stage with a brake in the center is sandwiched by two balloons. The doughnut-shaped balloon that is made of silicone rubber film with thickness of 0.5 mm has the outer diameter of 48 mm and the inner diameter of 26 mm. The theoretical generated force from the sectional area of the balloon with the supply pressure of 500 kPa is about 639 N. The size of the actuator is small, it has the outer diameter of 60 mm and the length of 50 mm. The mass of the actuator without tube is about 180 g.

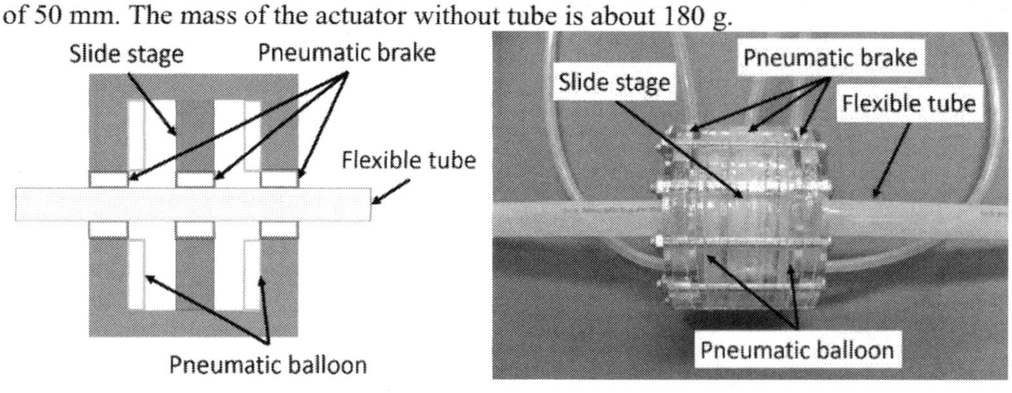

Figure 2. Construction and view of tested flexible linear stepping actuator.

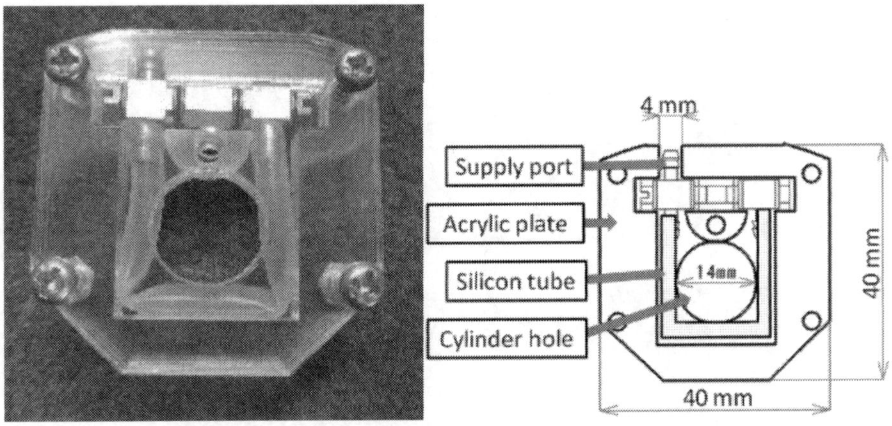

Figure 3. View and construction of pneumatic driven brake

4 Operating principle of flexible linear stepping actuator

Figure 4 shows the operating principle of the tested actuator. The operating principle is as follows: First, the both side brakes are driven to hold the tube for 0.2 s as shown in Fig.4 (1). In the condition, the left side balloon is driven for 0.15 s and the moving stage moves toward right as shown in Fig.4 (2). When the moving stage reaches at the right wall (input pressre is given for 0.15 s), the brake on the stage is driven to hold the tube as shown in Fig.4 (3). After that, the left balloon and both side brakes are released. At the same time, the right side balloon is driven for 0.15 s, the tube that is held by the brake on the moving stage is pushed toward left as shown in Fig.4 (4). By repeating this procedure, the tube can move toward left every a certain stroke ; the width of the balloon chamber(5mm).

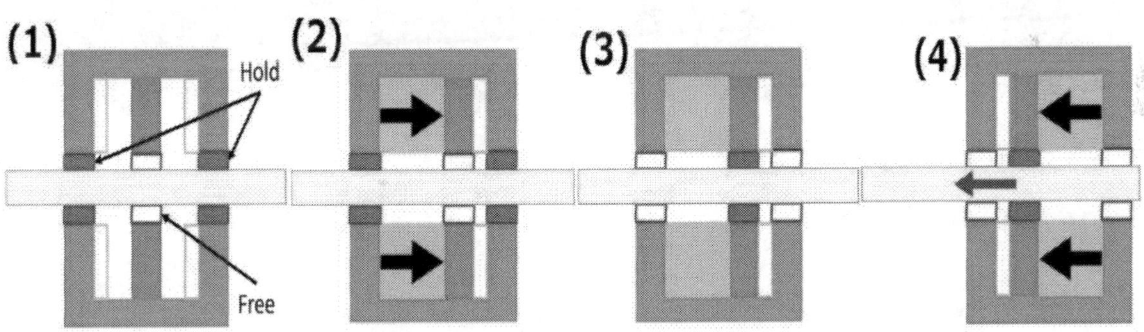

Figure 4. Operating principle of tested linear stepping actuator.

5 Control system and experimental result

Figure 5 shows the schematic diagram of the driving control system of the tested actuator. The system consists of the tested flexible linear stepping actuator, an embedded controller (Renesas Co. Ltd., SH7125) and four on/off control valves (Koganei Co. Ltd., G010E1). Valves 1 and 2 are used to drive both side balloons, respectively. Valve 3 is used to drive the brake on the moving stage. Valve 4 can drive both side brakes. These valves are controlled by the embedded controller through I/O ports.

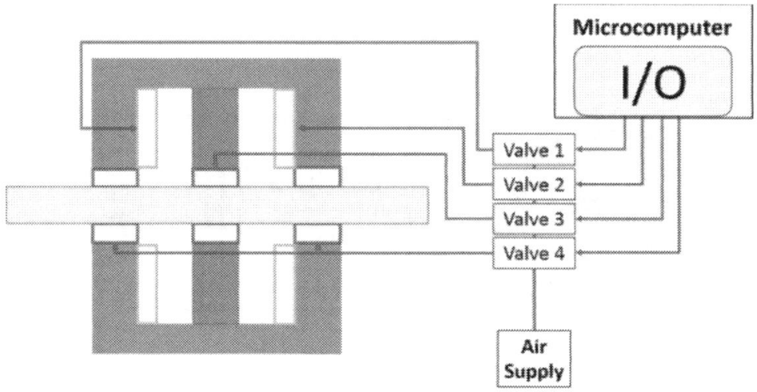

Figure 5. Schematic diagram of control system of tested linear stepping actuator.

Figure 6 shows the transient view of the movement of the tested actuator when the moving pattern of the brakes and balloons are given mentioned above. From Fig.6, it can be seen that the actuator can push the tube with step stroke of 5 mm every 0.7 s. The speed of the tested actuator is about 5.4 mm/s. The improvement of the moving speed is our future work.

Figure 6. Transient view of movement of the tested actuator.

6 Conclusions

As a flexible actuator with larger force and longer stroke, the flexible linear stepping actuator driven by pneumatic balloons and brakes was proposed and tested. The moving test using various moving pattern was carried out. As a result, it can be found that the tested actuator can move with maximum speed of 5.4 mm/s in experiment. It can be confirmed that the actuator can move every step that is about 5 mm.

Acknowledgements

This research was suppoted by the Ministery of Education, Culture, Sports, Science and Technology of Japan through a Financial Assistance for QOL Innovative Reseearch (2012-2016).

References

1. Ministry of Internal Affairs and Communications, Statistics Bureau, Statistics, Population Estimates, Result of the Population Estimates, Monthly Report. Result of the Population Estimates (online), Available: http://www.stat.go.jp/english/data/jinsui/tsuki/ (2015)

2. M. Ishii, K. Yamamoto, K. Hyodo, Stand-Alone Wearable Power Assist Suit -Development and Availability-. *Journal of Robotics and Mechatronics*, Vol.17, No.5, 575–583 (2005)
3. J. Piquion, A. Nayar, A. Ghazaryan, R. Papann, W. Klimek, and R. Laroia, Robot-assisted gynecological surgery in a community setting. *Journal of Robotics and Surgery*, Vol.3, Issue 2, 61–64 (2009)
4. T. Noritsugu, M. Takaiwa, D. Sasaki, Development of Power Assist Wear Using Pneumatic Rubber Artificial Muscles, *Journal of Robotics and Mechatronics*, Vol. 21, No. 5, 607-613(2009)
5. H. Kobayashi, T. Shiban, Y. Ishida, Realization of all 7 motions for the upper limb by a muscle suit, *Journal of Robotics and Mechatronics*, Vol. 16, 504-512(2004)
6. Y.Nagata ed., Soft Actuators –Forefront of Development-, NTS Ltd.(2004) (in Japanese)
7. T. Akagi, S. Dohta, Development of a Rodless Type Flexible Pneumatic Cylinder and Its Application, *Transactions on Robotics and Automation of the JSME (C)*, Vol. 73, No. 731, 2108–2114(2007) (in Japanese)

MATEC Web of Conferences **51**, 02005 (2016)
DOI: 10.1051/matecconf/20165102005

Development and Analysis of Flexible Thin Actuator with a Built-in Fluid Pressure Source

Shinji Senzaki [1], Tetsuya Akagi [1,a], Shujiro Dohta [1], and Yuto Fujiwara [1]

[1] *Department of Intelligent Mechanical Engineering, Okayama University of Science*

1-1, Ridai-cho, Kita-ku, Okayama, 700-0005, Japan.

Abstract. A flexible thin actuator using gas-liquid phase-change of a low boiling point liquid that can generate large force was proposed and tested in the previous study. The tested actuator is an envelope-type actuator that is made of laminating plastic sheets, low boiling point liquid and a flexible heater. In this paper, the analytical model of the flexible thin actuator was proposed and tested. The system parameters of the actuator were also identified. As a result, it was confirmed that the proposed analytical model can predict the behaviour of the tested actuator.

1 Introduction

Due to the aging of Japanese society and the decreasing birth rate, an important problem to keep a quality of life (QOL) for the elderly has occurred[1]. Especially, the decrease in physical ability of the elderly will be concerned to increase in nursing care task. Therefore, the national budget for social welfare will be increased. In order to solve this problem, that is, to improve a quality of life (QOL) for the elderly and the disabled and to decrease the budget for nursing care, a simple rehabilitation device to keep their physical ability might be one of solutions. The actuator used in such a rehabilitation device requires lightweight, softness, small size and larger generated force that can support human body[2,3]. Therefore, many researchers are studying wearable actuators such as a McKibben artificial muscle[4,5]. The McKibben actuator can generate larger force that is more than 300 N by using the actuator whose diameter of 10 mm in the initial condition. However, the actuator can not work with long stroke. Maximum stroke is one fourth of the original length of the actuator. To get larger force by using other pneumatic actuator, larger sectional area of the actuator or higher pressure supplied to the actuator is needed. In ideal, a thin actuator that can generate larger force and larger displacement is required. Therefore, in our previous study[6], the envelope-type actuator driven by pneumatic pressure was proposed and tested as a flexible thin actuator. The actuator can be used as a rehabilitation device for shoulder by being sandwiched it in armpit. In addition, the flexible thin actuator using gas-liquid phase-change of low boiling point liquid was also proposed and tested[7]. It was also confirmed that the tested actuator could be driven silently. In the next step, it is necessary to develop the method to estimate the characteristics of the actuator to design it so as to get the suitable behaviour. In this paper, an analytical model of the tested actuator is proposed. The identified parameters and the comparison

[a] Corresponding author : akagi@are.ous.ac.jp

of the dynamic characteristics between the theoretical and the experimental results of the actuator will be also described.

2 Flexible thin actuator

In the pervious study, the thinner and lighter wearable actuator was proposed and developed[6]. Figure 1 shows the construction and the view of flexible thin actuator that is an envelope-type actuator. The actuator consists of four sheets of paper and plastic laminate films as shown in Fig.1. The left figure in Fig.1 shows the shape of paper A and B. The middle figure shows the sectional structure of the actuator. The paper B works as a gusset in the actuator. The size of plastic laminate film is larger than the size of paper A and B. The plastic laminate films are bonded each other at the margin of the films with the exception of paper area. Then, the length of L_A, L_B and H in paper A and B are 90 mm, 110 mm and 40 mm, respectively. From Fig.1 on the right, it can be seen that the thickness of the actuator was greatly deformed about 39 times from the initial thickness of 1.5 mm with no supplied pressure to maximum thickness of 58.4 mm. By using typical laminate films and paper that can be easily got on the market as a stationery, the actuator can be quickly and easily fabricated. The time for production of the actuator is several minutes.

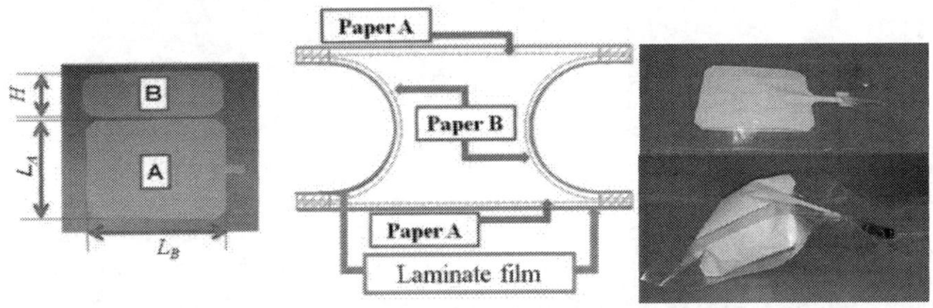

Figure 1. Contraction and view of the flexible thin actuator that is envelope-type actuator.

Figure 2 shows the relation between the expanded length (displacement) and the generated force of the actuator. From Fig.2, it can be seen that the maximum generated force of 657 N could be obtained in the case of supplied pressure of 100 kPa. From these results, the following experimental relation between the displacement and the generated force of the actuator can be obtained.

$$F = A_0 P\,(1\text{-}X / X_M) \tag{1}$$

where, F, A_0, P, X and X_M mean the generated force, equivalent sectional area, supplied pressure, displacement and maximum displacement of the actuator, respectively. In Fig.2, the solid lines show calculated results using Eq.(1) with A_0=0.8$L_A L_B$=0.00792 m^2 and X_M=2.9L_C=0.058 mm. From Fig.2, it can be seen that the proposed experimental relation expressed by Eq.(1) about the generated force and displacement of the actuator can predict well the static characteristics of the tested actuator.

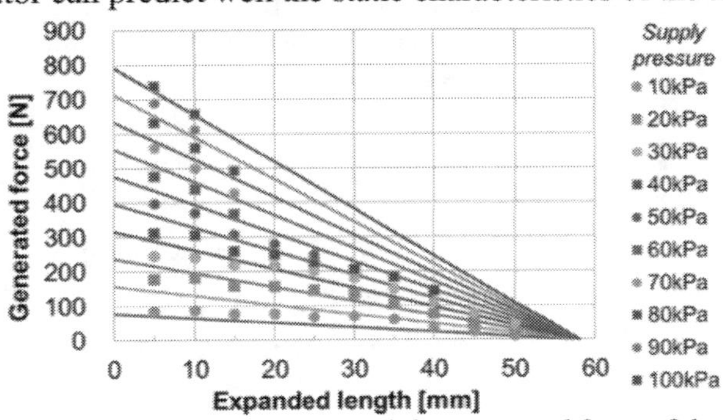

Figure 2. Relation between the expanded displacement and the generated force of the actuator.

3 Actuator with built-in fluid pressure source

3.1 Construction of thin actuator with built-in pressure source

In order to develop the actuator driven by the gas-liquid phase change of the low boiling point liquid, the heat of evaporation and the boiling point of the liquid must be low from the view point of the energy consumption and the response for heater. In addition, from the view point of working liquid in a rehabilitation device, it must have safety, non-toxicity to the human body and harmless to the environment. Therefore, we select a functional liquid (3M™ Co. Ltd., Novec™7000) as a working liquid. The liquid is used as a cleaning liquid of substrate in the electronic circuit. The boiling point is 34 degrees centigrade. The heat of evaporation is 142 kJ/kg. This value means about one sixteenth of that of that of case using the water (kJ/kg). The liquid is nonflammable and nonpoisonous fluid virtually. The high volume resistivity shows that the liquid can contact the electric heater directly during the high applied current.

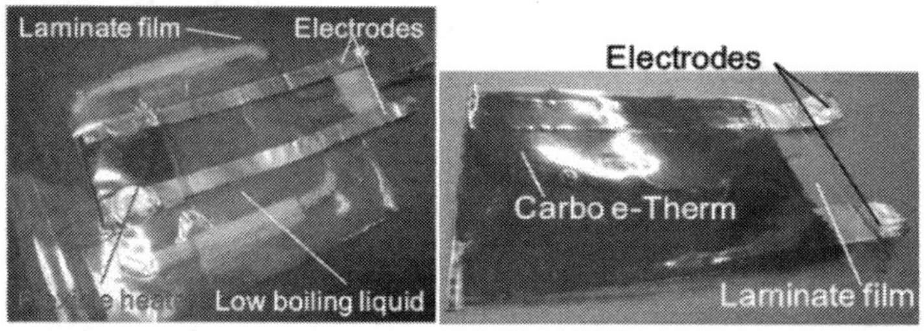

Figure 3. Construction of the flexible thin actuator with low boiling point liquid and flexible heater.

Figure 3 shows the construction of the flexible thin actuator using gas-liquid phase-change of low boiling point liquid and a flexible heater. The flexible heater consists of a copper tape as an electrode, a functional paint (Future Carbon GmbH Co. Ltd., Carbon e-Therm PUR-120) [8] and plastic laminate films. The size of the heater as shown in Fig.3 on the left is 60 x 65 mm. The thickness of the heater is 0.5 mm. The heater has flexibility. The flexible heater is installed into the envelope-type actuator with the low boiling point liquid. The liquid is used as a cleaning liquid for the electronic circuit board. The low boiling point liquid with volume of about 50 cm^3 is also poured into the actuator. To apply the voltage from the outside of the actuator, two copper tapes were connected with the heater. The size of the actuator is 110 mm in length, 90 mm in width and 6 mm in thickness. The total mass of the actuator including the liquid is 80 g.

3.2 Fundamental characteristics

Figure 4 shows the transient view of the movement of the tested actuator when the input voltage of 10V was applied. Each figure shows the view of the actuator when 5, 10, 15 seconds from the beginning of starting to apply the voltage. It can be seen that the actuator reached at maximum displacement within 15 seconds. After stopping to apply the voltage, the actuator was contracted according to the decreasing the temperature of the liquid by natural cooling. It was confirmed that the actuator did not make any sound while being driven.

Figure 4. Transient view of movement of the tested actuator when the input voltage of 10V is applied (Time:5s, 10s, 15s).

Figure 5 shows the experimental set up for measuring the displacement of the actuator and the transient response of the displacement of the tested actuator. In the experiment, the displacement of the actuator was measured by using the leaser displacement sensor (Optex FA Co. Ltd., CD33-120NV).The input voltage of 8 V was applied for 20 seconds.

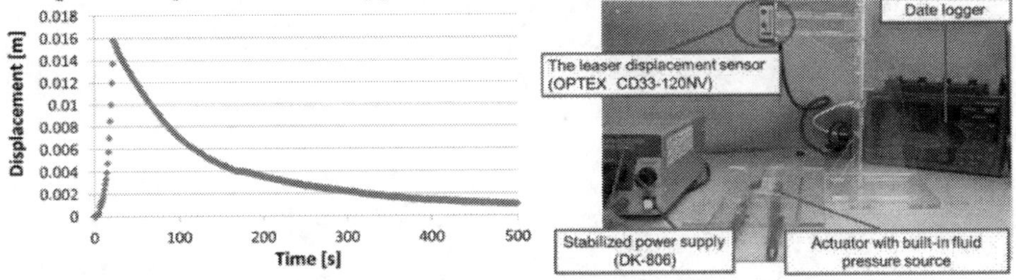

Figure 5. Transient response of the displacement of the actuator.

4 Analytical model of the actuator

In order to estimate the dynamics of the actuator, it is necessary to make an analytical model of gas-liquid phase change of low boiling point liquid. Therefore, the following energy equilibrium between the inner thermal energy of the actuator and the energy given by the heater and the atmosphere is considered.

$$m_G\{(T_b - T_a)\alpha + L\} = k_w \int W_{in} dt - \int A_h U (T_b - T_a) dt \tag{2}$$

where, U, k_w, L, m_G, W_{in} and α are the convective heat transfer coefficient, the efficiency of the heater, the latent heat of vaporization (=142 kJ/kg), the mass of the gas, the heating value of the heater (=16 W) and the specific heat (=1300 J/kgK) of the low boiling point liquid, respectively. T_a, T_b and T_0 mean the absolute temperature of the atmosphere (=296 K), the boiling point and the initial temperature of the liquid, respectively. Then, the left side term shows the energy required to make the liquid boiled. The right side terms express the energy from the heater and the air cooling, respectively. From Clausius-Clapeyron equation, the boiling point T_b of the liquid in the actuator is changed according to the inner pressure (=101.3 kPa) of the actuator P_i and can be given by the following equation.

$$T_b = \frac{T_{ba} L}{L - R T_{ba} \ln(P_i / P_a)} \tag{3}$$

where, P_a, R and T_{ba} show the atmospheric pressure, the gas constant (=287 J/(kg・K)) and the boiling point at the atmospheric pressure (=307 K),respectively. In Eq. (2), A_h is the equivalent contacting area with the gas and given by

$$A_h = L_B(2L_A + 4L_C)\frac{V_{Ga}}{V_{Ga} + V_L} \tag{4}$$

In Eq. (2), U means convective heat transfer coefficient and is given by

$$U = \frac{1}{\frac{1}{h_g} + \frac{t_w}{\lambda} + \frac{1}{h_a}} \tag{5}$$

where, t_w, h_g, h_a and λ show the thickness, thermal conductivity of the actuator's material, gas and atmosphere solidification heat transfer coefficient, respectively.
Then, it is assumed that the pressure of the gas is equal to the atmospheric pressure by changing the inner volume of the actuator. The volume V_{Ga} of the gas that is vaporized from the liquid is given as follows.

$$V_{Ga} = \frac{m_G}{M_{La}} M_{Ga} \tag{6}$$

where, M_{La} and M_{Ga} show the molar mass of the liquid (=0.02006 kg/mol) and gas (=22.4x10^{-3} m^3/mol) at the atmospheric pressure, respectively.

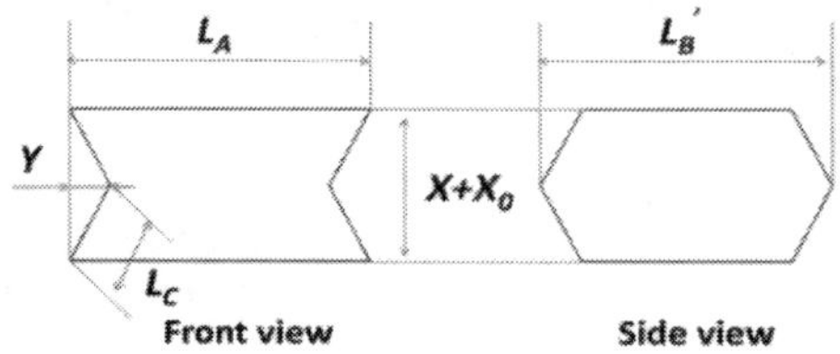

Figure 6. Geometric relationship of the actuator

Figure 6 shows the geometric relationship of the actuator. From Fig.6, the relation between the volume and the displacement of the actuator x can be obtained as follows.

$$(V_{Ga}+V_L)=L_A\left\{\frac{\pi-4}{4}(x+x_0)^2+L_B'(x+x_0)\right\}-YL_B'(x+x_0) \tag{7}$$

where, L_B' and Y are given by following equations.

$$L_B'=L_B+\frac{2-\pi}{4}(x+x_0) \tag{8}$$

$$Y=\sqrt{L_C^2-\left(\frac{x+x_0}{2}\right)^2} \tag{9}$$

The volume V_L of the liquid is given by

$$V_L=\frac{m_{all}-m_G}{\rho} \tag{10}$$

where, ρ, m_G and m_{all} mean the density of the liquid, mass of gas and whole mass, respectively. By using Eqs. (2) to (10), the behavior of the actuator when a heat is applied can be calculated.
Table 1 shows the identified parameters of the actuator. Most parameters are measured and investigated from the specification. The parameter of h_g and k_w are identified so that the calculated result of the behavior of the actuator agrees with the experimental result.

Table 1. Identified parameters of the flexible thin actuator using the low boiling point liquid.

Parameter	value	Parameter	value	Parameter	value
A_h(m^2)	variable	t_w(m)	0.0003	λ(W/mK)	0.33
α(J/kgK)	1300	m_G(kg)	variable	Ta(K)	296
h_a(W/m^{2}K)	3 , 5	M_{Ga}(m^3/mol)	22.4*10-3	Tb(K)	variable
h_g(W/m^{2}K)	6000	M_{LA}(kg/mol)	0.02006	Tba(K)	307
kw	0.55	Pa(kPa)	101.3	Vga(m^3)	variable
L_A(m)	0.09	Pi(kPa)	variable	Win(W)	16
L_B(m)	0.11	R(J/mol·K)	287	x(m)	variable
L_C(m)	0.02	L(kJ/kg)	142	ρ (kg/ m^3)	1400

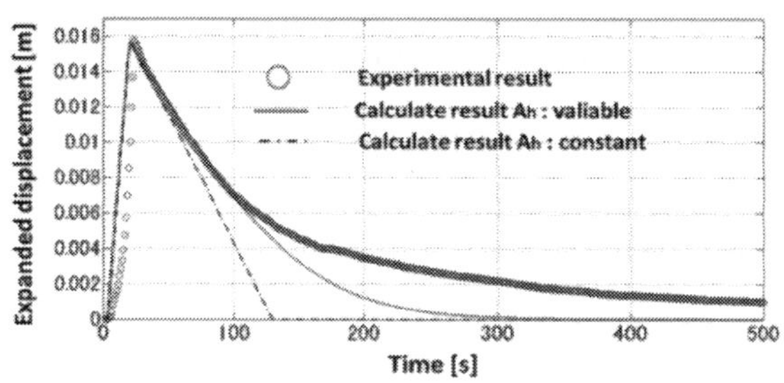

Figure 7. Transient response of the displacement of the actuator

Figure 7 shows the transient response of the measured and calculated displacement of the actuator by using the model and identified parameters. The experimental condition is same as the previous as shown in Fig.5. In Fig.7, symbols and lines show the experimental and calculated results, respectively. The solid line is the calculated result using the model in the case that the equivalent contacting area with the gas A_h is variable. The broken line shows the calculated result in the case when A_h is constant. It can be seen that the calculated result in the case that A_h is variable agrees with the experimental results compared with the case when A_h is constant. The difference between the experimental and calculated results was also found in the result after 100 s from the beginning. This is because the gas solidification heat transfer coefficient dose not change according to the condensation. Therefore, it is necessary to improve the analytical model as our further work. However, the calculation using the proposed model can mostly predict the behaviour of the actuator. It can be concluded that the proposed model and the identified parameters are valid to estimate the performance of the actuator theoretically.

5 Conclusions

This study that aims to develop and analyse the wearable thin actuator driven by low boiling point liquid for portable rehabilitation device can be summarized as follows. The analytical model of the actuator that includes the gas-liquid phase-change and the energy equilibrium between the inner thermal energy of the actuator and the given energy by the heater and the atmosphere was proposed and tested. The system parameters of the actuator were also identified. As a result, the calculated displacement of the actuator using the proposed model and the identified parameters can predict the behaviour of the tested actuator. It is confirmed that the proposed model is valid to estimate the dynamics of the tested actuator theoretically.

Acknowledgements

This work was partially supported by the Ministry of Education, Culture, Sports, Science and Technology of Japan through a QOL Innovative Research Program (2012-2016) and Grant-in-Aid for Scientific Research (C) (Subject No. 24560315).

References

1. Ministry of Internal Affairs and Communications, Statistics Bureau, Director-General for Policy Planning (Statistical Standards) & Statistical Research and Training Institute, Statistical handbook of Japan 2013 [Online].
2. T. Akagi, S. Dohta, M. Ihara, Improvement of McKibben Artificial Muscle with Long Stroke Motion and Its Application, *Journal of System Design and Dynamics*, Vol.4, No.4, 538-551(2010).
3. Y. Nagata ed., *Soft Actuators -Forefront of Development-*, NTS Ltd., 291- 335(2014)
4. H. Kobayashi, T. Shiban, Y. Ishida, Realization of all 7 motions for the upper limb by a muscle suit, *Journal of Robotics and Mechatronics*, Vol.16, No.5, 504-512(2004).
5. T. Noritsugu, M. Takaiwa, D. Sasaki, Development of Power Assist Wear Using Pneumatic Rubber Artificial Muscles, *Journal of Robotics and Mechatronics*, Vol.21, No.5, 607-613,(2009.)
6. Y. Tsuji, S. Dohta, T. Akagi S. Noguchi, Development of Envelope-type Pneumatic Thin Actuator, Proc. FLUCOME 2013, OS4-03-2pdf, 1-7,(2013)
7. 3M™ Novec™ 7000 Engineered Fluid (Catalog) [Online]
8. T. Akagi, S. Dohta, S. Fujimoto, Y. Tsuji, Y. Fujiwara , Development of Flexible Thin Actuator Driven by Low Boiling Point Liquid, *International Journal of Materials Science and Engineering* Vol. 3, No. 1, March 55-58,(2015)

MATEC Web of Conferences **51**, 02006 (2016)
DOI: 10.1051/matecconf/20165102006

Development of Pipe Holding Mechanism for Pipe Inspection Robot Using Flexible Pneumatic Cylinder

Kyujun Choi, Shujiro Dohta, Tetsuya Akagi[a] and Shota Ninomiya

Department of Intelligent Mechanical Engineering, Okayama University of Science
1-1, Ridai-cho, Kita-ku, Okayama, 700-0005, Japan

Abstract. A pipe inspection robot is useful to reduce the inspection cost. In the previous study, a novel pipe inspection robot using a flexible pneumatic cylinder that can move forward along to the pipe by changing the robot's body naturally was proposed and tested. In this paper, to improve its mobility for a corner of a pipe, the thin pipe holding mechanism using pneumatic bellows was proposed and tested. As a result of its driving test, the holding performance of the mechanism was confirmed.

1 Introduction

In Japan, it is found that the corresponding to 5 % of the total length of water supply pipe had been already past the statutory useful life of 40 years [1]. Water pipe lines are very complex and includes quite a lot of corners and joints. The cost of the inspection of pipe will be reduced by using inspection robot [2-4]. In the case of pipe inspection, the robot is required higher mobility. In ideal, it is more desirable that the shape of the robot changes naturally because of the lower energy consumption and decreasing the time for traveling. In our pervious study, a pipe inspection robot that consists of a flexible sliding mechanism using the flexible pneumatic cylinder that can travel in the narrower space smoothly by changing the body according to the shape of the pipe had been proposed and tested [5-7]. In the next step, it is necessary to improve its mobility for corners or joints. In this paper, the thin pipe holding mechanism using pneumatic bellows is proposed and tested so as not to prevent the smooth movement in a complex bending pipe. The holding performance of the tested mechanism is also investigated.

2 Pipe inspection robot using flexible pneumatic cylinder

Figure 1 shows the double ball type flexible pneumatic cylinder developed in our previous study [8]. As shown in Fig. 1(a), the cylinder consists of a flexible tube as a cylinder, two steel balls (Φ9 mm) as a cylinder head, and a slide stage. The slide stage has two brass rollers (Φ4 mm) set on the inner bore of the stage to press and deform the tube. The tube between two steel balls is pinched by the slide stage. When the pressure is applied to one side of the cylinder, the slide stage moves forward and backward. The slide stage can move even if the cylinder bends. The flexible cylinder can be used by two ways as shown in Figs.1(b) and (c). Figure 1(b) shows an ordinary rod-less type cylinder: the

[a] Corresponding author : akagi@are.ous.ac.jp

slide stage moves on the curved tube while the cylinder tube is fixed. On the other hand, the cylinder tube can move through the slide stage when the slide stage is fixed as shown in Fig.1(c). This usage can be easily realized because the cylinder tube is more light-weight compared with a usual metallic cylinder. In this study, both operations are used properly in a sliding mechanism.

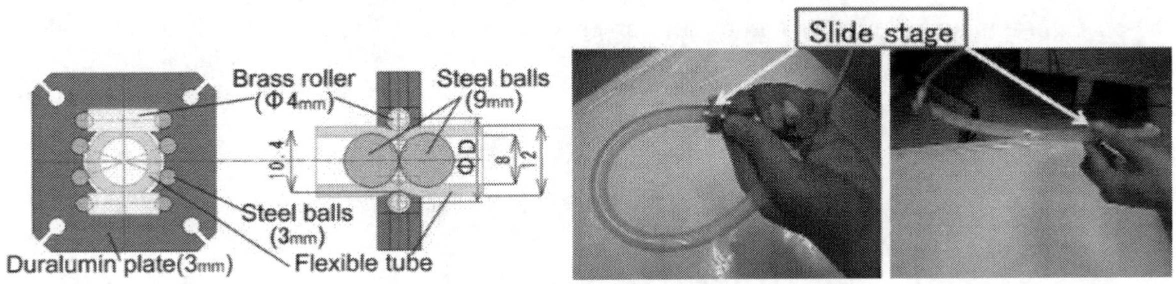

(a) Construction of the cylinder. (b) Cylinder tube is fixed. (c) Slide stage is fixed.

Figure 1. Construction and operation of flexible pneumatic cylinder.

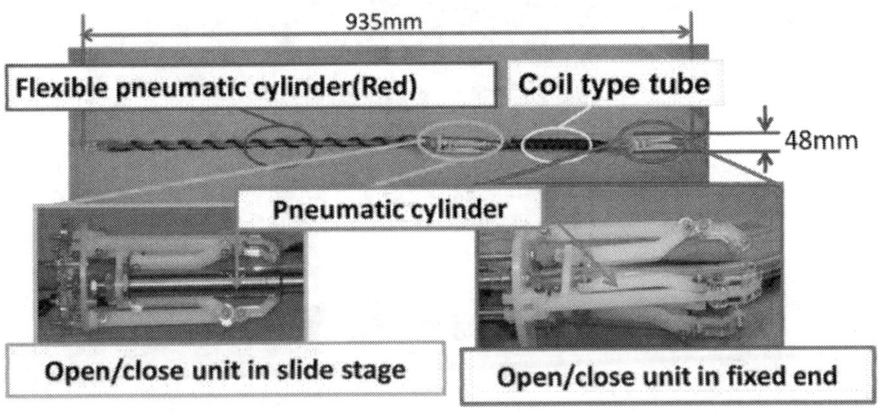

Figure 2. Flexible sliding mechanism for inspection robot.

Figure 2 shows a flexible sliding mechanism using a flexible pneumatic cylinder and two open/close units developed in the previous study [6]. The coil type of pneumatic supply tube is used as a supply pipe to the open/close unit on the slide stage and to the top end chamber of the cylinder. The open/close unit can expand the claw toward radial direction using the typical pneumatic cylinder. The flexible pneumatic cylinder is located into the coil tube. By this method, the smart configuration of air supply pipes from the end of the mechanism can be realized. Figure 3 (a) shows the operating principle of the sliding mechanism. The operating principle is as follows. First, the open/close unit at the end of the mechanism opens so that the claws can hold the pipe (2). Next, the right side chambers of the flexible cylinder are pressurized, then the slide stage with the open/close unit can slide toward the top end of the cylinder (3). When the slider reaches at the end of the cylinder, the open/close unit opens to hold the pipe (4). Then, the open/close unit at the end of mechanism closes (5), and the left side chambers of the cylinders are pressurized. At the same time, the flexible cylinder with end open/close unit can move forward (6). By repeating these operations from (2) to (6), the mechanism can move forward as an inchworm. Figure 3 (b) shows the view of movement of the tested inspection robot using the sliding mechanism when the robot passes through the pipe with the inner diameter of 100 mm. The size of the robot is 1.1 m in length and 48 mm in outer diameter. The whole mass of the robot including the pneumatic driving system is 0.78 kg. The motion for passing through pipe is as follows. First, from the far front of pipe tee, the bending unit of the robot was bent toward the desired direction (2). While keeping the bending motion, the robot moved forward. Then, the acrylic cover was inserted into the desired pipe. The robot can move forward along to the pipe by changing the robot's body naturally. Figure 3(b) (3) shows the case when the robot passes through the pipe elbow.

In the same manner as the case of the pipe tee, by bending toward the desired direction in the far front of the corner, the robot can easily move at the corner. The moving speed passing through both corners is almost same speed of moving straight (about 140mm/s). However, it was also observed that the open/close unit was often caught on the corner while bending. Therefore, a thinner pipe holding mechanism will be required.

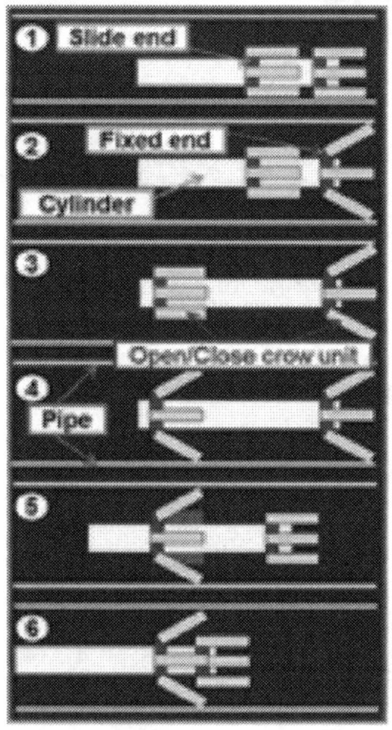

(a) Operating principle.

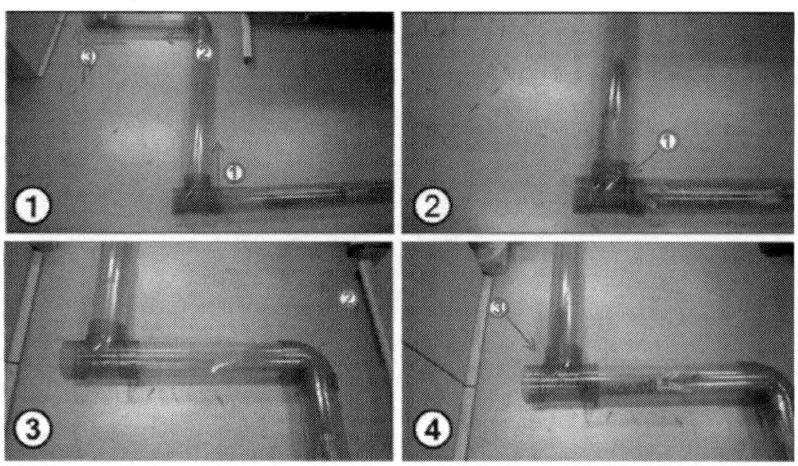

(b) View of movement of inspection robot in complex pipe.

Figure 3. Operating principle of sliding mechanism and movement of inspection robot in complex pipe.

3 Thinner pipe holding mechanism

Figure 4(a) shows the view of the thinner pipe holding mechanism proposed in this study. The mechanism consists of three acrylic claws that can expand toward radial direction and three bellows. Each claw is located every 120 degs. from the center of the mechanism. Each bellow is also set between the claw and the slide stage. The size of the mechanism is 72 mm in outer diameter and 31 mm in length. The mass of the mechanism is very small, that is 75 g.

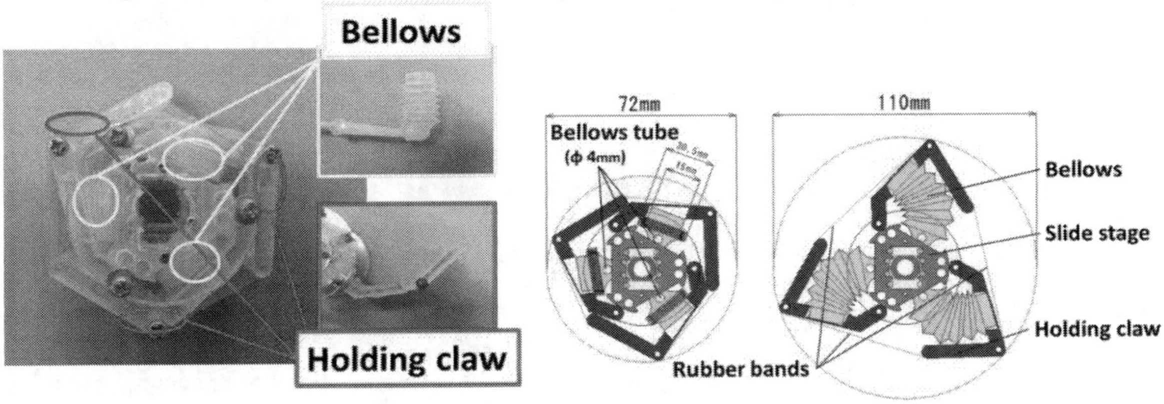

(a) View of the mechanism. (b) Operating principle of the mechanism.

Figure 4. Proposed thinner pipe holding mechanism.

Figure 4(b) shows the operating principle of the holding mechanism. The operating principle is as follows. When the supply pressure is applied to three bellows, the bellows expands towards the radial direction. The maximum outer diameter of 110 mm can be obtained when the supply pressure of 200 kPa is applied. When the supply pressure is released, the mechanism returns to its original shape by restoring force of rubber bands that cover the claws. By using this device, the mechanism can hold the pipe surely. Figure 5 shows the view of operation of the tested pipe holding mechanism. As a result of pipe holding test, the mechanism in the pipe with inner diameter of 100 mm can hold a load of 19.6 N. It is considered that this holding ability is sufficient to be applied for the pipe inspection robot. The travelling test of the proposed inspection robot in a pipe with corners is our future work.

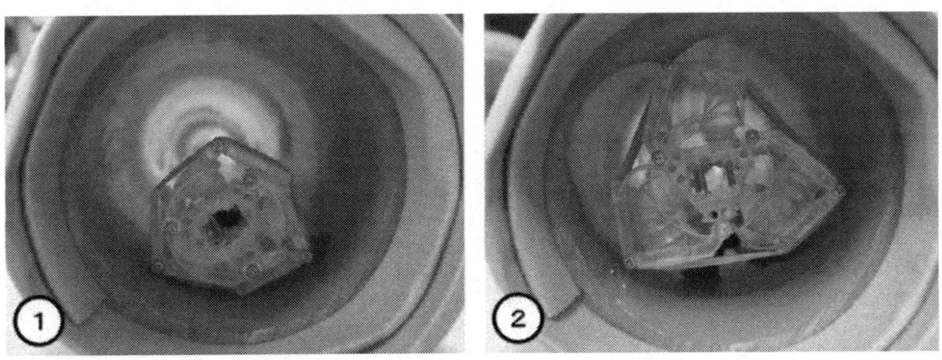

Figure 5. Operation of the tested pipe holding mechanism in a pipe.

4 Conclusions

In order to improve the mobility of the pipe inspection robot using flexible pneumatic cylinder when passing through corners, the thinner pipe holding mechanism using pneumatic bellows was proposed and tested. The holding performance of the mechanism was also investigated. As a result, the mechanism can hold a load of 19.6 N. It can be confirmed that the mechanism is useful for the pipe inspection robot because of its thinner body and sufficient holding ability.

Acknowledgements

This research was supported by the Ministry of Education, Culture, Sports, Science and Technology of Japan through a Financial Assistance Program of QOL Innovative Research (2012-2016).

References

1. http://ja.wikipedia.org/wiki/%E6%B0%B4%E9%81%93%E7%AE%A1.
2. J. Yuan et.al., Development of an inspection robot for long distance transmission pipeline on-site overhaul, *Int. J. Industrial Robot*, Vol. 36, pp.546-550 (2009)
3. E. Dertien et.al., Development of an inspection robot for small diameter gas distribution mains, *Proc. 2011 IEEE Int. Conf. R&A*, pp.281-294 (2011)
4. K. Nagatani et.al., Development of a Transformable Mobile Robot with a Variable Wheel Diameter, *J. Robotics and Mechatronics*, Vol.19, No.3, pp.252-257 (2007)
5. T. Kawasaki et.al., Development of a Flexible Sliding Mechanism Using Flexible Pneumatic Cylinder for Rescue Robot, *Proc. FLUCOME2011*, 41.pdf, pp.1-6 (2011)
6. T. Akagi et.al., Development of sliding type inspection robot using flexible pneumatic cylinder, *Int. J. Advanced Mechatronic Systems*, Vol.5, pp. 87-94 (2013)
7. T. Akagi et.al., Pipe Inspection Robot Using Flexible Pneumatic Cylinder Built-in Pneumatic Driving System, *Proc. Int. Forum on Systems and Mechatronics*, pp. 1-6 (2014)
8. T. Akagi et.al., Development of Wearable Pneumatic Actuator and Multiport Pressure Control Valve, *J. Robotics and Mechatronics*, Vol.17, No.5, pp. 529-536 (2005)

MATEC Web of Conferences **51**, 02007 (2016)
DOI: 10.1051/matecconf/20165102007

Characteristics of Linerboard and Corrugated Medium Paper Made from Durian Rinds Chemi-mechanical Pulp

Shaiful Rizal Masrol[1,2,a], Mohd Halim Irwan Ibrahim[2,3], Sharmiza Adnan[4], Muhammad Syauqi Asyraf Ahmad Tajudin[3], Radhi Abdul Raub[3], Siti Nurul Aqma Abdul Razak[3] and Siti Nur Faeza Md Zain[3]

[1]*Faculty of Engineering Technology, Universiti Tun Hussein Onn Malaysia (UTHM), Malaysia*
[2]*Advanced Manufacturing and Materials Center (AMMC), ORICC, Universiti Tun Hussein Onn Malaysia(UTHM), Malaysia*
[3]*Faculty of Mechanical and Manufacturing, Universiti Tun Hussein Onn Malaysia (UTHM), Malaysia*
[4]*Pulp and Paper Laboratory, Biomass Technology Programme, Forest Products Division, Forest Research Institute Malaysia (FRIM)*

Abstract. In recent years, there has been tremendous growth of interest in the utilization of non-wood based material to support the insufficient raw materials availability for the pulp and paper industry. Therefore, this study was conducted to investigate the characteristics of 120 gsm linerboard and corrugated medium paper made from unbeaten durian rinds chemi-mechanical (CMP) pulp. The linerboard and corrugated medium paper were prepared and tested according to Malaysian Standards / International Organization for Standardization (MS ISO) and Technical Association of the Pulp and Paper Industry (TAPPI) standard methods. Based on the burst index (2.03 $kPa.m^2/g$), RCT (1.97 $N.m^2/g$) and CMT (129N), the linerboard and corrugated medium from CMP durian rind pulp have shown a good potential as an alternative raw material for papermaking and comparable with other types of wood and non-wood based papers as well as current commercial papers.

1 Introduction

Nowadays, insufficient supply of wood-based raw materials for packaging industries due to environmental issues leads to tremendous interest in the utilization of non-wood based materials in papermaking. In Malaysia, almost 50% of total paper consumption comes from packaging papers [1, 2]. The country is a net importer of pulp and paper in order to satisfy the domestic consumption [3]. Eighteen(18) out of 20 Malaysian paper mills utilize 95% waste paper and old corrugated carton (OCC) as their main raw material, with the exception of Kimberly-Clark (M) Bhd. and Sabah Forest Industries that use 80% imported virgin pulp and 100% mixed tropical hardwood pulp, respectively [4]. In short, the paper and paperboard industries in Malaysia utilize imported virgin long-fiber pulp and secondary fibers, such as the old corrugated container (OCC) and waste papers to produce packaging papers [5]. In order to reduce the dependence on imported pulps, various researches have been conducted on the utilization of agricultural wastes as raw material alternatives for packaging paper such as kenaf [1, 2, 6], coir [5], oil palm empty fruit bunch [7, 8], Semantan bamboo [9] and oil

[a] Corresponding author : srizal@uthm.edu.my

palm male flower spikes [10,11]. Malaysia has an abundance of agro waste materials which have not been fully utilized [12].

Malaysia is one of the major fresh and frozen durian exporters besides Thailand and Indonesia [13–15]. In Malaysia, based on a study by Amid and Mirhosseini [16] stated that the rind constitutes in average about 45% for each mature fruit, Aditiya et al. [15] have calculated that durian rind waste generates about 133,688 to 171,303.9 metric tons from the year of 2006 until 2013 [15]. According to Ministry of Agriculture and Agro-based Industry Malaysia (MOA) [17], there were 376,565 metric tons of durian fruits were produced in 2014. Thus, massive amounts of the durian peels waste are disposed of, causing a severe problem in the community [18]. Therefore, the utilization of the durian rinds waste as pulp and paper alternative raw material would be beneficial. However, based on the authors' best knowledge there were limited research attentions and published works on the utilization of durian rinds waste as an alternative papermaking material.

The author previous study by Masrol, Ibrahim and Adnan [19] has successfully produced durian rinds pulp via chemi-mechanical pulping (CMP). In order to extend the study on utilizing durian rinds waste as packaging alternative raw material, this study was conducted to investigate the potential of durian rinds CMP pulp as linerboard and corrugated medium paper.

2 Experimental

2.1 Raw material preparations

Durian rinds were collected from a local durian flesh manufacturer at Batu Pahat, Johor, Malaysia. Only the rinds of D24 durian have been collected to ensure the accuracy and repeatability of results. Firstly, durian rinds were cleaned from the residual aril and dirt. Then, the cleaned durian rinds were sliced, spine removed, cubed and dried naturally according to a procedure developed by Masrol et al. [19]. Finally, the dried durian rinds were stored inside air-tight storage to prevent moisture and fungus infestation.

2.2 Chemi-mechanical pulping process

Durian rinds pulp was produced by a chemi-mechanical pulping process according to pulping conditions conducted by Masrol et al. [19]. Dried durian rinds were treated with 10% (based on raw material oven-dry (o.d.) weight) solution of Sodium Hydroxide (NaOH) at room temperature for 2 hours. Then, the treated durian rinds were refined by Sprout-Waldron refiner mechanical pulping (RMP) machine to produce pulp. Refiner mechanical pulping (RMP) is preferred because of its low production, labor and capital costs [20]. After the RMP process, screening was carried out using PTI Sommerville Fractionators according to TAPPI T 275 standard with a slot size of 0.15 mm. After the screening, a spin-dry Neng Shin extractor was used to remove excess water and the screened pulp was dispersed into smaller size using a Hobart mixer. Finally, durian rinds CMP pulps were sealed in polyethylene bags and stored inside chillers at a temperature of 6°C before laboratory hand sheets preparation.

2.3 Laboratory hand sheet preparation

Laboratory hand sheets with a grammage of 120±3 g/m^2 were produced from unbleached and unbeaten durian rinds chemi-mechanical pulp using a semi-automatic sheet machine (British Handsheet Machine) based on MS ISO 5269-1: 2007 "Pulps – Preparation of Laboratory Sheets for Physical Testing – Part 1: Conventional Sheet-Former Method (ISO 5269-1: 2005, IDT)". The hand sheets were dried and conditioned at 23 ± 1 °C and 50 ± 2.0% relative humidity (RH) according to TAPPI T 402 sp-03 "Standard Conditioning and Testing Atmospheres for Paper, Board, Pulp Handsheets and Related Products" and MS ISO 187: 2001 "Paper, Board and Pulps - Standard

Atmosphere for Conditioning and Testing and Procedure for Monitoring the Atmosphere and Conditioning of samples (ISO 187: 1990, IDT)" for at least 24 hours before further evaluation.

2.4 Characteristics' tests

The characteristics' tests (a–h) listed in Table 1 were conducted according to ISO 5270: 2012 "Pulps - Laboratory sheets determination of physical properties" with accordance to Malaysian Standards / International Organization for Standardization (MS ISO). The test for CMT, RCT and $Cobb_{30}$ used for packaging grade paper (Table 1) were carried out in this study. All tests were performed in a controlled temperature and humidity environment as stipulated in TAPPI T 402 sp-03 "Standard Conditioning and Testing Atmospheres for Paper, Board, Pulp Handsheets and Related Products" and MS ISO 187: 2001 "Paper, Board and Pulps - Standard Atmosphere for Conditioning and Testing and Procedure for Monitoring the Atmosphere and Conditioning of samples (ISO 187: 1990, IDT)". The sampling was conducted according to TAPPI T400 sp-02 "Sampling and Accepting a Single Lot of Paper, Paperboard, Containerboard, or Related Product" and MS ISO 186: 2002, IDT "Paper and Board - Sampling to Determine Average Quality".

Table 1. Characteristics' tests list and standards

No.	Test	MS ISO / TAPPI
a.	Grammage	MS ISO 536: 2001: Paper and Board –Determination of Grammage
b.	Thickness	MS ISO 534: 2007, IDT: Paper and Board – Determination of Thickness, Density and Specific Volume
c.	Brightness	MS ISO 2470-1: 2010: Paper, Board and Pulps – Measurement of Diffuse Blue Reflectance Factor-Part 1: Indoor Daylight Conditions -ISO Brightness-First revision (ISO 2470-1: 2009, IDT)
d.	Opacity	MS ISO 2471: 2010: Paper, Board and Pulps – Determination of Opacity (Paper Backing) – Diffuse Reflectance Method - First revision (ISO 2471: 2008, IDT)
e.	Tensile	MS ISO 1924-2: 2010: Paper and Board – Determination of Tensile Properties – Part 2 – Constant Rate of Elongation Method - 20 mm/min – First revision (ISO 1924-2: 2008, IDT)
f.	Tearing	MS ISO 1974: 1999: Paper – Determination of Tearing Resistance – Elmendorf Method – Second Revision (ISO 1974:1990, IDT)
g.	Bursting	MS ISO 2758: 2007: Paper-Determination of Bursting Strength (ISO 2758: 2001, IDT)
h.	Folding	MS ISO 5626: 1999: Paper – Determination of Folding Endurance
i.	RCT	TAPPI T 818: Ring Crush of Paperboard
j.	CMT	TAPPI standard T 809 om-99: Flat crush of corrugating medium (CMT test)
k.	$Cobb_{30}$	MS ISO 535: 2001: Determination of water absorptiveness – Cobb test

2.4.1 Ring crush test (RCT) and corrugated medium test (CMT)

The ring crush test (RCT) can be defined as an edgewise compression strength test of corrugated boards which is intended to characterize a paperboard's resistance to crushing by in-plane compressive forces. RCT is used for test liner and other papers used in packaging manufacture [21]. The test specimen with a dimension of 12.7 mm width x 152.44 mm length is cut by a strip punch as shown in Figure 1a and inserted into the sample holder (Figure 1b). Samples were tested using the L & W Crush Tester to measure the compression force required before the test specimens buckle as shown in Figure 1c.

CMT can be defined as a test to measure the flat compression strength of a laboratory fluted strip of a corrugating medium and can be used as a basis for judgment of fabrication efficiency. The CMT permits the evaluation of corrugating medium before it is fabricated into combined board [22]. The test specimen with the size of 12.7mm wide x 152.4 mm long is cut by a special die cutter and formed into a flute shape using Concora Medium Fluter in Figure 2a. The rack and comb inside the fluter machine work to perfectly shape the sample into flute shape as shown in Figure 2b. Samples were tested using the same L & W Crush Tester machine shown in Figure 1c.

Figure 1. Ring crush test (RCT) procedure: (a) strip punch; (b) sample holder; and (c) L&W crush tester

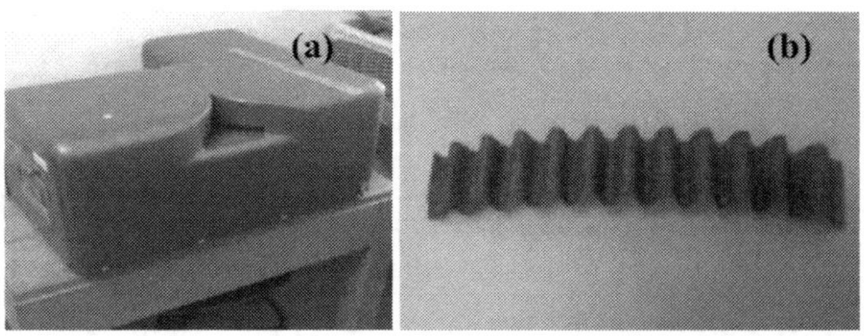

Figure 2. Corrugated medium test (CMT): (a) concora medium flutter; and (b) corrugated medium flute

2.4.2 Water absorption test ($Cobb_{30}$ test)

The purpose of this method is to determine the quantity of water absorbed by a linerboard in a specified time. Distilled water is used for the test to avoid any substance factor that can affect the absorption characteristics of the paper tested. The weight of the sample is measured before and after the test (Figure 3a and Figure 3d). The samples were tested for water penetration for at 20±1 seconds within 30±1 seconds total test time as shown in Figure 3c.

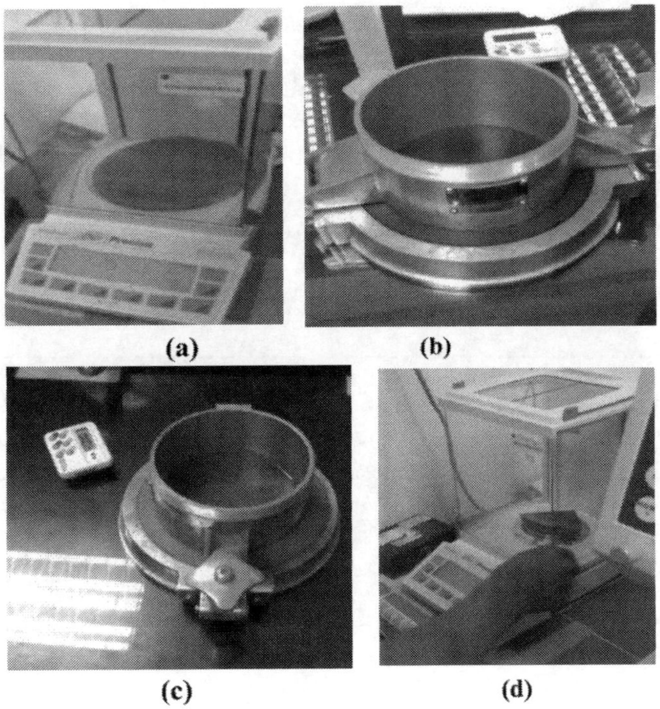

Figure 3. $COBB_{30}$ test procedure: (a) weight before; (b) Cobb test device; (c) sample holder; and (d) weight after

3 Results and Discussion

3.1 Laboratory hand sheet characteristics

Table 2 shows the physical and optical characteristics of 120 gsm linerboard produced from durian rinds chemi-mechanical pulp (CMP). The number of folds for 120 gsm linerboard (37) was higher compared to 15 by previous 60 gsm paper [19]. For optical characteristics, ISO brightness (13.20%) and opacity (97.73%) are similar with Masrol et al. [19]. The result of $Cobb_{30}$ test (225 g/m^2) shows that water penetrates durian rinds CMP linerboard too fast. This result could be improved by the addition of a sizing agent such as starch during the paper sheet preparation in the future.

Table 2. Physical and optical characteristics of durian rinds linerboard

Characteristics	Value
Grammage (g/m^2)	122
Apparent bulk density(g/cm^3)	0.73
Double Fold, No	37
ISO Brightness (%)	13.20
ISO Opacity (%)	97.73
$Cobb_{30}$ (g/m^2)	225

3.2 Paper mechanical properties comparison with other raw materials

Table 3 shows the comparison of tensile, tear and burst indices between durian rinds CMP pulp and other types of materials reported by previous researchers. Results indicate that there is no significance difference in mechanical characteristics (tensile, tear and burst indices) between 120 gsm and 60 gsm reported by Masrol et al. [19]. Durian rinds unbeaten CMP pulp's tensile index is higher than empty fruit bunch (EFB) [8], *Acacia mangium* [8], *Eucalyptus globulus* [8], rapeseed [23], Semantan bamboo [24] and coir CMP [25]. For tear index, unbeaten durian rinds CMP recorded a higher value than *Acacia mangium* [8], *Eucalyptus globulus* [8] and rapeseed [23]. Durian rinds CMP also recorded a better burst index than empty fruit bunch (EFB) [8], *Eucalyptus globulus* [8], rapeseed [23], Semantan bamboo [24] and coir CMP [25]. The results show that durian rinds CMP is comparable with other types of raw materials for papermaking. However, mechanical treatments such as beating or refining is recommended for unbeaten durian rinds CMP pulp in order to enhance the pulp and paper characteristics [5, 11, 24].

Table 3. Mechanical properties of durian rinds CMP compared with another types of linerboard

Material	Tensile Index (N.m/g)	Tear Index ($mN.m^2/g$)	Burst Index ($kPa.m^2/g$)
Durian CMP	36.48	6.72	2.03
Kenaf KHK[1]	71.03	13.38	4.18
Kenaf KLK[1]	78.84	15.11	5.12
Kenaf Bast[1]	40.83	20.68	2.66
Kenaf Core[2]	83.09	5.87	6.09
[b]Kenaf KHK[2]	78.21	14.35	5.21
[b]Kenaf KLK[2]	86.10	14.36	5.90
[b]Kenaf Bast[2]	72.47	23.95	4.70
[b]OCC[2]	42.27	7.02	2.79
[b]USWK[2]	95.18	12.49	7.23
[a]Coir Soda-AQ[5]	44.66	11.76	4.57
EFB[8]	21.37	5.85	1.41
Acacia mangium[8]	2.53	3.28	6.76
Eucalyptus globulus[8]	24.84	4.33	0.13
Rapeseed[23]	25.57	3.70	1.24
[a]Semantan bamboo[24]	32.54	13.45	1.54
Coir CMP[25]	18.33	7.05	1.85
Bagasse[26]	43.60	5.47	2.63

a= optimum pulp; b : beaten pulp

3.3 Comparison of burst index, RCT and CMT with other types of paper

Table 4 shows that RCT results of durian rinds 120 gsm linerboard and corrugated medium made from CMP pulp are comparable with other paper types and commercial linerboards.

Linerboards with a high apparent density indicates a good bonding ability between fibers during paper formation [5]. In this study, the durian rinds linerboard produced has an apparent density of 0.73 g/cm^3 which is similar with other non-wood based and commercial linerboards (Table 4). Apparent density is one of the most significant properties of paper and influences almost all mechanical, physical, and electrical properties [8].

In industrial practices, burst strength indirectly measures the internal resistance of corrugated box container to any kind of force applied [5]. The burst index of durian rinds CMP linerboard recorded a value of 2.03 $kPa.m^2/g$ which is higher than some commercial linerboards' minimum requirement such as GSPP IL3, GSPP IL4, GSPP IL5, GSPP IL6, Pascorp IL1, Pascorp IL4, Pascorp IL4 and Pascorp IL6 reported by Main et al. [5]. The burst index for durian rinds linerboard is also higher than commercial scale linerboards developed by Rushdan [8] except for SANYEN 6 and SANYEN 13.

Unbeaten durian rinds CMP linerboard recorded RCT value of 1.58 kN/m and 1.97 $N.m^2/g$ for RCT index. For RCT index, durian rinds CMP linerboard recorded a higher value than the minimum requirement of commercial linerboards by GSPP and Pascorp as reported by Main et al. [5]. RCT of durian rinds CMP linerboard is higher than Kenaf bast [1], OCC [2] and overall commercial scale linerboards developed by Rushdan et al. [8].

Durian rinds CMP linerboard CMT (129N) value is higher than coir [5, 27] and OCC [28] linerboards. The CMT index of durian rinds linerboard (1.06 $N.m^2/g$) is also slightly higher than Semantan Bamboo [9], Kenaf [9] as well as commercial scale SANYEN 40 linerboard [8]; and a bit similar with CF *Mischantus* pulp [29].

Therefore, based on the result of burst index, RCT and CMT, durian rinds CMP pulp has shown a good potential as an alternative material for linerboard and corrugated medium

4 Conclusion

In this study, durian rinds chemi-mechanical pulp (CMP) linerboard and corrugated medium flutes were successfully produced and analyzed. The results demonstrated that durian rinds chemi-mechanical pulp is suitable for linerboard and corrugated medium applications. The properties of durian rinds CMP linerboard and corrugated medium produced are comparable to other types of wood and non-wood raw materials as well as commercial linerboards specifications produced in Malaysia. However, continuous improvement is necessary in order to achieve optimum quality for CMP pulp produced from durian rinds. In future, optimization of CMP pulping variables (active alkali percentage, digestion time and beating revolutions) is required to obtain the optimum pulp that can produce durian rinds CMP paper with the best characteristics and quality. Furthermore, the addition sizing agents should be considered to improve durian rinds linerboard water absorptiveness. Beating or refining process also needs to be applied to enhance the characteristics of the CMP durian rinds pulp and paper. The findings of this research would be beneficial to the pulp and paper industry as well as the agriculture, namely durian based, industry.

Table 4. Characteristics of durian rinds CMP compared to various type linerboards and corrugated mediums

Material	G (g/m^2)	BI ($kPa.m^2/g$)	Bulk (cm^3/g)	Bulk (g/cm^3)	RCT ($N.m^2/g$)	RCT (kN/m)	CMT (N)	CMT ($N.m^2/g$)
100 % Durian	122	2.03	1.38	0.73	1.97	1.58	129	1.06
Kenaf core[1]	150	4.18	-	0.716	-	2.87	-	-
Kenaf KHK[1]	150	5.12	-	0.590	-	1.85	-	-
Kenaf KLK[1]	150	2.66	-	0.620	-	2.25	-	-
Kenaf bast[1]	150	6.09	-	0.502	-	1.43	-	-
[b]Kenaf KHK[2]	150	5.21	-	0.651	-	2.25	-	-
[b]Kenaf KLK[2]	150	5.90	-	0.672	-	2.68	-	-
[b]Kenaf bast[2]	150	4.70	-	0.594	-	2.61	-	-
[b]OCC[2]	150	2.79	-	0.533	-	1.32	-	-
[b]OSWK[2]	150	7.23	-	0.723	-	3.14	-	-
Coir[5][27]	120	4.57	-	0.71	1.76	-	87	-
GSPP IKL1[5]	150	≥ 3.12	-	0.68	≥ 1.39	-	-	-
GSPP IKL2[5]	175	≥ 3.12	-	0.73	≥ 1.47	-	-	-
GSPP IKL3[5]	275	≥ 2.85	-	0.79	≥ 1.71	-	-	-
GSPP IKL4[5]	165	≥ 2.70	-	0.72	≥ 1.58	-	-	-
GSPP IL1[5]	150	≥ 2.09	-	0.68	≥ 1.13	-	-	-
GSPP IL2[5]	180	≥ 2.09	-	0.72	≥ 1.27	-	-	-
GSPP IL3[5]	275	≥ 1.80	-	0.79	≥ 1.38	-	-	-
GSPP IL4[5]	140	≥ 1.80	-	0.79	≥ 1.04	-	-	-
GSPP IL5[5]	170	≥ 1.80	-	0.74	≥ 1.03	-	-	-
GSPP IL6[5]	250	≥ 1.80	-	0.81	≥ 1.03	-	-	-
Pascorp IL1[5]	135	≥ 2.00	-	0.71	≥ 1.11	-	-	-
Pascorp IL2[5]	140	≥ 2.09	-	0.70	≥ 1.14	-	-	-
Pascorp IL3[5]	150	≥ 2.09	-	0.71	≥ 1.13	-	-	-
Pascorp IL4[5]	165	≥ 2.00	-	0.72	≥ 1.21	-	-	-
Pascorp IL5[5]	170	≥ 2.00	-	0.71	≥ 1.26	-	-	-

Pascorp IL6[5]	175	≥ 2.00	-	0.70	≥ 1.30	-	-	-
SANYEN 6[8]	123.58	2.54	1.49	-	-	1.4	-	1.6
SANYEN 13[8]	121.90	2.30	1.51	-	-	1.5	-	1.5
SANYEN 20[8]	118.39	1.92	1.55	-	-	1.4	-	1.5
SANYEN 26[8]	118.77	1.95	1.58	-	-	1.3	-	1.6
SANYEN 40[8]	121.45	1.55	1.56	-	-	1.1	-	1.0
SANYEN 50[8]	114.87	1.59	1.87	-	-	1.2	-	1.3
TRI025[8]	122.53	1.71	1.93	-	-	1.2	-	1.2
TRIO45[8]	124.62	1.67	1.82	-	-	1.1	-	1.2
TRIO50[8]	122.55	1.75	1.76	-	-	1.0	-	1.2
PASCORP20[8]	113.21	1.40	1.68	-	-	1.1	-	1.3
PASCORP25[8]	112.05	1.25	1.82	-	-	1.1	-	1.3
100% Old OCC[8]	121.90	1.54	1.79	-	-	1.3	-	1.3
Bamboo[9]	120		-	-	-	-	-	1.04
Softwood[9]	120		-	-	-	-	-	1.78
Kenaf[9]	120		-	-	-	-	-	0.98
Cardoon pulp[21]	130		-	-	-	-	235	-
Waste paper pulp[21]	130		-	-	-	-	145	-
Hardwood NSSC[21]	130		-	-	-	-	200	-
Wheat straw[28]	120		5.7	-	-	-	330	-
OCC[28]	120		1.3	-	-	-	103	-
Wheat Straw+OCC[28]	120		4.0	-	-	-	260	-
Semi-Chemical[30]	151-172		-	-	-	-	421	-
Recycled medium[30]	91-111		-	-	-	-	187	-
Mischantus pulp[29]	-	2.3	-	-	-	-	-	1.54
CF *Mischantus* pulp[29]	-	1.67	-	-	-	-	-	1.065

G= grammage; BI = Burst Index; Bulk: Apparent bulk density; RCT: ring crush test; CMT= flat crush test for corrugating medium

Acknowledgements

The authors would like to thank Universiti Tun Hussein Onn Malaysia for funding this project under UTHM Short Term Grant (STG Vot. 1333) and Ministry of Higher Education Malaysia under SLAI scholarship scheme. The authors would also like to acknowledge Pulp and Paper Laboratory, Biomass Technology Programme, Forest Products Division, Forest Research Institute Malaysia for research facilities and support

References

1. A. A. Mossello, J. Harun, H. Resalati, R. Ibrahim, P. Md Tahir, S. R. Fallah Shamsi, and A. Z. Mohamed, Soda-anthraquinone pulp from Malaysian cultivated kenaf for linerboard production. BioResources. **5(3)**, 1542–1553 (2010).
2. A. A. Mossello, J. Harun, R. Ibrahim, H. Resalati, S. R. Fallah Shamsi, P. Md Tahir, and M. N. Mohad Yusof, Evaluation of linerboard properties from Malaysian cultivated kenaf soda-anthraquinone pulps

versus commercial pulps. BioResources. **5(3)**, 1595–1604 (2010).

3. P. Singh, O. Sulaiman, R. Hashim, L. C. Peng, and R. S. Singh, Using biomass residues from oil palm industry as a raw material for pulp and paper industry: potential benefits and threat to the environment. Environment, Development and Sustainability, **15(2)** , 367–383 (2013).
4. L. Jasmani, S. Mahmudin, and R. Ibrahim, Recycled paper strengthened with Kenaf. *FRIM in Focus: Waste to Wealth* (Forest Research Institute Malaysia, Kepong, Selangor, p. 10, Sep-2009).
5. N. M. Main, R. A Talib, R. Ibrahim, R. Abdul Rahman, and A. Z. Mohamed, Linerboard made from soda-anthraquinone (Soda-AQ) treated coconut coir fiber and effect of pulp beating. BioResources. **10(4)**, 6975–6992 (2015).
6. A. A. Mossello, J. Harun, H. Resalati, R. Ibrahim, S. R. Fallah Shamsi, and P. Md Tahir, New approach to use of kenaf for paper and paperboard production. BioResources **5(4)**, 2112–2122 (2010).
7. R. Ibrahim, Structural, mechanical and optical properties of recycled paper blended with oil palm empty fruit bunch. Journal of Oil Palm Research. **15(2)**, 28–34 (2003).
8. I. Rushdan, J. Latifah, W. K. Hoi, and M. Y. Mohd Nor, Commercial-scale production of soda pulp and medium paper from oil palm empty fruit bunches. Journal of Tropical Forest Science. **19(3)**, 121–126 (2007).
9. N. H. Mohd Hassan, S. Muhammed, and R. Ibrahim, Properties of corrugated paper from recycled paper blended with Semantan bamboo. Australian Journal of Basic and Applied Sciences. **8(9)**, 113–117 (2015).
10. S. R. Masrol, M. H. I. Ibrahim, S. Adnan, M. S. S. Amir Shah, N. M. Main, M. F. Esa, and M. H. Othman, Soda Anthraquinone pulping of oil palm male flower spikes. Applied Mechanics and Materials. **660**, 373–377 (2014).
11. S. R. Masrol, M. H. I. Ibrahim, S. Adnan, M. S. S. Amir Shah, N. M. Main, M. F. Esa, and M. H. Othman, Effect of beating process to soda anthraquinone pulp of oil palm male flower spikes fibre. Applied Mechanics and Materials. **773–774**, 158–162 (2015).
12. Daud, Z., Mohd Hatta, M. Z., Mohd Kassim, A. S. and Aripin Mohd, A. Suitability of Malaysia's pineapple leaf and napier grass as a fiber substitution for paper making industry, *EnCon 2013, 6th Engineering Conference, "Energy and Enviroment"*. Kuching, Sarawak. 2-4 July (2013).
13. S. Hokputsa, W. Gerddit, S. Pongsamart, K. Inngjerdingen, T. Heinze, A. Koschella, S. E. Harding, and B. S. Paulsen, Water-soluble polysaccharides with pharmaceutical importance from Durian rinds (*Durio zibethinus Murr.*): isolation, fractionation, characterisation and bioactivity. Carbohydrate Polymers. **56(4)**, 471–481 (2004).
14. M. Unhasirikul, N. Naranong, and W. Narkrugsa, Reducing sugar production from durian peel by hydrochloric acid hydrolysis. World Academy of Science, Engineering and Technology. **6(9)**, 394–399 (2012).
15. H. B. Aditiya, W. T. Chong, T. M. I. Mahlia, A. H. Sebayang, M. A. Berawi, and H. Nur, Second generation bioethanol potential from selected Malaysia's biodiversity biomasses: A review. Waste Management, **47(Part A)**, 46–61 (2016).
16. B. T. Amid and H. Mirhosseini, "Optimisation of aqueous extraction of gum from durian (*Durio zibethinus*) seed: A potential, low cost source of hydrocolloid. *Food Chemistry*, **132(3)**, 1258–1268 (2012).
17. Ministry of Agriculture and Agro-Based Industry Malaysia, *Agrofood Statistics 2014* (Information Management and Statistics Section, Policy and Strategic Planning Division, Putrajaya, 2015).
18. B. H. Hameed and H. Hakimi, Utilization of durian (*Durio zibet*hinus Murray) peel as low cost sorbent for the removal of acid dye from aqueous solutions. Biochemical Engineering Journal. **39(2)**, 338–343 (2008).
19. S. R. Masrol, M. H. I. Ibrahim, and S. Adnan, Chemi-mechanical pulping of durian rinds. *Procedia Manufacturing*. **2**, 171–180 (2015).
20. R. Ibrahim, Moulded products from oil palm pulp. *FRIM in Focus: Waste to Wealth*. (Forest Research Institute Malaysia, Kepong, Selangor, p. 11, Sep-2009).
21. S. Abrantes, M. E. Amaral, A. P. Costa, and A. P. Duarte, Cynara cardunculus L. alkaline pulps: alternatives fibres for paper and paperboard production. Bioresource Technoloy. **98(15)**, 2873–2878 (2007).
22. A. Ghasemian, M. Ghaffari, and A. Ashori, Strength-enhancing effect of cationic starch on mixed recycled and virgin pulps. Carbohydrate Polymers. **87**, 1269–1274 (2012).
23. I. Gonzalez, M. Alcalá, G. Arbat, F. Vilaseca, and P. Mutjè, Suitability of rapeseed chemithermomechanical pulp as raw material in papermaking. BioResources. **8(2)**, 1697–1708 (2013).
24. N. H. Mohd Hassan, S. Muhammed, and R. Ibrahim, Properties of *Gigantochloa scortechinii* paper enhancement by beating revolution. Journal of Tropical Resources and Sustainable Science. **2**, 59–67 (2014).
25. S. Mohamad Jani and I. Rushdan, Effect of bleaching on coir fibre pulp and paper properties. Journal of Tropical Agricuture and Food Science. **42(1)**, 51–61 (2014).

26. P. Sheikhi, G. Asadpour, S. Zabihzadeh, and N. Amoee, An optimum mixture of virgin bagasse pulp and recycled pulp (OCC) for manufacturing fluting paper. Bioresources. **8(4)**, 5871–5883 (2013).
27. N. M. Main, R. A Talib, R. Abdul Rahman, A. Z. Mohamed, R. Ibrahim, and S. Adnan, Effect of amphoteric and cationic polyacrylamide on the structural and strength properties of coir paper. Procedia Manufacturing. **2**, 28–34 (2015).
28. F. Sarkhosha and T. Mohammad, Soda-AQ pulping of wheat straw and its blending effect on old corrugated cardboard (OCC) pulp properties. Tappsa Journal. **March**, 35–39 (2009).
29. F. Marín, J. L. Sánchez, J. Arauzo, R. Fuertes, and A. Gonzalo, Semichemical pulping of Miscanthus giganteus. Effect of pulping conditions on some pulp and paper properties. Bioresource Technology. **100(17)**, 3933–3940 (2009).
30. S. Adamopoulos, E. Martinez, and D. Ramirez, Characterization of packaging grade papers from recycled raw materials through the study of fibre morphology and composition. Global NEST Journal. **9(1)**, 20–28 (2007).

MATEC Web of Conferences **51**, 02008 (2016)
DOI: 10.1051/matecconf/20165102008

Development of Electric Wheelchair with Input of Force Feedback Joystick

Keisuke SUGIHARA[1], Shinsaku FUJIMOTO[1, a] and Koji YOSHIDA[1]

[1]Okayama University of Science, Engineering Graduate Course,

1-1 Ridaicho, Kita-ku, Okayama, JAPAN

Abstract. The purpose of this study is to realize the high function of an electric wheelchair with the consideration of the driving environment. The orientation sensor is used to comprehend the driving environment. The estimated method can be obtained unknown parameters of the driving environment by the orientation sensor. The developed electric-wheelchair operates by the force-feedback joystick. When the marketed wheelchairs drive on the slope or the side-slope, the wheelchairs flow to the slope lower direction. In this case, the wheelchairs might go to roadway side in JAPAN, and such situations are very dangerous. The force-feedback joystick is developed to avoid dangerous situation. The force-feedback joystick is combined with the orientation sensor to the force feedback of joystick. The effectiveness of the force-feedback joystick is confirmed by comparison experiments with the control performance of the wheelchair without the force feedback. Therefore, this study is expected to be useful to social welfare in the future.

1 Introduction

This study focuses on an electric wheelchair which is a welfare apparatus for mobility support of a physical handicapped person. The purpose of this study is to develop the electric wheelchair which can reduce the burden of user and caregiver by the consideration of the driving environment as such the slope or the bad road.

The general operation methods of the electric wheelchair[1] are a hand-rim, joystick and the handle. This study discusses about an electric wheelchair with the joystick type. When the marketed electric-wheelchairs drive on the slope or the side-slope, wheelchairs flow to the slope lower direction (unintentional direction). In this case, wheelchairs might go to the roadway side, and such situations are very dangerous. Thus, it is necessary to consider the driving environment for the electric wheelchair. Therefore several kinds of inertia sensors measure the dynamic slope-angles, while moving the wheelchair. In order to avoid dangerous situations, the proposed electric-wheelchair is equipped the force-feedback (hereafter FFB) joystick system[2][3].

Figure 1 shows the control method of the FFB joystick. On the Figure 1, the blue arrow is the direction of the desired movement, a red arrow is the direction of the force-feedback, and a blue dashed line is the straight direction. The pale joystick is the position of center (neutral position). The

[a] Corresponding author : fuji@are.ous.ac.jp

dark joysticks is the user's mistake operation. This study is based on the control concepts as shown in Figure 1. The user avoids the mistake operation and danger by FFB joystick, according to the direction of the slope and the acceleration (and velocity) of the wheelchair. It is dangerous to operate the joystick to the straight direction on the side-slope[3]. If the user operates it to the slope direction as shown in Figure 1(a) and (d), the wheelchair suddenly accelerates. In addition, if the user gives the quantity of excessive operation to the slope direction (See Fuigure 1 (b) and (c)), the wheelchair becomes the same situation. Therefore, the joystick suppresses the mistake operation by the force feedback (direction of red-arrow). Since the proposed FFB joystick should not suppress the desired movement for the user, the maximum magnitude of the force feedback is 3N. However, the maximum magnitude must be adjusted by each user.

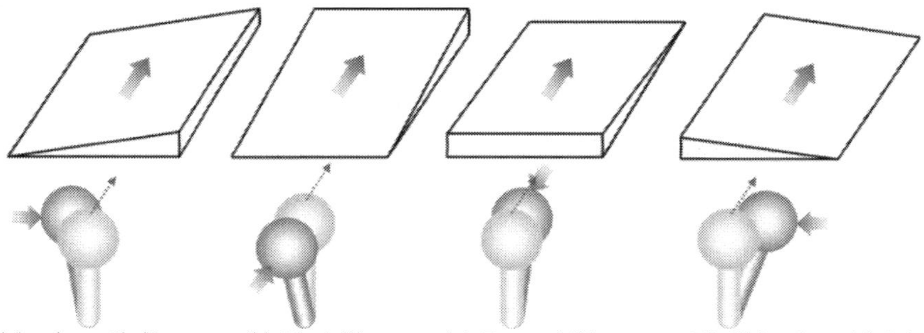

(a) Side-slope (left) (b) Up-hill (c) Down-hill (d) Side-slope(right)

Figure 1. Concepts of force-feedback.

2 Structure of FFB joystick[4]

The FFB joystick is developed as the same excursion of both axes (See Figure 2). When wheelchair users release the hand from a joystick by the operating mistake, the linear spring is attached to return to an origin (neutral position) of each axis. In addition, DC motor and potentiometer are equipped to axes of x-axis and y-axis. Therefore, x and y positions of the joystick can be measured by two potentiometers.

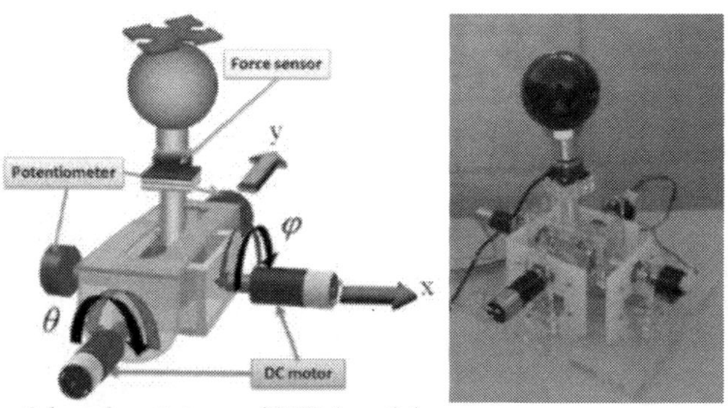

Figure 2. Structure FFB joystick and prototype of FFB joystick

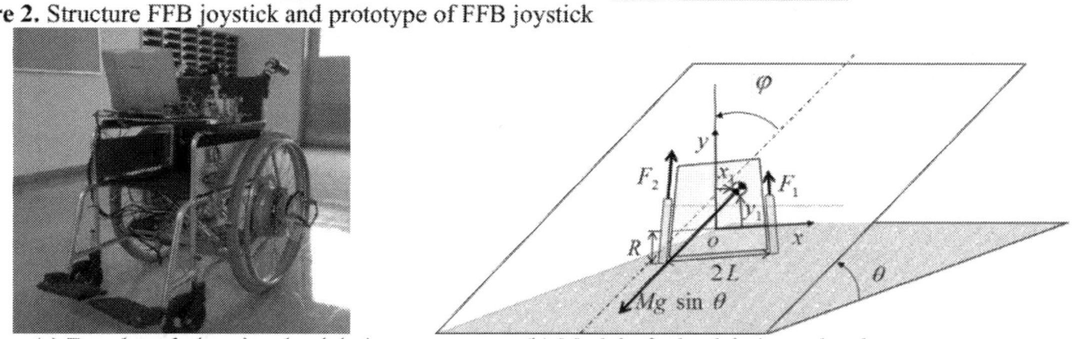

(a) Developed electric-wheelchair (b) Model of wheelchair on the slope

Figure 3. Developed electric-wheelchair and model of wheelchair on the slope

Figure 3(a) shows the developed electric-wheelchair. When the marketed electric-wheelchair drives on the side-slope, the wheelchair goes to the slope direction against user's will. This case is caused by the difference(error vector) between the user's joystick operation(vector of desired movement) and the wheelchair's movement vector. Hereupon, the control error is modified by the FFB joystick in this study. In addition, the developed electric-wheelchair compensates the gravity term by slope angles as shown in Figure 3(b). Then, if the gravity compensation[5] is insufficient by changing mass of wheelchair, the FBB joystick is also used to compensate the modelling error.

The gravity terms of both wheels (F_1:Right, F_2:Left) are the following equations.

$$F_1 = Mg\sin\theta\{\ \sin\varphi - \frac{(L+y_1)\sin\varphi - x_1\cos\varphi}{2L}\} \tag{1}$$

$$F_2 = Mg\sin\theta\{\ \frac{(L+y_1)\sin\varphi - x_1\cos\varphi}{2L}\} \tag{2}$$

where M is the total mass of wheelchair, g is acceleration of gravity, R is wheel radius, $2L$ is the distance between the axes, θ is angle of slope, φ is direction angle of wheelchair and x_1,y_1 are positions of the center of gravity, respectively. The upper equation is the gravity compensation of each wheel.

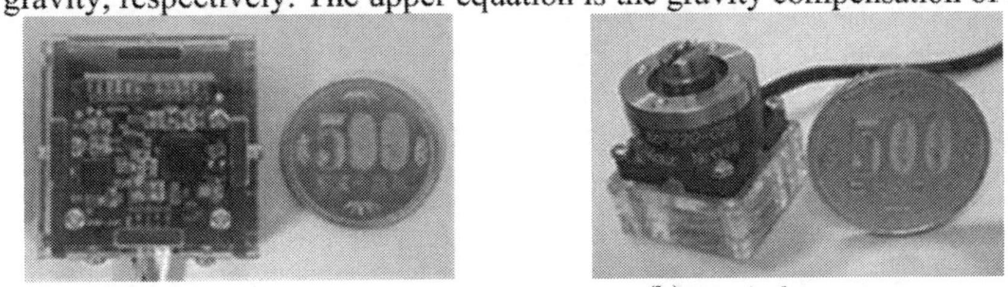

(a) Orientation sensor (b) 3-axis force sensor

Figure 4. Orientation sensor (Roll Pitch Yaw angle) and 3-axis force sensor for FFB joystick

This study adopted the orientation sensor (CH Robotics: UM7-LT orientation, size: 27×26×6.6mm, mass: 3.5g) to perform gravity compensation (See Figure 4(a)). This sensor can be measured the dynamic slope-angles(θ and φ) in real time. The gravity compensation of equations (1) and (2) is calculated by using the sensor data.

In addition, Figure 1 and Figure 4(b) show the 3-axis force sensor for FFB joystick. 3-axis force sensor (Nippon Liniax Co. Ltd.: MFS20-025, size: 20×20×18mm, mass: 26g, Moment Mx, My: 0.25Nm, Force Fz: 50N) is attached to the central part of the joystick in order to measure the operating moments (about x and y axes) and force (z-axis).

The sensor can be also measured the operating force of z-axis. In other words, this sensor is used to determine that the user is holding the joystick. In addition, if the wheelchair passes on the slope (or if the speed of wheelchair is too fast), FFB joystick presents the force-feedback to the user. And the proposed wheelchair is controlled by the joystick such that the user is not dangerously.

3 Force-Feedback Method

The positions (x and y) of the FFB joystick system can be expressed in following equation easily. Figure 5 shows the joystick coordinate system.

$$\begin{cases} x = L\cos\varphi\sin\theta = f_1(\boldsymbol{q}) \\ y = -L\sin\varphi\cos\theta = f_2(\boldsymbol{q}) \end{cases} \quad \because \boldsymbol{q} = [\varphi \quad \theta]^{\mathrm{T}} \tag{3}$$

where L is the length of the joystick, φ and θ are rotary angles of x-axis and y-axis.

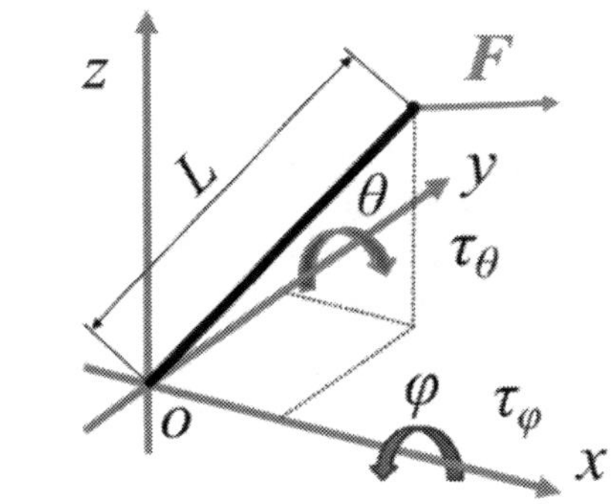

Figure 5. Joystick coordinate system.

The equation (3) is differentiated at time, and tip velocity of joystick is given by:

$$\begin{bmatrix} \dot{x} \\ \dot{y} \end{bmatrix} = \boldsymbol{J}(\boldsymbol{q})\dot{\boldsymbol{q}} = \begin{bmatrix} -L\sin\varphi\sin\theta & L\cos\varphi\cos\theta \\ -L\cos\varphi\cos\theta & L\sin\varphi\sin\theta \end{bmatrix} \begin{bmatrix} \dot{\varphi} \\ \dot{\theta} \end{bmatrix} \tag{4}$$

where $\boldsymbol{J}(\boldsymbol{q})$ is Jacobian matrix.

In addition, linear spring is attached to the edge of each axis in order to return to neutral positions. Therefore dynamic models of the prototype joystick are given by:

$$\begin{cases} I_\varphi \ddot{\varphi} + D_\varphi \cdot \dot{\varphi} + K_\varphi \cdot \varphi = \tau_{M\varphi} - \tau_\varphi \\ I_\theta \ddot{\theta} + D_\theta \cdot \dot{\theta} + K_\theta \cdot \theta = \tau_{M\theta} - \tau_\theta \end{cases} \tag{5}$$

where I_* are the inertia moments, D_* are coefficients of viscous friction and K_* are spring constants, respectively. ($\because * = \varphi$ or θ)

Assume that the operation of the joystick is in a static state, from principle of virtual work, the torques of motors can be expressed as:

$$\begin{bmatrix} \tau_{M\varphi} \\ \tau_{M\theta} \end{bmatrix} = \begin{bmatrix} -L\sin\varphi\sin\theta & -L\cos\varphi\cos\theta \\ L\cos\varphi\cos\theta & L\sin\varphi\sin\theta \end{bmatrix} \begin{bmatrix} f_x \\ f_y \end{bmatrix} + \boldsymbol{K} \begin{bmatrix} \varphi \\ \theta \end{bmatrix} \tag{6}$$

Hence, upper equation holds:

$$\boldsymbol{\tau}_M = \boldsymbol{J}^{\mathrm{T}}(\boldsymbol{q}) \cdot \boldsymbol{F} + \boldsymbol{K} \cdot \boldsymbol{q} \qquad \because \boldsymbol{K} = diag(K_\varphi \quad K_\theta) \tag{7}$$

If the external force $\boldsymbol{F}$ is given by the designer and angular vector $\boldsymbol{q}$ can be measured, we can derive $\boldsymbol{\tau}_M = [\tau_{M\varphi}\ \tau_{M\varphi\theta}]^{\mathrm{T}}$ (motor torques). Therefore, the FFB joystick can present the force feedback to the user. In addition, if the relation between the force sensor and motor torques can calibrate, note that the force-feedback with the expensive force sensor is not necessary for FFB joystick.

4 Verification Experiment of Force-Feedback Method

The verification experiment is performed to confirm the effectiveness of force-feedback method. The experimental method is as follows. When the user operated joystick to x and y axes, 3-axis force sensor is measured the feedback forces (Desired forces:2N) of the user. And, the user is presented the force by the FFB joystick.

However, 3-axis force sensor is used only the force sensing. In other words, the force sensor isn't necessary for the force-feedback.

Figure 6 shows the experimental results of the proposed method(Equation (7)), desired forces (2N) are shown to the user. This result is confirmed the effectiveness of proposed force-feedback method.

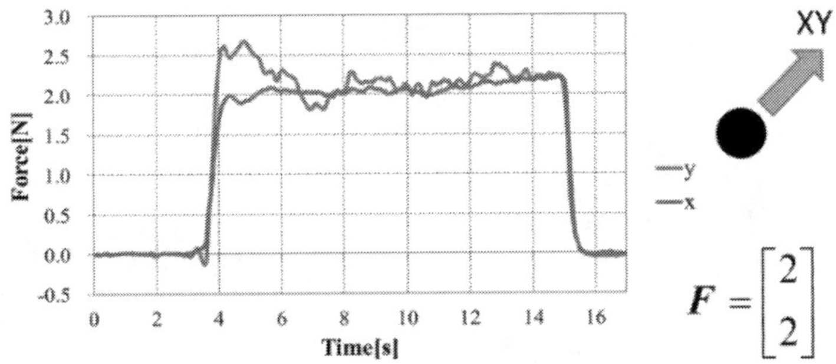

Figure 6. Experimental results of the Force-Feedback Method(Proposed method)

5 Measurement of Orientation Sensor

Figure 7 shows the result which the orientation sensor was measured on the side-slope. Red line is the direction of movement(Sagittal plane). Blue line is right and left direction(Lateral plane). These lines denote angles of inclination. The orientation sensor could measure the inclination angle of about 5 degrees in the case of the side-slope. From Figure 7, a little vibration occurred in direction of movement. The inclination angle(Red line) was measured the angular error within ±1 degree. However, the angle of the wheelchair oscillates actually in the direction of movement by means of the acceleration (and deceleration) of the wheelchair. The right and left direction was measured about 5 degrees. Therefore, the orientation sensor is able to be applied to an electric wheelchair system.

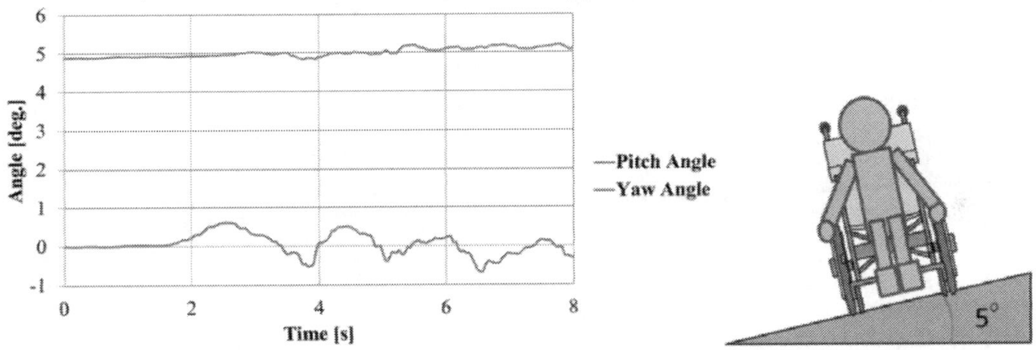

Figure 7. Measurement results of the side-slope(Slope angle: θ=5 degrees)

6 Driving Experiment

This chapter is shown by the experimental results about the effectiveness of the FFB joystick. This experiment was performed on the side-slope of 2 degrees. First, the wheelchair without the force-feedback was operated on the side-slope. Next, the wheelchair with the force-feedback was operated in same environment. The user generally operates the joystick in the forward direction. But, the user might modify the direction of the FFB joystick by the force-feedback.

The line shows the driving path of the wheelchair as shown in Figure 8. Figure 8(a) shows the driving path of the wheelchair that was equipped the normal joystick without FFB. Figure 8(a) shows that the wheelchair(Conventional wheelchair) greatly flowed to the slope direction(Left side). In case of this experiment, the user stopped the wheelchair at around 2 meters because the user was dangerous. Figure 8(b) shows the driving path of the wheelchair that equipped the FFB joystick. Since the behavior of the wheelchair is different from the direction vector of the joystick as soon as the wheelchair starts, the movement direction is modified by the force-feedback. Thus, a wheelchair with the FFB joystick didn't flow to the slope direction(Left side). When the FFB joystick was used, the position error was about 20 cm. But, the position error of the conventional wheelchair was 60 cm. Therefore, the effectiveness of the FFB joystick was confirmed by these experiments on the side-slope.

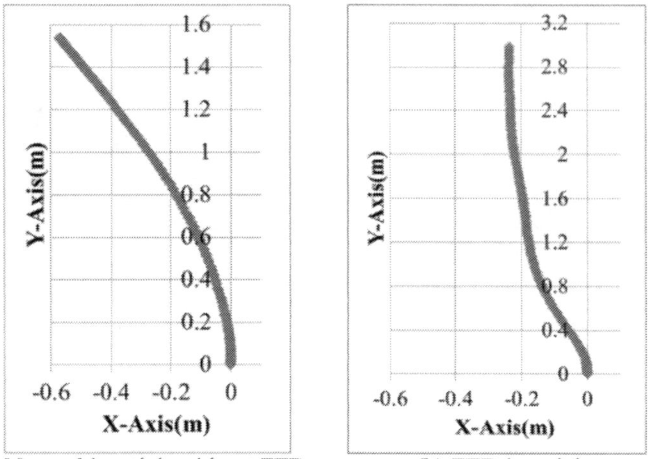

(a) Normal joystick without FFB (b) FFB joystick

Figure 8. Comparison with the driving paths of the wheelchairs without FFB and with FFB

7 Conclusions and Future Work

This study focused on the electric wheelchair of the joystick input and the FFB joystick was produced. On the other hands, the sensor system was constructed to measure the driving environment.

As a result, the dynamic orientation could be measured when a target object (wheelchair) moved. The FFB joystick was produced and force-feedback method was proposed. In addition, the effectiveness of the FFB joystick was confirmed by experimental results on the side-slope.

In future work, this study will propose the method of force-feedback for the slope (up and down hill), an evaluation method of the joystick and an adaptive control design with consideration of the driving environment.

Acknowledgements

This work was supported by Grant-in-Aid for Scientific Research (C) 25350696. In addition, we are deeply grateful to Professor M. Kurisu (Tokyo Denki University) about that we could receive precious time and advice about software development the orientation sensor (UM-7).

References

[1] S. Watanabe, Development of Power Assisted Wheelchair Easy to Drive on Slope, University of Tokyo, Graduation thesis (2004) (in Japanese)

[2] V. Vashista, X. Chen and S. K. Agrawal, Adult-Human Learning on a Robotic Wheelchair Using a Force Feedback Joystick, *Proceedings of the 1st International and 16th National Conference on Machines and Mechanisms (iNaCoMM2013)*, 794-799(2013)

[3] M.A.H. Abdelkader, G. Bourhis and B. Cherki, Haptic Feedback and Motor Disability: A Preliminary Study, *HuMaN 07 proceedings*, Timimoun(2007)

[4] Brubaker CE, McLaurin CA, McClay IS, Effects of side slope on wheelchair performance, Journal of Rehabilitation Research and Development, 23-2 , 55-57(1986)

[5] K. Sugihara, S. Fujimoto K. Yoshida, Development of Electric Wheelchair with Input of Force Feedback Joystick, *Proceedings of the 24th Annual Conference of the SICE Chugoku Chapter*, 90-91 (2015) (in Japanese)

MATEC Web of Conferences **51**, 02009 (2016)
DOI: 10.1051/matecconf/20165102009

Level-Ground Walking for 3D Quasi-Passive Walker with Flat Feet - Lateral-plane Input using McKibben-Type Artificial Muscle -

Akihiro Yamamoto[1] , Shinsaku Fujimoto[1,a] and Tetsuya Kinugasa[1]

[1]*Okayama University of Science, Engineering Graduate Course*

1-1 Ridai-cho, Kita-ku, Okayama, Japan

Abstract. Currently, many bipedal robots have been proposed to realize the high energy efficiency walking. The passive dynamic walking does not require control input. Generally, a foot of passive dynamic walking robot is an arc foot. In this paper, it is intended to establish a control method and control mechanism to achieve energy efficient and stable gate. Therefore, we developed 3D quasi-passive walker with flat feet driven by an antagonistic pneumatic artificial muscle. An antagonistic mechanism is constituted by a pair of McKibben muscle. And an antagonistic pneumatic system is used as joint actuators of linkage mechanisms which control the torque, joint stiffness and position simultaneously. Finally, this report shows that the 3D quasi-passive walking in the level ground can realize by the swinging (simple) input of the frontal direction, and the stride of the robot is proportional to lateral-plane input.

1 Introduction

An energy efficiency and locomotion such as human [1] are very important issues in a biped robot walking. Recently, a passive dynamic walking has received considerable research attention. The investigation of passive dynamic mechanism can help to solve both issues. The idea of passive dynamic walking was pioneered by McGeer [2]-[4] who studied the so-called compass gait biped, a two DOF walker that can achieve passive dynamic walking down shallow inclines powered only by gravity. Most previous studies of passive dynamic walker with arc-shaped feet (no feet) have been constrained by the locomotion of the sagittal plane [5][6]. Since the ground-direction contact occurs only at a point, the arc-shaped feet provide insufficient friction against ground. Passive walkers with arc-shaped feet are unstable in the upright state.

Wisse et al.[7]-[9] studied a 2D passive walker with flat feet and ankle springs by utilizing the equivalence between the torsional spring and the arc-shaped feet. Therefore, the arc-shaped feet can be replaced by flat feet that are mounted on ankles with torsional spring stiffness.

In this paper, we develop the thee-dimensional quasi-passive walker with flat feet which is driven by antagonistic pneumatic artificial muscle.

This paper is organized as follows. Chapter 2 explains the principle of 3D quasi-passive walking, and antagonism pneumatic artificial muscle. Chapter 3 then presents the structure of quasi-passive

[a] Corresponding author : fuji@are.ous.ac.jp

dynamic walker, followed by the method of our experiment and its results in Chapter 4. Finally, the paper concludes with the results of the 3D quasi-passive walking in the level ground.

2 Relations of Arc foot and Equivalent spring constant

Wisse et al. showed that the rigid arc-shaped feet can be replaced by flat feet that are mounted on ankles with a torsional spring stiffness. The spring stiffness has a similar effect as the foot radius; it reduces the sensitivity to disturbances due to friction (slipping) and collisions, and thus improves the disturbance handling. Therefore, 3D passive dynamic walking with flat feet are walking forms that extend the lateral motion to 2D passive dynamic walking in sagittal plane as indicated in Figure 1. They studied a 2D passive walker with flat feet and ankle springs by utilizing the equivalence between the torsional spring and the arc-shaped feet.

A typical antagonistic pneumatic mechanism consists of a pair of McKibben artificial muscles [10] as shown in Figure 2. The developed walker [11] has the equivalent characteristic of torsional spring to the walking robot by equipping antagonism pneumatic artificial muscle. It is confirmed that sustainable 3D quasi-passive walker can walk on a slope. In this study, we measure the natural frequency, the ankle rigidity and the stride of the walking robot based on 3D quasi-passive gait on the slope.

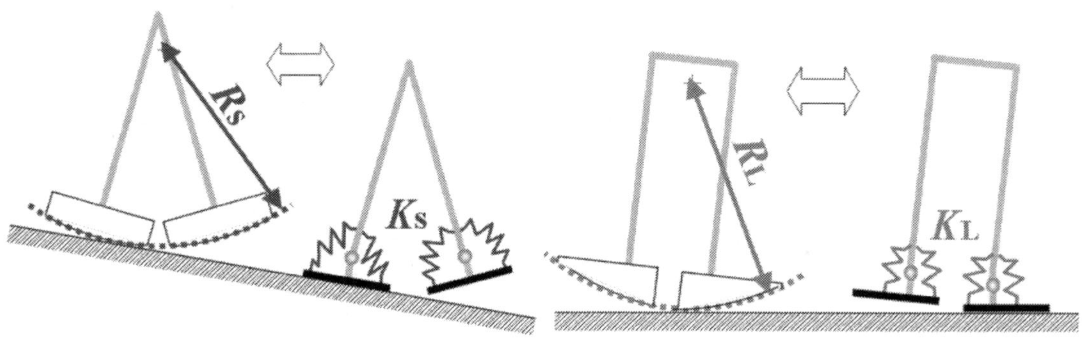

(a) Sagittal plane (b) Lateral plane

Figure 1. 3D passive dynamic walking with flat feet.

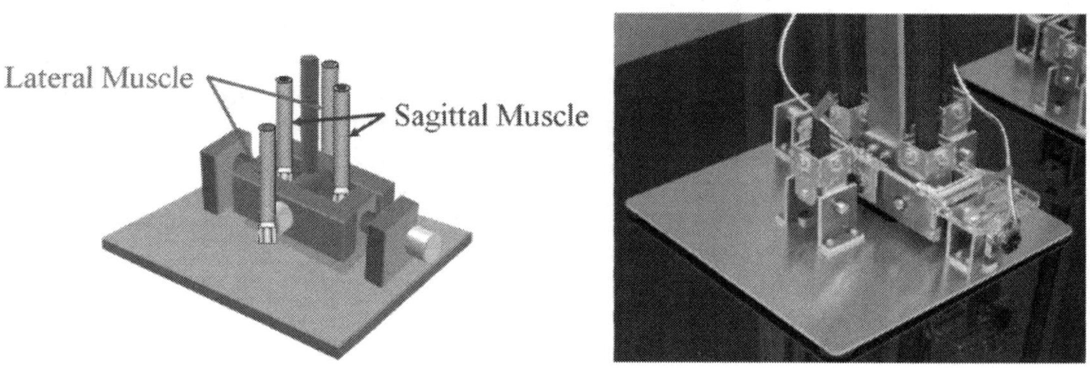

Figure 2. Structure of ankle joint.

3 Structure of Bipedal Robot

Figure 2 and Figure 3 showed whole structure and dimensions of the quasi-passive dynamic walker (TENBU 3). TENBU produced in this study is equipped the McKibben artificial muscle at sagittal and lateral direction of the ankle joint. The biped walking robot is equipped with micro-computer

(Renesas Co. Ltd.,H8/3664), battery, air tank , On/Off valves and wireless module(ZigBee). Passive dynamic walking is a gait form that depends on the dynamic characteristics of biped walking robot. It is desirable that there is no external cable(umbilical cable) as much as possible. Thus, "TENBU" robot is a self-contained robot. The trial number of the walking experiment in level ground is about 5 times by using two air tanks(air capacity : 4 liter). Circular rubber sheet with the force sensor is attached to the soles of the feet to increase friction on the ground.

The micro-computer adjusts the internal pressure of artificial muscle by controlling the On/Off valve. The embedded controller that is the micro-computer gets the output voltage from six potentiometers and eight force sensors. Then, the sampling period of the On/Off valve control was about 30 ms. However, the period of the sampling data is 70 ms by using the wireless ZigBee module(57600 bps).

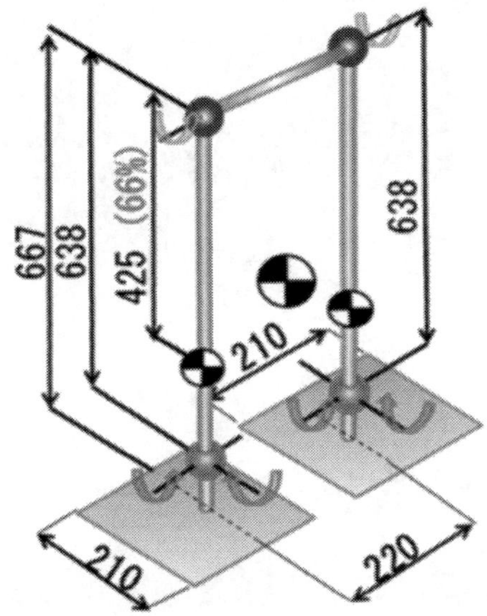

Figure 3. Structure and size of quasi-passive dynamic walker.

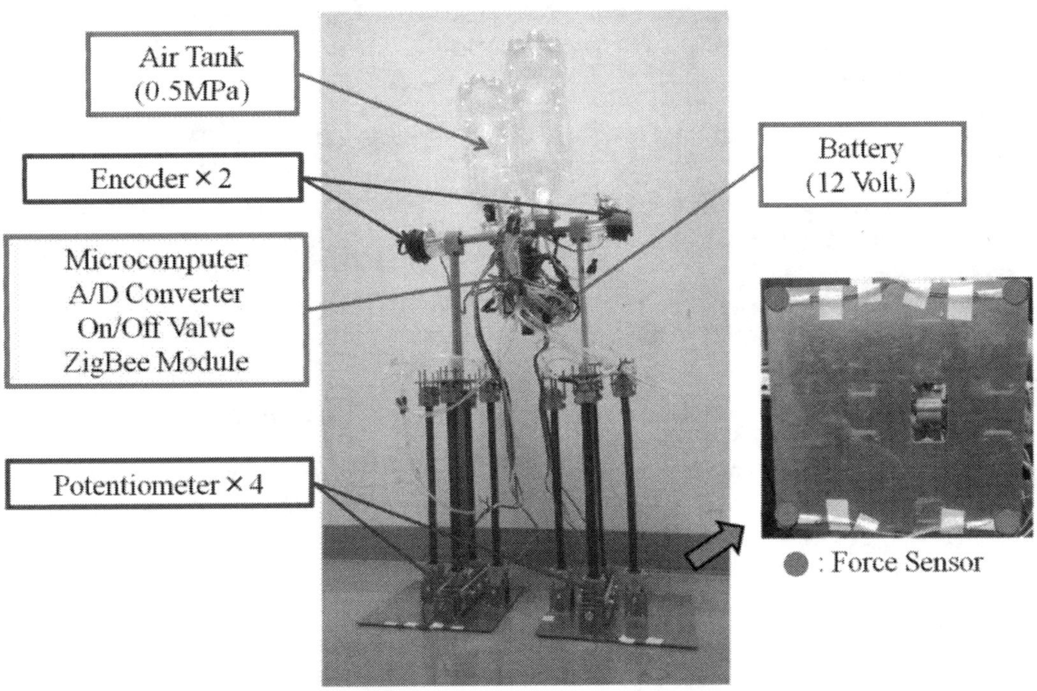

Figure 4. Quasi-passive dynamic walker (TENBU 3).

4 The experiment of 3D quasi-passive walking

In this chapter, we carried out the experiment of 3D quasi-passive walking on the slope to confirm the effectiveness of the produced ankle experimentally and to acquire basic walking experiment in the level ground by lateral-plane input based on the walking data in the slope.

4.1 Quasi-passive walking on the slope

In this paragraph, we state the 3D quasi-passive walking experiment on the slop. Figure 5 shows the most common experimental result in which the 3D quasi-passive walker walked on the slope.

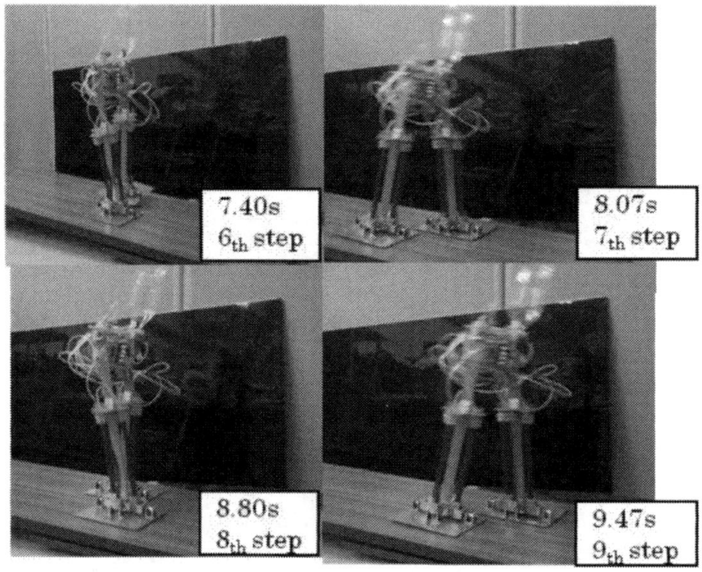

Figure 5. 3D quasi-passive walking on the slope.

The experimental conditions of the passive walking are as follows. But, valve controller is only control to keep the internal pressure of the McKibben artificial muscles to a constant value to maintain ankle rigidity constantly.

1. Angle of inclination of the slope : 4.0[deg.]
2. Internal pressure of artificial muscles : Sagittal plane direction : 0.20[MPa]
   Lateral plane direction : 0.15[MPa]

As results of 3D quasi-passive walking experiment, the following gait data were obtained.
The sustained walking are possible under this condition:

1. Average step : 149[mm]
2. Average walking period : 1.36[s] (Half period : 0.68[s])
3. Average walking speed : 220[mm/s]

4.2 The experiment of 3D quasi-passive walking in the level ground

One of the purposes of this study includes that they realize 3D quasi-passive walking in the level ground. Therefore, we examined level ground walk by swing motion input (Lateral plane direction) using the quasi-passive walking without the knee as the first step in this paper.

We let fluctuate internal pressure P_R and P_L of the antagonism pneumatic artificial muscle attached to lateral plane direction of each ankle joint as following equation.

$$P_R = P_{AMP}\,sin(\omega t) + P_0 \qquad : [\mathrm{Pa}]\ (\mathrm{Right}) \tag{1}$$

$$P_L = P_{AMP}\,sin(\omega t - \pi) + P_0 \qquad : [\mathrm{Pa}]\ (\mathrm{Left}) \tag{2}$$

where P_{AMP} is amplitude, ω is angular frequency and P_0 is initial pressure. The initial pressure set up the sagittal plane muscle 0.20 MPa and lateral plane muscle 0.15 MPa from the experimental results of the passive walking on the slope. The angular frequency ω is set approximately 1.36s (Half period : 0.68s) like the walking period of the 3D quasi-passive walking on the slope. In addition, the slope is inclined approximately 4 degrees.

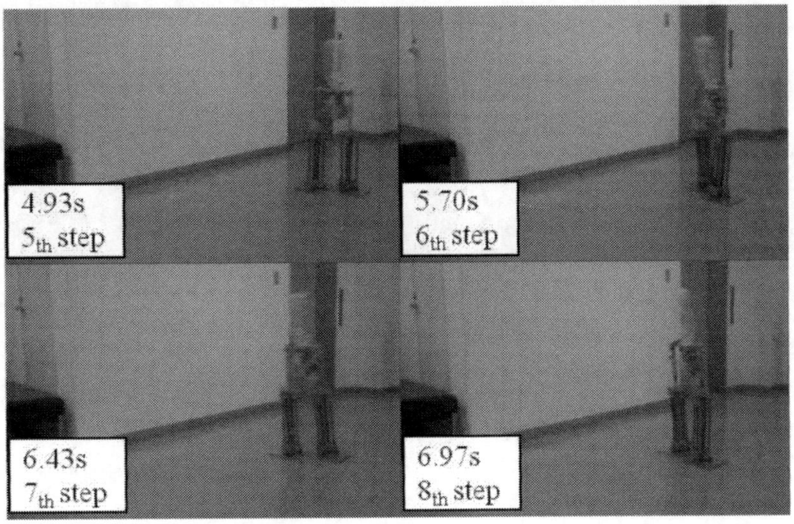

Figure 6. 3D quasi-passive dynamic walking (Level-ground).

Figure 6 shows the most common experimental results in which 3D quasi-passive walking in the level ground. TENBU was able to realize 3D quasi-passive walking in the level ground though it was only gave very simple lateral-plane input.

Figure 7 shows the relations between the stride and every half period. Half period that TENBU can walk from 0.3s to 0.5s. In addition, we obtained the following equation that approximates to the liner function by the least-squares method from Figure 6.

$$b = 320 \cdot T_P - 70.4 \text{ [mm]} \tag{3}$$

where b is stride, T_P [s] is half period. Therefore, it is thought that the stride and the period of lateral-plane inputs is proportion.

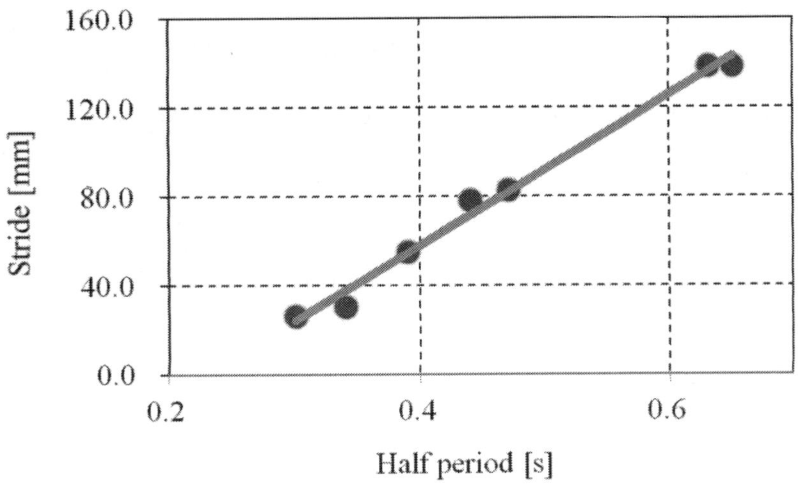

Figure 7. Relation between stride and half period.

5 Conclusions and Future Work

This paper designed and produced the bipedal robot which had the antagonism pneumatic artificial muscle with flat feet. And we carried out 3D quasi-passive walking experiment. In addition, we carried out 3D quasi-passive walking experiment in the level ground by simple lateral-plane input. As a result, we were able to get the resulting knowledge.

1. The quasi-passive dynamic walker was able to be realized 3D quasi-passive walking in the level ground by means of giving simple lateral-plane input.
2. The proportion relations between the half period of lateral-plane input and the walking stride were discovered.

As the future work, we are going to conduct the gait analysis based on COP (Center Of Pressure) ROS (Roll Over Shapes) using encoder, potentiometer and power sensor.

References

[1] S. Collins, A. Ruina, R. Tedrake and M. Wisse : Efficient Bipedal Robots Based on Passive-Dynamic Walkers, *Science*, **307**, 1082–1085 (2005)
[2] T. McGeer : Passive Dynamic Walking :*Simson Fraser University Burnaby British Columbia, CSS-ISS* TR88-02 (*Technical report*), 1-51 (1988)
[3] T. McGeer : Passive Dynamic Walking, IJRR, **9**, 62-82 (1990)
[4] T. McGeer : Passive Walking with Knees, *Proceedings of the IEEE International Conference on Robotics and Automation*, **3**, 1640-1645 (1990)
[5] A. Goswami, B.Thuilot and B. Espiau : A study of the passive gait of a compass-like biped robot symmetry and chaos, IJRR, **17**, 1282-1301 (1998)
[6] M. Garcia, A. Chatterjee, A. Ruina and M. Coleman : The Simplest Walking Model: Stability, Complexity, and Scaling, J BIOMECH ENG-T ASME, **120**,281-288 (1998)
[7] M. Wisse at all : Ankle springs instead of arc-shaped feet for passive dynamic walkers, *IEEE-RAS International Conference on Humanoid Robots*,110-116 (2006)
[8] T. Kinugasa, K. Yoshida, K. Kotake, K. Fujimura, H. Tanaka and K. Ogawa : 3D Passive Walker with Ankle Springs and Flat Feet , JRSJ, **27**, 91-94 (2009) (in Japanese)
[9] T. Narukawa, K. Yokoyama, M. Takahashi and K. Yoshida : Design and Construction of a Simple 3D Straight-Legged Passive Walker with Flat Feet and Ankle Springs, JSDD, **3**, 1 (2009)
[10] S. Fujimoto, T. Ono, K. Ohsaka and Z. Zhao : Modeling of Artificial Actuator and Control Design for Antagonistic Drive System, Transactions of JSME (C), **73** , no.730, 1777-1785 (2007) (in Japanese)
[11] S.Fujimoto, T.Kinugasa K.Yoshida and H.Watanabe: 3D Quasi-passive Walking of Bipedal Robot with Flat Feet -Quasi-passive Walker Driven by Antagonistic Pneumatic Artificial Muscle-, IJAMechS, **5**, 95-104 (2013)

MATEC Web of Conferences **51**, 02010 (2016)
DOI: 10.1051/matecconf/20165102010

Development of Active Orthosis for Lumbago Relief-Improvement of Pneumatic Textile Actuator for Orthosis -

Shinsaku FUJIMOTO [a], Tetsuya AKAGI and Akihiro YAMAMOTO

Department of Inteligent Mechnical Engineering, Okayama University of Science,

1-1 Ridaicho, Kita-ku Okayama, JAPAN

Abstract. It is important to develop the orthosis which improves the Quality of Life (QOL) and maintains health conditions. As one of the treatment methods done to lumbago(low back pain), the waist fixation method with the spinal brace or the orthosis is prescribed. A waist active orthosis implemented with pneumatic flexible actuators have been developed. However, several problems of the previous actuator were that the strain and the generated force were small for the orthosis.Thus, this paper proposesthe improved actuator for the orthosis. Theimproved actuator is modeled and the reliability of static and dynamic model is validated through experiment. As a result, it was confirmed that the improved actuator had the strain of 2 times and the generated force of 1.3 times, in comparison with the previous actuator. And the dynamic model including volume of actuatorcould be represented by a second-order form with a dead time.

1 Introduction

It is important to improve the QOLand maintain the health condition for elderly people. Our special attention is directed to lumbago because many people feel a back pain in daily life. As one of the treatment methods done to lumbago, the waist fixation method[1] with a spinal brace or an orthosis is performed. Generally, an orthosis is wrapped around the waist made of cloth and the rubber textile. However, the existing orthosis have the three difficult and important issues.

1) Problem difficult to meet on-demand requirements2) Problem to suppress blood stream
3) Problem difficult to customize

In order to solve these issues, a waist active orthosis implemented with pneumatic flexible actuators is developed and pressure control method of pneumatic flexible actuators is established. However, the customizing issues have not been solved.In this study, firstly, the prototype of pneumatic textile actuator (hereafter called PTA) has been developed. But several problems of the previous actuator were that the strain and the generated force were small for the orthosis.The previous actuatoris indirectly driven by a McKibbenactuator. Thus, the improved actuator changes to the driving method that can be contracted directly by a silicone tube. Therefore the improved actuator is modeled and the reliability of static and dynamic model is validated through experiment.

---

[a] Corresponding author : fuji@are.ous.ac.jp

2 Structure of Improved PTA

The previous PTA has a structure which inwrought with a long McKibben-type actuator[2] into two soft cloths (See Figure 1(a)). When the compressed air is inject into the supply port, the McKibben-type actuator contracts to an axial direction. As the result, the PTA shrinks indirectlyby the seam constraint. Thus, two problems of the previous PTAwere that the strain and the generated force were small for the orthosis.

On the other hand, the improved PTA has a structure which inwrought with a silicone tube into two soft cloths as shown in Figure 1(b).

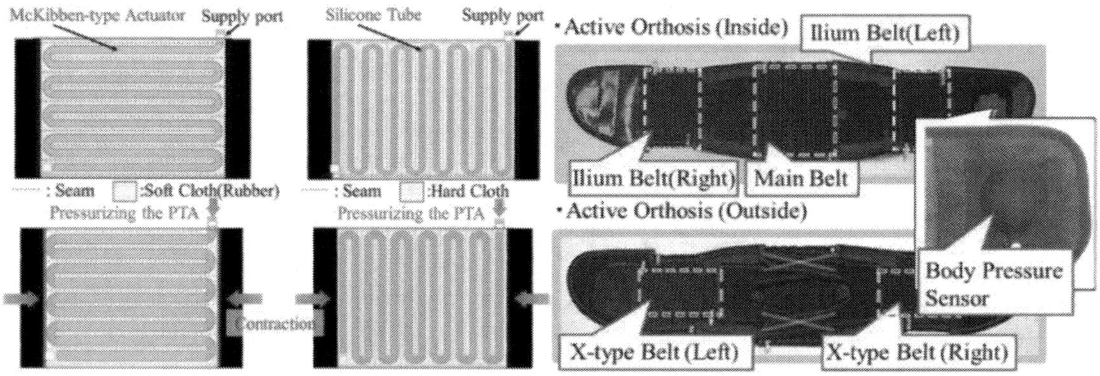

(a) Previous PTA. (b) Improved PTA. (c) Prototype of Active Orthosis.

Figure 1. Previous PTA and improved PTA.

Figure 1(c) shows the prototype of the active orthosis for the lumbago relief. The proposed active orthosis consists of five PTAs (=belts) and the body pressure sensor. The size of the orthosis is 970×200×7mm, andthe mass is 300g. The active orthosis has a double structure. One structure is the body-orthosis that is configured in the main belt and the ilium belts (right and left). Another structure is two X-type belts.

Figure 2(a) shows the operating principle of PTA.When the compressed air is injected into the supply port, the silicone tubeexpands a radial direction. As the result, the PTA shrinks directlyby the constraints of the seam constraint and the hard cloth.

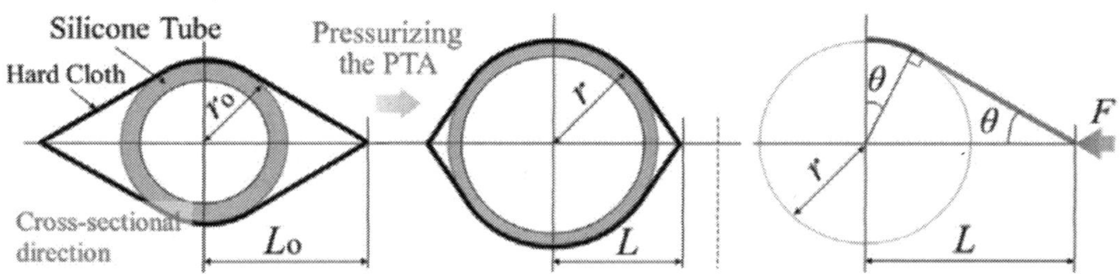

(a)The operating principleof improved PTA. (b) Static model.

Figure 2. The operating principle and static model of improved PTA.

3 Modeling of Improved PTA

This chapter will be described about the construction methods of a static model and a dynamic model. And theorthoses control valvesystem is modeled and the reliability of exhaust and supply model is validated through experiment.

3.1 Static Model of Improved PTA

From the view of energy conservation, the input work of the improved PTA should equal the output work which a system includes force dissipation or force loss F_{diss}. Suppose that the actuator is in this ideal condition, a following equation of static model holds.

$$P \cdot dV = -(F + F_{diss}) \cdot dL \tag{1}$$

where P is the internal gauge pressure,dV is the volume change,F is the force generated by the PTA, and dL *is* the axial displacement, respectively.

When the clothes of PTA are made with the material which can't shrink, the half pitch length b of the seam doesn't change. The outside radius r of the silicone may be expressed as

$$r = b(\theta + \cot\theta)^{-1} \tag{2}$$

where θ is the angle between the seam axis and the silicone (See Figure 2(b)).

The inner volume V of silicone is given by

$$V = l \cdot \pi \, r_S^2 \tag{3}$$

where r_S is the inside radius and l is length of the silicon, respectively.

Assume that the outside radius r change but the cross-sectional area S_O and the length l of silicone don't change. The following equation is expressed by

$$V = l\,(\pi r^2 - S_O) \because S_O = \pi(r^2 - r_S^2) = Const. \tag{4}$$

Eq.(2) is substituted for eq.(4). Then, the inner volume V can be rewritten as

$$V = l\,\{\pi b^2(\theta + \cot\theta)^{-2} - S_O\} \tag{5}$$

On the other hand, since the force term $F + F_{diss}$ of eq.(1) is given by

$$F + F_{diss} = -P\frac{dV}{dL} = -P\frac{\partial V}{\partial \theta} \cdot \frac{\partial \theta}{\partial L} \tag{6}$$

Hence, the generated force F is

$$F = 2\pi lbP\left\{\frac{\cos\theta}{\theta(\theta + \cot\theta)}\right\} - F_{diss} \tag{7}$$

where each partial differentiation with eq.(6) is

$$\frac{\partial V}{\partial \theta} = \frac{2\pi lb^2 \cot^2\theta}{(\theta + \cot\theta)^3}, \quad \frac{\partial L}{\partial \theta} = -\frac{b\theta\cos\theta}{(\theta\sin\theta + \cos\theta)^2} \tag{8}$$

The length L is the involute function with respect to θ. Thus, the angle θ is the nonlinear inverse-function with respect to L. However, the inverse function of an involute function can't be solved analytically. In this paper, this inverse function will be solvednumerically.

Therefore, to solute the nonlinear function $f(\varepsilon)$ that is the first term of right side in eq.(7), the relation between the function $f(\varepsilon)$ and the strain ε is calculated by the numerical solution such as Newton method.

$$f(\varepsilon) = \frac{\cos\theta}{\theta(\theta + \cot\theta)} \because \varepsilon = \frac{L - L_O}{L_O} \tag{9}$$

where L_O denotes an initial length of PTA.

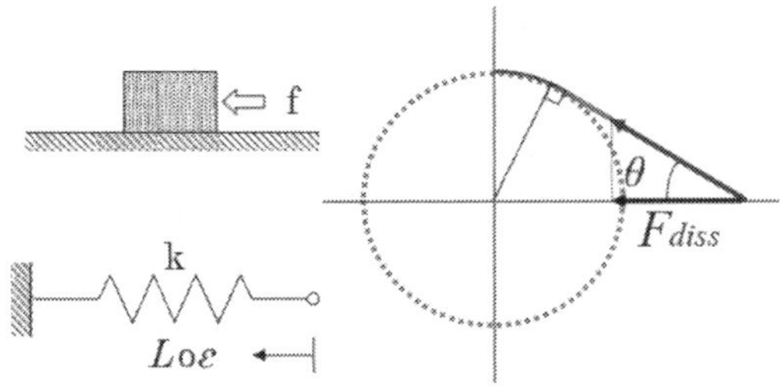

Figure 3. Concept of dissipation forceF_{diss}.

If a strain of the PTA is held in less than 25 percent, the function $f(\varepsilon)$ with respect to strain ε is approximated by following first-order function.

$$f(\varepsilon) = \frac{\cos\theta}{\theta(\theta + \cot\theta)} \equiv \alpha \cdot \varepsilon + \beta \tag{10}$$

where α and β are characteristics parameters of PTA.

Eq.(10) is substituted for eq.(7). The generated force F is given by

$$F = 2\pi lbP(\alpha \cdot \varepsilon + \beta) - F_{diss} \tag{11}$$

Then, the dissipation force is discussed. In this paper, the dissipation force F_{diss} includes a static frictional force and the force loss of an elastic force which occur to the silicone and the cloth. Figure 3 shows the concept of dissipation force F_{diss}.

The static model of the improved PTA with the dissipation force is expressed by

$$F = 2\pi lbP(\alpha \cdot \varepsilon + \beta) - (kL_0\varepsilon + f)\cos\theta \tag{12}$$

where k is an elastic coefficient and f is a static frictional force.

Furthermore, we discuss the relation between strain ε and cosine function that is the second term of right side in eq.(12). If a strain of the PTA is held in less than 20 percent, the function $\cos(\theta)$ with respect to strain ε is approximated by following first-order function.

$$\cos\theta \equiv \gamma \cdot \varepsilon + \eta \tag{13}$$

where γ and η are the approximate coefficients of the cosine function.

Eq.(13) is substituted for eq.(12). Therefore, the static model of the improved PTA is given by

$$F = P(\alpha_1\varepsilon + \alpha_0) - (\beta_2\varepsilon^2 + \beta_1\varepsilon + \beta_0) \tag{14}$$

where coefficients $\alpha_1 = 2\pi lb\alpha$, $\alpha_0 = 2\pi lb\beta$, $\beta_2 = kL_o\gamma$, $\beta_1 = kL_o\eta + f\ \gamma$ and $\beta_0 = f\ \eta$ are PTA characteristics constants, respectively. Clearly, the force F depends on both pressure P and strain ε. When pressure increases, the silicone tube expands radially and PTA shortens in length to generate a radial contraction force.

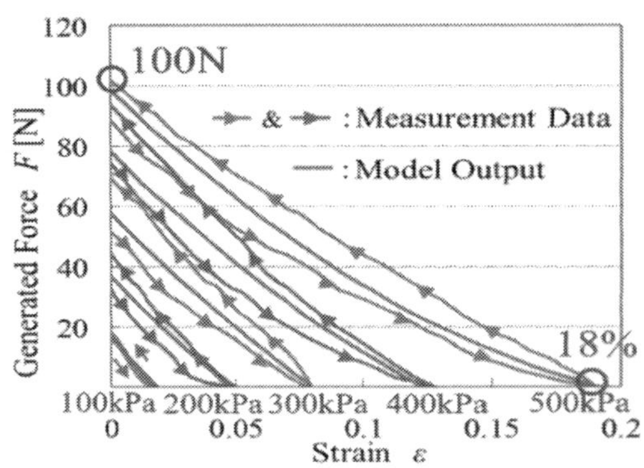

Figure 4. Varification experiment of static model (Results of measurement and model output signals).

In order to prove versatility of equation (14), the comparison was done between the measurement data and force model. It can be seen in Figure 4 that both generated force $F$and strain εincrease with pressure. The static force model of the improved PTA is that it is located in the average of the hysteresis loop. Therefore, the accurate fitting is demonstrated in Figure 4. And ,The improved actuator had the strain performance of 2 times and the generated force of 1.3 times, in comparison with the previous actuator.

3.2 Three Modes of Tested Valve

Figure5 shows the operating principle of the orthosis controlvalve. The tested valve consists oftwo on/off-type control valves (SMC Co. Ltd.,S070C- SDC-32) that both output ports are connected each other. One valveis used as the supply valve, and another is used as the exhaust valve. Two valves can adjust output flowrate like a variable fluid resistance by means of the fast switching. The size of the on/off valve is 36×14.5×7.2mm, andthe mass is 5g. The total mass of the orthosis control valve including the controller (Micro-computer: Renesas Co. Ltd., H8/3664F) is very light, that is about 220g.The orthosis controlvalve has three modes (See Figure 5(a),(b) and (c)) as follows:

(a) Supply Mode: The exhaust valve is the off state. And,the pressure of the tank (PTA) can be adjusted by the fast switching of the supply valve.
(b) Exhaust Mode: Conversely, the supply valve is the off state. And, the pressure of the tank (PTA) can be adjusted by the fast switching of the exhaust valve.
(c) Hold Mode: When both valves are the off states, the pressure of the tank (PTA) is kept a constant pressure.

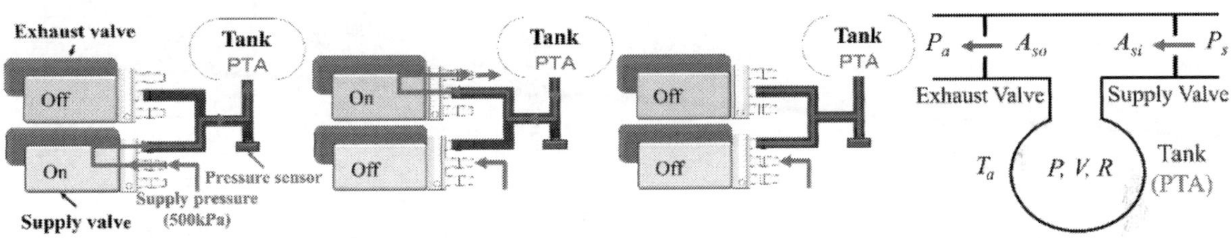

(a) Supply Mode (b) Exhaust Mode (c) Hold Mode (d) Analytical model

Figure 5. Three Modes and analytical modelof the orthosis controlvalve.

3.3 Dynamic Model of Improved PTA[3] [4]

Figure 5(d) shows the analytical model of the orthosis control valve. The mass flow rate of supply valve$Q_i$and the exhaust valve$Q_o$are given as follows.

$$Q_i = A_{si} P_s \sqrt{\frac{2}{RT_a}} g(z_i), \quad z_i = \frac{P}{P_s} \quad : \text{Supply State} \tag{15}$$

$$Q_o = A_{so} P_a \sqrt{\frac{2}{RT_a}} g(z_o), \quad z_o = \frac{P_a}{P} \quad : \text{Exhaust State} \tag{16}$$

where R and $T_a$mean a gas constant and an absolute temperature, respectively. The function $g(z)$that expresses the state of flow is given as follows.

$$g(z) = \sqrt{\frac{\kappa}{\kappa - 1}\left(z^{2/\kappa} - z^{(\kappa+1)/\kappa}\right)} \quad [0.528 \le z \le 1] \ \text{Subsonic Flow} \tag{17}$$

$$g(z) = \sqrt{\frac{\kappa}{\kappa + 1}\left(\frac{2}{\kappa + 1}\right)^{2/(\kappa-1)}} \quad [0 \le z \le 0.528] \ \text{Choked Flow} \tag{18}$$

where κ means a specific heat ratio (=1.4). The pressure P in the volume$V$ of PTA is given by next equation.

$$\dot{P} = \frac{dP}{dt} = \frac{\kappa R T_a}{V}(Q_i) \quad : \text{Supply State} \tag{19}$$

$$\dot{P} = \frac{dP}{dt} = \frac{\kappa R T_a}{V}(-Q_o) \quad : \text{Exhaust State} \tag{20}$$

The systemsof Eqs.(19) and (20) are non-linear systems with respect to the input of sectional areaA_{s*}(*=i or o). Here, the simple idea is to approximate a non-linearsystem by a linear one (around the pressure point $P=P_E$ (z= 0.528))

$$\dot{x} = -\frac{2.70\times10^2 \cdot A_{so}}{V}x - \frac{4.61\times10^4}{V}v \qquad \because x = P_E - P, v = A_o \quad : \text{Supply State} \tag{21}$$

$$\dot{x} = -\frac{2.78\times10^2 \cdot A_{si}}{V}x + \frac{4.68\times10^4}{V}v \qquad \because x = P_E - P, v = A_i \quad : \text{Exhaust State} \tag{22}$$

In the linearization, the atmosphericpressure $P_a$of 101.3 kPa, the room temperature $T_a$of 298 K, the gas constant $R$of 287 J/kg/K, the supply pressure $P_s$of 500 kPa were used. And the sectional area of the supply port A_{si}(=2.02×10^{-7} m^2) and thearea of exhaust port A_{so}(=2.30×10^{-7} m^2)were adopted the values of catalog specification (on/off valve). The obtained linear system can be expressed by the first-order transfer function. The pole of first-order system depends on the volume V of PTA. The smallervolume of PTA, the faster response speed becomes.

On the other hand, suppose that the sectional area$v=A_*$ of the on/off valve is opened (or closed) slowly.Suppose that the switching area A_*(*=i or o) of valve is approximated by dead time L(=3 ms) and time constant T_o (=2 ms) of a primary delay system.

$$v(s) = A_*(s) = \frac{A_{s*}}{T_o s + 1} \cdot e^{-Ls} \cdot u(s) \qquad \begin{cases} u(s) = 1 & : \textit{Supply Mode} \\ u(s) = -1 & : \textit{Exhaust Mode} \\ u(s) = 0 & : \textit{Hold Mode} \end{cases} \quad \text{New Input} \tag{23}$$

From Eqs.(21)-(23), the transfer function of on/off valve system with volume of PTA is given by a second-order form and a dead time.The transfer function of the control valve system with volume of PTA is given by a second-order form with a dead time[5].To validate the reliability of the linear model, the verification experiment of supply (500kPaG) and exhaust (0 kPaG) motion was performed on conditions of different volumes. Figure 6(a) shows the result (bode diagram) of system identification using the experimental device. And,Figure 6(b) shows the experimental result and output of the proposed model at the volume V=7 ml. From Figure6(b), it can be seen that the output results (rise action and fall action) using the proposed model agree well with the experimental result.The proposed model was a very simple model, but it could be confirmed that the actual valve system including volume can be represented by means of a second-order form with a dead time.

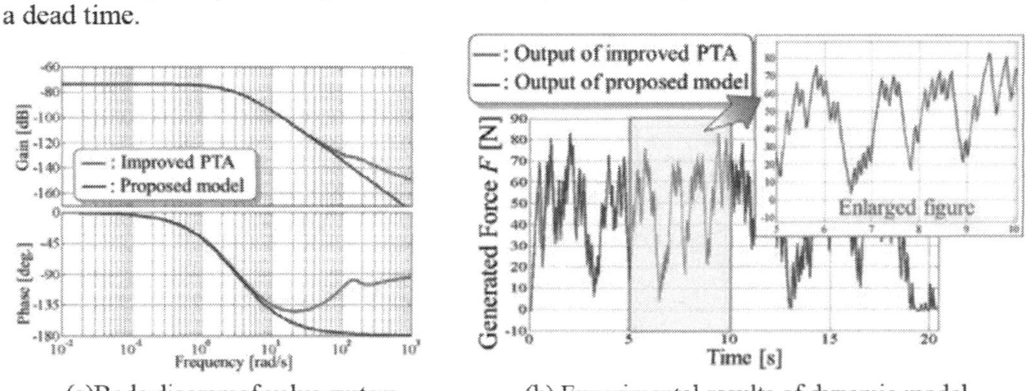

(a)Bode diagramof valve system. (b) Experimental results of dynamic model

Figure 6. Results of verification experiment.

4 Conclusions

This study was aimed to develop the PTA of the orthosis for lumbago relief and the resultingknowledges are summarized as follows:

1) The improved PTAwas modeled and the reliability of static and dynamic model was validated through experiment.
2) The improved actuator had the strain performance of 2 times and the generated force of 1.3 times, in comparison with the previous actuator.
3) The dynamic model including volume of PTA could be represented by means of a second-order form with a dead time.

Acknowledgments

The authors would like to acknowledge financial support for this workprovided by QOL (Quality of Life) Innovative Research (2012–2016)from the Ministry of Education, Culture, Sports, Science and Technology of Japan.

References

1. S. Schulte, H. Rossler, P. Ziehe and F. Higuchi:Bracing and supporting of the lumbar spine, Prostheses and Orthotics International, **6**, **3**,139-146(1982)
2. H. F. Schulte: The characteristics of the McKibben Artificial Muscle, The Application of External Power in Prosthetics and Orthotics, National Academy of Sciences–National Research Council,94-115(1961)
3. R. B. van Varseveld and G. M. Bone: Accurate Position Control of a Pneumatic Actuator Using On/Off Solenoid Valves, IEEE/ASME Transactions onmechatronics, **2**,**3**, 195-204(1997)
4. Y.Moriwake,T. Akagi, S.Dohta and F.Zhao: Development of Low-cost Pressure Control Type Quasi-servo Valve UsingEmbedded Controller,*International Symposium on Robotics and Intelligent Sensors*,493-500(2012)
5. N.Abe: The Limiting Optimal Regulator for Systems with Time-delay Considering Disturbance Decoupling, Mem. Inst. Sci., Tech., Meiji Univ., **39**, **1**,1-6 (2000) (in Japanese)

MATEC Web of Conferences **51**, 02011 (2016)
DOI: 10.1051/matecconf/20165102011

EEG Controlled Wheelchair

Sim Kok Swee[1,a], Kho Desmond Teck Kiang[1] and Lim Zheng You[1]

[1] *Faculty of Engineering and Technology, Multimedia University, Malaysia*

Abstract. This paper describes the development of a brainwave controlled wheelchair. The main objective of this project is to construct a wheelchair which can be directly controlled by the brain without requires any physical feedback as controlling input from the user. The method employed in this project is the Brain-computer Interface (BCI), which enables direct communication between the brain and the electrical wheelchair. The best method for recording the brain's activity is electroencephalogram (EEG). EEG signal is also known as brainwaves signal. The device that used for capturing the EEG signal is the Emotiv EPOC headset. This headset is able to transmit the EEG signal wirelessly via Bluetooth to the PC (personal computer). By using the PC software, the EEG signals are processed and converted into mental command. According to the mental command (e.g. forward, left...) obtained, the output electrical signal is sent out to the electrical wheelchair to perform the desired movement. Thus, in this project, a computer software is developed for translating the EEG signal into mental commands and transmitting out the controlling signal wirelessly to the electrical wheelchair.

1 Introduction

At the present time, the amount of disabled person is increasing annually in Malaysia. By referring to the statistical data that obtained from the Department of Social Welfare Malaysia, The information of the statistical data shows that the amount of physical disabled person are rising from year 2008 to 2011[1]. According to the statistics data, there are total of 657,152 people with disabilities registered in Malaysia calculated from 2009 until 2013[2] [3] [4] [5] [6].

According to the statistical data in 2013, there are 162,215 persons suffered from physical disability among 494,074 number of disabled person [6]. The percentage of physically disabled persons among all the disabled person is 33%, which is the second majority of the disabled person [6]. Most of these physically disabled persons are facing the difficulty to move around freely.

Since the physically disabled person is one of the major contribution to total number of disabled persons, this indicates that the amount of disabled persons who lost their mobility are substantial. Wheelchair is the most common device that used to provide mobility for the physically disabled person. However, most of the wheelchair today especially the affordable manual wheelchair requires human power to maneuver. Even for the electrical wheelchair, it still requires user's finger to move the joystick or press the button in order to control the movement of the electrical wheelchair [7]. In this way, some users who completely lost their hands or those who having the difficulty to control their hands such as Poliomyelitis patients are not able to navigate the wheelchair movement. Hence, they are incapable to move around by themselves. With the intention to resolve this issue, another

[a] Corresponding author : kssim@mmu.edu.my

constructive way is by using the brain to directly control the movement of the wheelchair. This method will allow most of the people to navigate the wheelchair by themselves. Therefore, this will bring an extremely high impact, especially for the disabled individuals who are not able to communicate physically [8].

The activities of the brain are required to be analyzed for implementing the Brain-computer interface (BCI) for the wheelchair. According to the technology nowadays. There are several methods that can be utilized to obtain the signals of brain activities. These methods are known as Single Photon Emission Tomography, Positron Emission Tomography [9], Functional Magnetic Resonance Imaging and Electroencephalogram (EEG). However, EEG is the most suitable method to be utilized for retrieving the signal of brain activity in order to implement the Brain-computer interface system. This is because the EEG device is portable and non invasive [10].

EEG is a method that captures electrical waveform from the brain. This method involves the EEG device that equipped with electrodes. After appropriately placed these electrodes on the scalp, the voltage fluctuations resulting from ionic current within the neurons of the brain are measured [11]. The electrical motivations resulted from the nerves of the brain are recorded using this method [12]. According to the human thoughts, different electrical waves are produced. Even for a muscle contraction, it will also generate a unique electrical signal [13]. By using the EEG device, these signals are captured and will be transmitted to the computer for further processing.

The patterns of the electrical waves that produced by the brain is distinctive due to different kind of thoughts. Therefore, during the EEG signal processing, the raw EEG signals are processed in a manner to differentiate the mental commands that are thinking by the user. The examples of these commands are forward, backward, left, right and stop [14]. After obtaining the mental command, it is used to produce the electrical control signals for controlling the movement of electrical wheelchair.

The method of translating the brain signals into controlling signals extends to the study of BCI [15]. BCI is a direct communication between the brain and a device that enables signals from the brain to direct some external activity. In this project, the signal from the brain is captured and transmitted to control the external device, which in this case is the electrical wheelchair.

2 Design and Implementation

2.1 Methodology

An electrical wheelchair is required for this project. The electrical wheelchair structure is built from scratch and the electronic circuits are designed in order to control the electric motor on the wheelchair. The electronic circuits include the microcontroller and motor driver. The connections of the electronic boards are connected as shown in the block diagram in Figure 1.

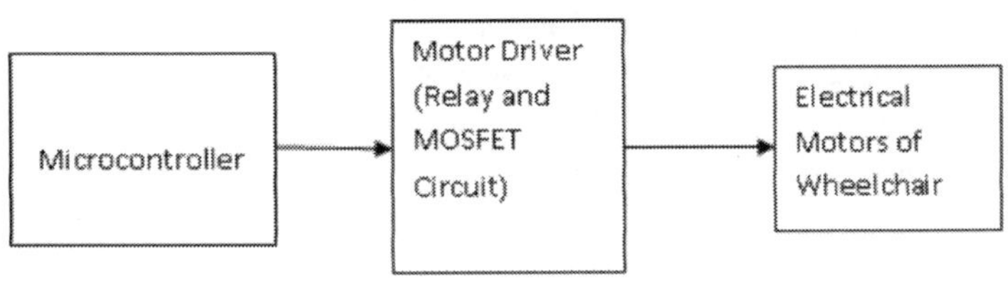

Figure 1. Electrical Board Connection in Electrical Wheelchair System

The whole BCI system consists of several devices. These devices are known as the Emotiv EPOC headset, laptop or PC with Windows operating system and electrical wheelchair with a microcontroller. The interconnection between the devices is illustrated in the Figure 2.

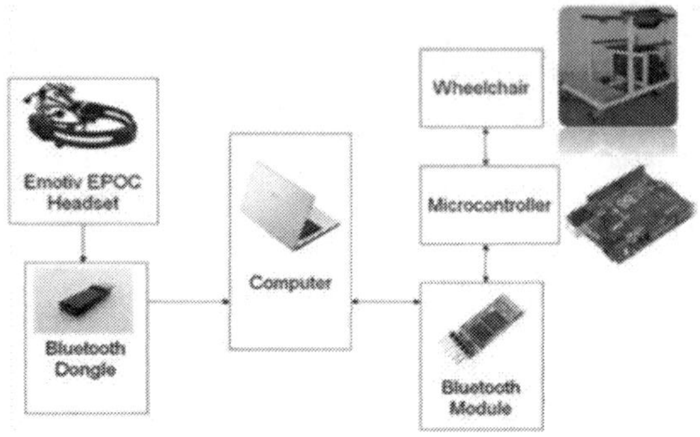

Figure 2. Block Diagram of Brainwave Controlled

2.2 Selection of EEG Device

With the intention to realize the Brain-computer interface (BCI), a device is required to capture the information of the brain activity. The method proposed for obtaining the brain activity signal is the Electroencephalogram (EEG). Therefore, it is necessary to choose the most suitable EEG device for this project. There are several EEG devices that can be found in the commercialized market. They are known as OpenEEG, Mattel Mindflex, Neurosky Mindset and Emotiv EPOC.

In order to implementing the BCI for the electrical wheelchair, there are few requirements that have to be fulfilled by the EEG device. First, the device must be portable and convenient. In this case, the OpenEEG is not suitable as the electronic boards involved are space consuming[16]. Beside OpenEEG device, the other devices are portable and wearable. Another requirement for this BCI is the software developer tool. As EEG is a complicated electrical waveform, it will be very time consuming for developing a software that is able to interpret and process these waveforms from fundamental. The provided developer tool is able to lighten the developer's burden. This is because the developer tool contains the basic algorithm to process the raw EEG waveform, such as Fast Fourier Transform. After undergo the fundamental signal processing, the developer is able to acquire the desired data based on the BCI purpose. In this case, there are only Neurosky Mindset and Emotiv EPOC provide developer tool. Therefore, Mattel Mindflex headset is not chosen for this project. In reviewing the Neurosky Mindset and Emotiv EPOC, the major difference between Emotiv EPOC and Mindset is the number of sensors and the location of the sensors. Emotiv EPOC has 14 sensors placed all over the head while the Mindset has only one sensor located on the forehead. With these sensors, Emotiv EPOC is able to detect the strength of signal in every different part of the brain. The Mindset only relies on one sensor to capture the EEG signal, which can only monitor the activity in one part of the brain. Therefore, Emotiv EPOC is able to provide more accurate data compare to the Neurosky Mindset. In addition, Emotiv EPOC is able to monitor the whole brain activity pattern. This feature is very useful for this project to differentiate the brain activity pattern that associates to the mental command which is thinking by the user. Therefore, Emotiv EPOC is the most ideal EEG device for this project.

2.3 Hardware

A wheelchair structure with size of 36 inches long and 24 inches wide is designed. The wheelchair is rear wheel differential drive. The structure of the wheelchair is constructed by using the material of steel.

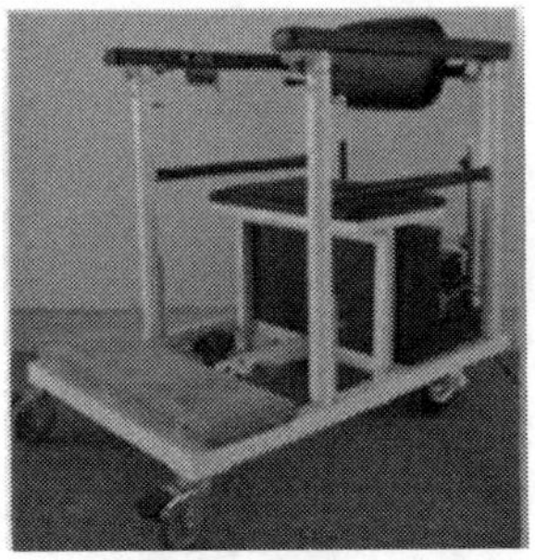

Figure 3. The Structure of Wheelchair.

In order to drive the wheelchair with a person on it, high torque motors are required. Therefore, scooter motors are chosen to drive the rear wheels of the wheelchair. The maximum voltage for this motor is DC 24 V. The power rating for the motor is 250 W. The speed of the motor is 3000 rpm and its rated torque is 0.80Nm. This motor draws maximum current up to 13.4 A.

Figure 4. Scooter Motor

As the scooter motor draws high current, a motor driver which is able to sustain high current is required for driving the motors. The scooter motor will draw high current up to 13.4 A, therefore the current rating for motor driver must exceed the maximum current of the DC motor. The chosen motor driver is high current motor driver. This motor driver can support DC 12 V to 30 V. It can support up to 160 A peak current and 60 A continuous current. Thus, this motor driver is very suitable for driving the scooter motor of the wheelchair. Figure 5 shows the high current motor driver.

Figure 5. High Current Motor Driver.

Arduino Uno microcontroller board is using ATmega328P microcontroller IC. This microcontroller has 14 digital I/O pins and 6 analog inputs. Within these 14 digital I/O pins, 6 of them can be directly configured as PWM output. It also has 16 MHz quartz crystal on this microcontroller board. This board also consists voltage regulator IC which can support input voltage from 6 to 20 V. The dimension of this microcontroller board is 68.6 mm x 53.4 mm. This microcontroller board is selected because it consists enough number of I/O pins to control the electrical wheelchair. The entire wheelchair has 4 switches as input. The motor driver as described earlier requires 4 output pins and 2

PWM pins as control signals. In this way, four analog pins are used for input switches, four digital output pins and two PWM pins are used for controlling the motor driver. Besides, this microcontroller is relatively small size, thus it is convenient to be mounted on the wheelchair. In addition, this microcontroller has the feature of plug and play; it only requires a USB type B cable for both programming and serial interface purpose. Figure 6 shows the Arduino Microcontroller Board that used as the main controller of the electrical wheelchair.

Figure 6. Arduino Uno

The wireless module used in this project is Bluetooth HC-06 module. HC-06 is a Bluetooth serial module that is used to convert the serial port into Bluetooth. This Bluetooth module is using Serial Port Protocol (SPP). It has 2.4 GHz radio transceiver and baseband. This module also works on low power operation, 1.8 to 3.6 V I/O. It also has a feature that it will auto-connect to the last device on power as default. HC-06 is used to connect with the Arduino Uno microcontroller. This allows the microcontroller board on the wheelchair to receive command signals wirelessly from the PC or smartphone. Figure 7 shows the HC-06 Bluetooth module.

Figure 7. HC-06 Bluetooth Module

Emotiv EPOC [17] headset is the EEG device manufactured by the company of Emotiv. This headset is a multi-channel with high resolution portable EEG device for the purpose of brain-computer interface (BCI). This headset contains 14 channels which means 14 electrodes to capture the EEG signal from the brain. It has the most electrodes among the commercialized EEG device in the market. This device uses a single ADC with sequential sampling method. The sampling rate is at 128 samples per second. Emotiv EPOC headset is built in with digital 5th order Sinc filter and digital notch filters at 50Hz and 60 Hz. The battery of the headset is rechargeable LiPoly battery and the battery life can lasts for 12 hours. The wireless connectivity of the headset is using Bluetooth. The feature of Emotiv EPOC headset is it is able to translate the brainwaves which in EEG form into the metrics that can be easily understood. These metrics are translated by using different detection suite and they are known as facial expression, performance metrics and mental commands. The Facial Expression detection suite can detect the real-time facial expression of the user. Performance metrics is used to analyze the user's real time emotional states and monitor the user's state of mind. The Mental Command detection suite is used to read in and interpret the user's thought and intent. The mental commands are push, pull, left, right, and etc. The Mental Command detection suite is very useful for interpreting the user's mental command in order to control the electrical wheelchair movement. Figure 8 shows the Emotiv EPOC headset.

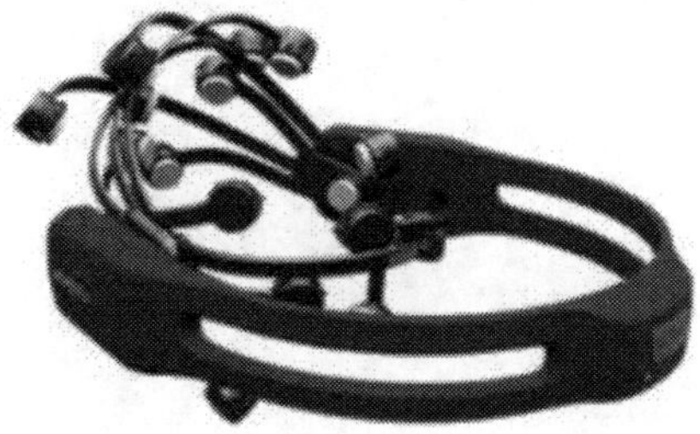

Figure 8. Emotiv EPOC Headset

2.4 Hardware

For the purpose of programming the Arduino Uno microcontroller, Arduino Integrated Development Environment (IDE) [18] is utilized to compile and upload the program code to the microcontroller. The advantages of this IDE are the provided software is open-source and only requires simple C language for the programming. Besides, it consists of various built-in functions which are very useful in most of the applications, for example analogWrite() which is the output PWM function.

The wheelchair is designed to be operated in two modes, which is the normal electrical mode that is triggered by switches and the brain controlled mode that is using the BCI method. The design of the program in flowchart is shown in Figure 9.

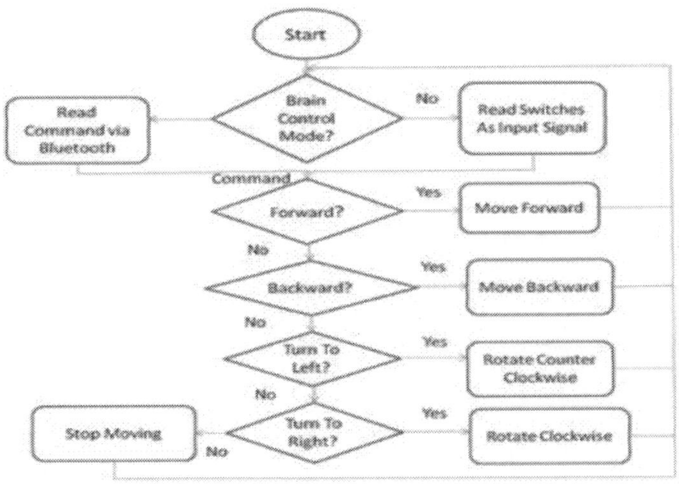

Figure 9. Flowchart of Microcontroller Program

Microsoft Visual C# [19] is the IDE (Integrated Development Environment) that supports Visual C# with a full-featured code editor and compiler. This IDE is used to write the Windows based application program for conducting several tasks. The first task is to receive the EEG data that transmitted out from Emotiv EPOC headset. Then, it processes the EEG signal. Lastly, it sends the electrical command signal out to the microcontroller by using Bluetooth wireless communication. The algorithm of the PC application in flowchart is shown in Figure 10.

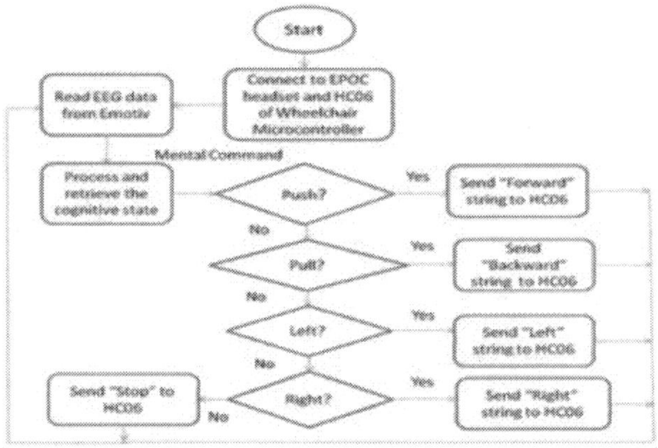

Figure 10. Flowchart of Microsoft Visual C# Application Program

The program uses Emotiv API (Application Program Interface) to interface with the Emotiv EPOC headset. The EmoEngine in this API is used to communicate with the Emotiv EPOC headset, receives pre-processed EEG data, execute post-processing on the data and convert the detection results into another structure which is known as Emostate. Emostate is a data structure that consists of the state of the Emotiv detections, which are the user's facial, emotional and cognitive state. By analyzing the cognitive state, it is able to determine the mental command which is thinking by the user. Output signal is generated as string according to these mental commands (e.g. left, right and etc.). Then, the output signal is sent out to the wheelchair's microcontroller via Bluetooth.

3 Results and Discussion

As result, an electrical wheelchair which is able to function as a common electrical wheelchair is developed.

Windows-based Application for this project is written as shown in figure below. This program is able to read and process the EEG data from the Emotiv EPOC headset into mental command. After processing, this program sends out the command signal to the connected HC-06 on the microcontroller board of the wheelchair. In the end, the electrical wheelchair is able to move according to the user's desire.

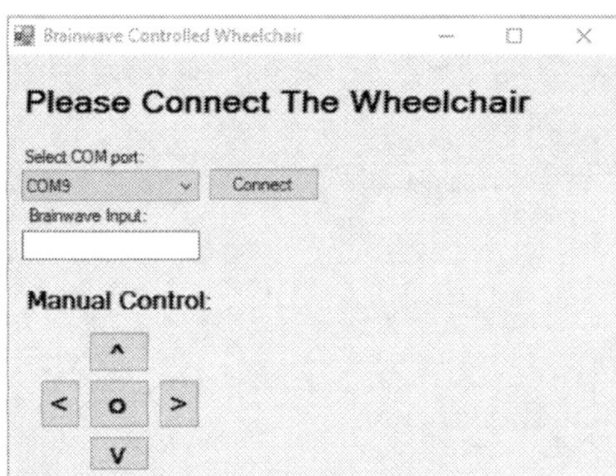

Figure 11. GUI for EEG Controlled Wheelchair

By using this application, the electrical wheelchair can be directly controlled by human brain. The results shows that the processed EEG data does not provide 100% accuracy according to the human mental command. However, it can achieve up to 90% of accuracy. The testing result for mental commands which conducted by 5 users are shown in table below:

Table 1. Test Result for Mental Command

Command/User	1	2	3	4	5	Total
Forward	5/10	6/10	8/10	8/10	4/10	31/50
Backward	9/10	8/10	9/10	9/10	7/10	42/50
Left	6/10	7/10	9/10	8/10	8/10	38/50
Right	8/10	9/10	7/10	9/10	8/10	41/50
Total	28/40	30/40	33/40	34/40	27/40	

Therefore, this shows that the BCI (brain-computer interface) for the electrical wheelchair is successfully established.

4 Conclusion

The main objective of this project is to construct an electrical wheelchair which can be directly controlled by the brain of the user. At the end of this project, an electrical wheelchair which can be operated in normal wheelchair mode and brainwave control mode is successfully developed. The brainwave controlled wheelchair can be controlled by most of the people. Besides, this project also contributes a new and effective solution for the physically disabled people or the patients who suffered from neuromuscular disorder to regain their mobility.

References

1. Public Service Department Malaysia, "What Is The Definition Of Senior Citizens In Malaysia?" [Online]. Available: http://www.jpapencen.gov.my/english/senior_citizen.html. [Accessed: 01-Dec-2013].
2. Department of Social Welfare Malaysia, "Bhg. B 10 - Pendaftaran Orang Kurang Upaya," 2009. [Online]. Available: http://www.jkm.gov.my/file/file/Statistic_2009/BHG B 10 - PENDAFTARAN ORANG KURANG UPAYA.pdf. [Accessed: 20-Jan-2016].
3. Department of Social Welfare Malaysia, "Bhg. B 10 - Pendaftaran Orang Kurang Upaya," 2010. [Online]. Available: http://www.jkm.gov.my/file/file/Statistic_2010/BHG B 10 - PENDAFTARAN ORANG KURANG UPAYA.pdf. [Accessed: 20-Jan-2016].
4. Department of Social Welfare Malaysia, "Bhg. B 10 - Pendaftaran Orang Kurang Upaya," 2011. [Online]. Available: http://www.jkm.gov.my/file/file/Statistic_2011/BHG B 10 - PENDAFTARAN ORANG KURANG UPAYA.pdf. [Accessed: 20-Jan-2016].
5. Department of Social Welfare Malaysia, "Bhg. B 10 - Pendaftaran Orang Kurang Upaya," 2012. [Online]. Available: http://www.jkm.gov.my/file/file/Statistic_2012/BHG B 10 - PENDAFTARAN ORANG KURANG UPAYA.pdf. [Accessed: 20-Jan-2016].
6. Department of Social Welfare Malaysia, "Bhg. B 10 - Pendaftaran Orang Kurang Upaya," 2013. [Online]. Available: http://www.jkm.gov.my/file/file/Statistic_2013/BHG B 10 - PENDAFTARAN ORANG KURANG UPAYA.pdf. [Accessed: 20-Jan-2016].
7. L. Bi, X. A. Fan, Y. Liu "EEG-Based Brain-controlled Mobile Robots: A survey", Human-Machine Systems, 43, 161-176, (2013).
8. A. Nijholt, D. Tan, G. Pfurtscheller, C. Brunner, J.del R. Mill´an, B. Allison, B. Graimann, F. Popescu, B. Blankertz, and K.-R. M¨uller, "Brain–computer interfacing for intelligent systems.", IEEE Intelligence System, 23, 72–79. (2008).
9. WebMD. "Positron Emission Tomography (PET)."[Online]. Available: http://www.webmd.com/cancer/lymphoma/positron-emission-tomography. [Accessed: 9-Sep-2014].

10. T. F. Collura. "EEG Frequency Bands." The Measurement, Interpretation, and Use of EEG Frequency Bands. (1997).
11. E. Niedermeyer & B. F L. Silva. "Electroencephalography: Basic Principles, Clinical Applications and Related Fields." (2005).
12. K. H. Solanki, H. Pujara. "Brainwave Controlled Robot.", International Research Journal of Engineering and Technology (IRJET), 2, 609 -612. (2015).
13. K. Yendrapalli , S. S. N. P. K. Tammana The "Brain Signal Detection for Controlling the Robot.", International Journal of Scientific Engineering and Technology, 3, 1280-1283. (2014).
14. B. Rebsamen, C. Guan, H. Zhang, C. Wang, C. Teo, M. H. Ang, Jr., and E. Burdet, "A brain controlled wheelchair to navigate in familiar environments.", IEEE Trans. Neural Syst. Rehabil, 18, 590–598. (2010).
15. K. R. M¨uller and B. Blankertz, "Toward noninvasive brain–computer interfaces," IEEE Signal Processing Magazine, 23, 125–128. (2006).
16. A. Griffith, "An Exploration of the OpenEEG Project", C.H.G.Wright's BioData Systems, 2006.
17. Emotiv, " Brain Computer Interface & Scientific Contextual EEG.", 2014. [Online]. Available: http://www.emotiv.com. [Accessed: 20-Jan-2016]
18. Arduino LLC, "Arduino Software IDE." [Online]. Available: http://www.arduino.cc. [Accessed: 27-Aug-2015].
19. "Microsoft Visual C#," Microsoft, 1989. [Online]. Available: http://msdn.microsoft.com [Accessed: 27-Aug-2015.

MATEC Web of Conferences **51**, 03001 (2016)
DOI: 10.1051/matecconf/20165103001

Experimental Investigation on the Performance of Grinding Assisted Electrochemical Discharge Drilling of Glass

Ladeesh V G[a], Manu R

National Institute of Technology Calicut, Mechanical Engineering Department, Kerala- 673601, India

Abstract. Grinding assisted electrochemical discharge drilling (G-ECDD) is a novel technique for producing micro and macro holes in brittle materials including advanced ceramics and glass, both efficiently and economically. G-ECDD involves the use of a rotating diamond core drill as the tool in a normal electrochemical discharge machine setup. The material removal happens by a combination of thermal melting due to electric discharges, followed by grinding action of diamond grits and chemical etching action. In this study, the effect of process parameters like voltage, duty cycle, cycle time and electrolyte concentration on material removed (MR) was investigated systematically using response surface methodology. Analysis of variance was performed to identify the significant factors and their percentage contribution. The most significant factor was found to be duty cycle followed by voltage, cycle time and concentration. A quadratic mathematical model was developed to predict MR. Tool wear was found for different frequencies and voltages. Higher tool wear was observed for high frequency above 5kHz pulsed DC supply at high voltage of 110V. Tool wear at the end face of the tool was found to be a significant problem affecting the tool life.

1 Introduction

Grinding assisted electrochemical discharge drilling (G-ECDD) is a hybrid machining technique which utilizes the combined effect of electrochemical discharges, grinding action of diamond coated/impregnated tool along with high temperature chemical etching action of the electrolyte for removing material from the workpiece. Here, the rotating tool will be polarized as the cathode and a steel plate will be maintained as the anode. Both the electrodes are dipped in an electrolyte and a pulsed DC voltage will be applied between the electrodes. The electrochemical reactions at the cathode liberates hydrogen bubbles which will combine to form a gas film at the critical voltage and the film separates the tool from the electrolyte. This develops high current density at the tool tip and causes ionization of the hydrogen gas inside the film. An avalanche of electrons which appear as an electric arc strikes the workpiece surface placed at the vicinity of discharge. The kinetic energy of electrons will be dissipated in the form of heat causing the melting of the workpiece. The rotating tool with diamond grits at its tip remove the molten material from the machining zone. This reduces the heat affected zone and increase the MRR. Only few researchers have conducted a systematic study of ECDM process using abrasive tools and there is a great need to unveil the full potential of G-ECDD through proper research strategies.

[a] Corresponding author : ladeesh.vg@gmail.com

Jain et al. [1] performed electrochemical discharge machining of alumina and glass using conventional cutting tool (CCT) and abrasive cutting tool (ACT). The material removed for different voltages and temperatures was found to be high for ACT when compared with CCT. Sanjay and Venkateswara [2] attempted drilling of alumina with a rotary abrasive electrode by ECDM. They found that the use of pulsed DC can improve the quality of the hole with a reduction in cracks. Liu et al. [3] used grinding aided ECDM for the machining of composites (alumina particles reinforced aluminium alloy) for the first time. They confirmed the material removal mechanism as a combination of ECDM and grinding action of the diamond core drill from SEM images. Sanjay and Venkateswara [4] made an attempt to make larger diameter hole in alumina using trepanning operation. An increase in the volume of material removed was observed with an increase in voltage, duty factor and conductivity of electrolyte. Peng and Liao [5] used traveling wire- ECDM for slicing glass rod and quartz bar. They found that frequency and duty ratio interact mutually and decides the energy density of the spark. A detailed study of current wave forms during ECDM was conducted and they identified two modes of current pulse (higher and smaller current impulses).

This study aims to reveal the performance of G-ECDD for machining glass and to find a practical solution to reduce the tool wear rate. A central composite design was used to develop the experimental runs. The machining parameters selected are voltage, duty cycle, cycle time and electrolyte concentration. A quadratic mathematical model was developed to predict material removed (MR). Additional experiments are performed to study the effect of voltage and frequency on tool wear rate.

2 Experiment Details

2.1 Experimental setup

All the experiments are performed using a modified CNC Router (Make- Newclear technologies, India: Model- TR 203) which is shown in Fig. 1. The experimental setup consist of an electrolyte bath which serves as the machining chamber, attachment for supplying power to rotating tool, a spring fed tool arrangement and a DC power supply. The spring fed tool arrangement was designed and fabricated so as to enable smooth feed of the tool and avoids the mechanical grinding action of diamond grits to predominate. This spring fed tool system ensures a proper balance of different material removal mechanisms in producing holes with minimal defects. A customized power supply capable of delivering pulsed DC with provisions for varying duty cycle and frequency was used for the experiments.

Figure 1. Integrated experimental setup for G-ECDD

The workpiece used was sodalime glass (Make- Saint gobain, India) of dimension 40(L) x 30(B) x 5(H) mm. The electrolyte used was potassium hydroxide (KOH). A thermostat with heating coil was used to maintain the temperature of electrolyte at 40° Celsius. Material removed was determined by measuring the difference of the weight of workpiece before and after machining. Tool wear was also determined in a similar manner by taking the weight of the tool before and after machining.

2.2 Design of experiment

The central composite rotatable design (CCRD) has been adopted to study the four process parameters (n=4). The CCRD design consist of 2^n factorial points, $2n$ axial points and 7 centre points which gives a total of 31 runs. A design that produces contours of constant standard deviation of predicted response is called rotatable design. A central composite design can be made rotatable by choosing the α value of axial points equal to $n^{1/4}$. Here the number of factors being 4, the value of $\alpha = 2$. Hence the coded values for different levels in CCRD are -2, -1, 0, 1, 2. The selected factors and their levels are shown in Table 1.

Analysis of variance (ANOVA) was conducted to identify the significant factors and the main interactions. ANOVA is a method of partitioning total variation into accountable sources of variation in an experiment. The technique can be used to interpret experimental data and make decision about the parameters under study

Table 1. Factors and their levels for CCRD

Symbol	Parameters	Units	Coded levels				
			-2	-1	0	1	2
A	Voltage	V	90	95	100	105	110
B	Duty cycle		0.4	0.5	0.6	0.7	0.8
C	Cycle time	s	0.0002	0.00065	0.00110	0.00155	0.002
D	Electrolyte ncentration	M	2	2.5	3	3.5	4

3 Results and Discussion

3.1 Results of CCRD

Experiments are conducted with central composite rotatable design and the results obtained are shown in Table 2. Randomization of the run order was performed for averaging out the effect of noise factors. A statistical software MINITAB-17 (trail version) was used for the analysis of the data. The main effects plot of the material removed is shown in Fig. 2. All the experiments were performed for a duration of two minutes and the weight of the workpiece before and after machining was taken to determine the material removed (MR).

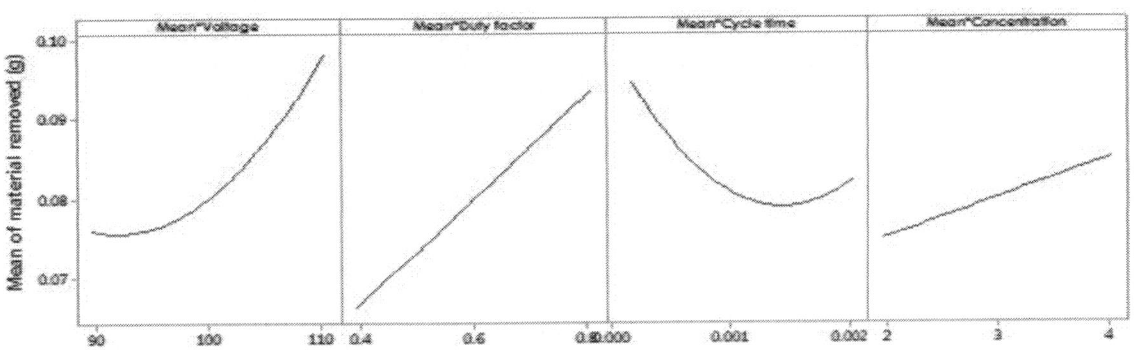

Figure 2. Main effects plot for material removed

As voltage increases, the material removal also increases. An increase in voltage will produce discharges of higher energy. This high energy discharges cause the melting of workpiece surface in the form of craters. The crater volume will be more for discharges at higher voltage and this will increase the overall material removed.

Duty cycle can be defined as the ratio of pulse-on time to the total cycle time (pulse-on + pulse-off time). Hence a higher value of duty cycle indicates a higher duration of pulse supply time in a cycle. An increase in duty cycle will deliver the discharge energy for more duration and hence cause more material to melt. This accounts for the rise in material removed with an increase in duty cycle.

Cycle time decides the frequency of sparking (frequency = 1/cycle time). At low cycle time, the frequency will be high and this produce large number of sparks at a particular location causing more material to be removed. When cycle time increases the frequency of sparking decreases causing less material removal.

Table 2. Results of CCRD

Standard order	Run order	A	B	C	D	Material removed (g)
1	1	95	0.5	0.00065	2.5	0.0747
2	6	105	0.5	0.00065	2.5	0.0857
3	11	95	0.7	0.00065	2.5	0.0777
4	18	105	0.7	0.00065	2.5	0.0887
5	25	95	0.5	0.00155	2.5	0.0577
6	30	105	0.5	0.00155	2.5	0.0788
7	7	95	0.7	0.00155	2.5	0.0891
8	19	105	0.7	0.00155	2.5	0.0918
9	2	95	0.5	0.00065	3.5	0.0792
10	12	105	0.5	0.00065	3.5	0.0902
11	8	95	0.7	0.00065	3.5	0.0912
12	26	105	0.7	0.00065	3.5	0.1032
13	13	95	0.5	0.00155	3.5	0.0692
14	20	105	0.5	0.00155	3.5	0.0812
15	3	95	0.7	0.00155	3.5	0.0852
16	27	105	0.7	0.00155	3.5	0.0962
17	21	90	0.6	0.0011	3.0	0.0769
18	14	110	0.6	0.0011	3.0	0.0948
19	31	100	0.4	0.0011	3.0	0.0654
20	9	100	0.8	0.0011	3.0	0.0908
21	22	100	0.6	0.0002	3.0	0.0941
22	28	100	0.6	0.002	3.0	0.0801
23	15	100	0.6	0.0011	2.0	0.0775
24	23	100	0.6	0.0011	4.0	0.0813
25	4	100	0.6	0.0011	3.0	0.0801
26	16	100	0.6	0.0011	3.0	0.0791
27	10	100	0.6	0.0011	3.0	0.0793
28	24	100	0.6	0.0011	3.0	0.0821
29	29	100	0.6	0.0011	3.0	0.0843
30	17	100	0.6	0.0011	3.0	0.0801
31	5	100	0.6	0.0011	3.0	0.0771

When concentration increases, the conductivity of the electrolyte increases and the electrochemical reaction is accelerated. This cause more number of hydrogen bubbles to be liberated at the tool surface and the gas film formation happens at a faster rate. This helps to utilize a greater portion of the energy supplied per pulse for arcing and hence the material removed will be more at higher electrolyte concentration. Thus the highest MR of 0.1032g was obtained for the combination of high voltage of 105V, high duty cycle of 0.7, low cycle time of 0.00065s and high concentration of 3.5M. The drilled holes using G-ECDD are shown in Fig. 3.

3.2 ANOVA and regression analysis

ANOVA results are given in Table 3. From the *F*- values, the most significant factor was found to be duty cycle followed by voltage, cycle time and concentration. The square effects of voltage and cycle time are found to be significant. The two- way interaction of duty cycle and cycle time was also found to be significant. A second order regression model was developed for material removed (MR) and is given by (1).

MR = 0.624 - 0.01185*Voltage - 0.0021*Duty factor - 64.4*Cycle time + 0.00493*Concentration + 0.000065 *Voltage*Voltage + 9560 *Cycle time*Cycle time + 61.5 *Duty factor*Cycle time (1)

Backward elimination regression was used to eliminate insignificant terms and make the model more efficient. The standard deviation of material removed is 0.00341 which is very less. The co-efficient of multiple determination (R^2) and adjusted R^2 for MR are 89.78 and 86.67 respectively which are high and indicate that the model terms are significant and the model can be used for predicting the response. Lack-of-fit was found to be insignificant which means that the model can fit the data well.

Table 3. ANOVA table for material removed

Source	Sum of Squares	Degrees of freedom	Mean Square	F Value	p-value Prob > F	Percentage Contribution (%)
Model	0.0023508	7	0.0003358	28.868271	< 0.0001	
A	0.0006813	1	0.0006813	58.567876	< 0.0001	26
B	0.00103	1	0.00103	88.540853	< 0.0001	39
C	0.0002003	1	0.0002003	17.219183	0.0004	8
D	0.0001458	1	0.0001458	12.534083	0.0017	6
BC	0.0001226	1	0.0001226	10.536101	0.0036	4.6
A*A	7.605E-05	1	7.605E-05	6.5375305	0.0176	2.9
C*C	0.0001094	1	0.0001094	9.4049748	0.0055	4.1
Residual	0.0002676	23	1.163E-05			
Lack of Fit	0.0002355	17	1.386E-05	2.596983	0.1215	
Pure Error	3.201E-05	6	5.335E-06			
Cor Total	0.0026183	30				

3.3 Effect of voltage and frequency on tool wear

The effect of voltage and pulse frequency on tool wear is shown in Fig. 4. An increase in voltage will deliver more current to the machining zone and causes ohmic heating of the tool and surrounding electrolyte. This local rise in temperature may deteriorate the bonding of diamond grits to the tool shank thus causing dislodging of grits and increase tool wear. The discharges occur randomly at the location of high electric field and the current flows to the discharge channel via the tool surface. At high frequency, the electric and magnetic field interacts. The sum of main current and eddy current rises on the electrode surface and the discharge probability from tool edge will be accelerated. This cause an increase in tool wear at the end face while using high frequency power supply when compared to low frequency.

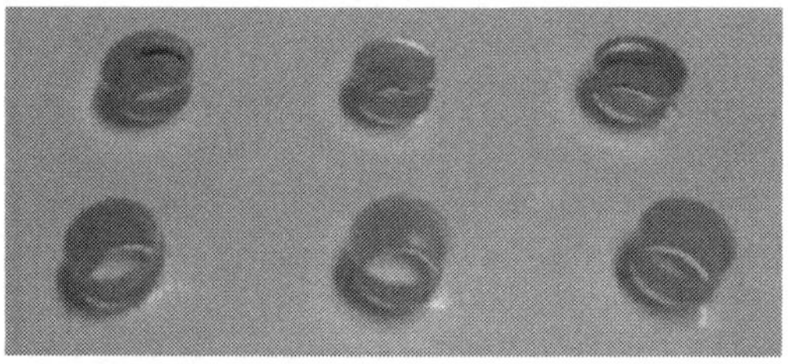

Figure 3. Machined holes of 6mm diameter

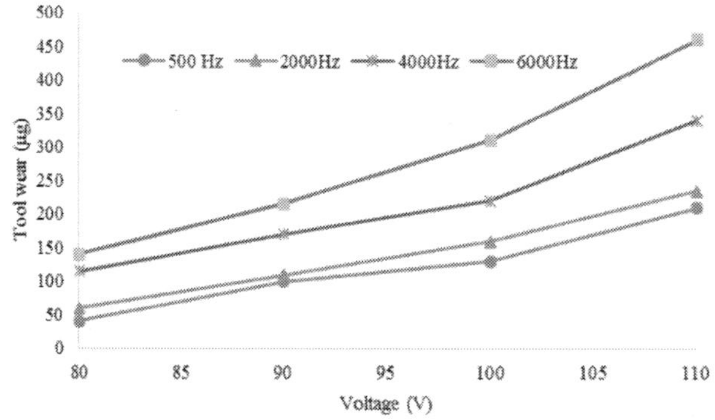

Figure 4. Effect of voltage and frequency on tool wear

4 Conclusions

The following conclusions are drawn from the above studies.

- Duty cycle and voltage are found to be the most significant factors that influence material removal. Cycle time and electrolyte concentration are the other main factors.
- The highest MR of 0.1032g was obtained for the combination of high voltage (105V), high duty cycle (0.7), low cycle time (0.00065s) and high concentration of 3.5M.
- The use of a low frequency (500Hz) and low voltage (80V) can significantly reduce the tool wear. The tool wear at the end face of the diamond core drill was found to be more predominant than the wear at the lateral face. Hence an increase in the thickness of diamond coating layer at the end face of the tool will increase the tool life.

References

1. V. K. Jain et al.," On the machining of alumina and glass," Int. J. of Mach. Tools & Manuf. 42, 1269-1276 (2002).
2. K. Sanjay et al., "The drilling of Al2O3 using a pulsed DC supply with a rotary abrasive electrode by the electrochemical discharge process," Int. J. Adv. Manuf. Technol. 39, 633–641 (2008).
3. J. W. Liu et al., "Grinding-aided electrochemical discharge machining of particulate reinforced metal matrix composites," Int. J. Adv. Manuf. Technol. 68, 2349–2357 (2013).
4. K. Sanjay et al., "Trepanning of Al2O3 by electro-chemical discharge machining (ECDM)," Int. J. of Mach. Tools & Manuf. 47, 2061–2070 (2007).
5. W. Y. Peng et al.,"Study of electrochemical discharge machining technology for slicing non-conductive brittle materials," J. of Mat. Pro. Tech. 149, 363-369 (2004).

MATEC Web of Conferences **51**, 03002 (2016)
DOI: 10.1051/matecconf/20165103002

PEMFC Performance with Metal Bipolar Plates Depending on the Channel Dimension

Kuikam Kwon[1], Doogon Kim[1], Yun Sik Kang[2], Sang Moon Kim[1], Sung Jong Yoo[2], and Chong Nam Chu[1,a]

[1]*Department of Mechanical and Aerospace engineering, Seoul National University, Seoul, Korea*
[2]*Fuel Cell Research Centre, Korea Institute of Science Technology (KIST), Seoul 136-791, Republic or Korea*

Abstract. Bipolar plates of a proton exchange membrane fuel cell (PEMFC) play an important role in removing liquid phase water as a by-product in order to facilitate the reaction between fuel and oxygen. A great amount of effort has been made to improve the performance of a fuel cell such as maximum current density or maximum power, by improving water removability of a bipolar plate. Most of the studies, however, are conducted numerically because of the complexity of analysing gas and liquid and the poor manufacturability of graphite bipolar plates. In this proceeding, we demonstrate that the performance of a PEMFC with metal bipolar plates can be enhanced by reducing the dimension of the channel. Bipolar plates were machined with stainless steel (type 316L) to have three different channel size (1000 μm, 500 μm and 300 μm) and the performance of each assembled cells were tested. As a result, the maximum power density and the maximum current density increased by 25%.

1 Introduction

A polymer exchange membrane fuel cell (PEMFC) is gaining significant attention in the various industry such as vehicle manufacturers, as eco-friendliness is becoming more and more important value nowadays. [1] A PEMFC emits pure water after converting electricity using hydrogen as a fuel, which is a representative merit to be nominated as a "green" energy device in the future. Accordingly, a great effort has been putting to develop related technologies for advanced PEMFCs worldwide. While a huge advance has been made in the boundary of academia, it still remains in question whether PEMFCs can be successfully commercialized, due to the high barrier of overall cost arising from expensive materials, complex manufacturing process, *etc.*

In order to meet the criteria to commercialize PEMFC, the cost of each part composing a cell should be minimized as much as possible while maintaining the overall performance. One of the effective ways is to replace the material of bipolar plates with economical conductive materials, *i.e.*, commercial metals. [2] Graphite has been widely used as a common material for bipolar plates in the laboratory scale; however, graphite for manufacturing bipolar plates in the mass production basis is not affordable due to the high cost of the material itself, as well as the high cost of machining flow channels. Therefore, a great number of studies have been reported to increase performance of fuel cell

[a] Corresponding author : cnchu@snu.ac.kr; chu7147@gmail.com

with metal bipolar plates fabricated with commercial materials such as stainless steel, copper alloys, *etc*.

There are several theoretical reports with computational results, arguing that the performance of a PEMFC with metal bipolar plates can be maximized by diminishing the dimension of flow channel below sub-millimetre scale; yet few experiments has been conducted to demonstrate and verify such result. [3-6] In case of using commercial metals for fabricating bipolar plates, it is inevitable to accompany overall performance decrease compared to using advanced carbon-based materials. One of the methods to minimize the decrement in the performance is to optimize the design of flow channels on the bipolar plates. While a number of studies reports that reducing flow channels below sub-millimetre scale can be a useful way to maximize fuel cell performance in a given condition, most of the studies mainly rely on theoretical simulation results due to the difficulty in machining small flow channels. Therefore, it is necessary to test the fuel cell performance having sub-millimetre-sized flow channels on metal bipolar plates in various conditions to support the preceding results arguing the effectiveness of smaller flow channels.

In this proceeding, we demonstrate that the performance of a PEMFC with stainless steel bipolar plates increases as the dimension of flow channel decreases.

2 Experimental

2.1 Bipolar plate fabrication

To investigate the relationship between the channel size of a bipolar plate and the performance of a cell, metal bipolar plates with different channel sizes were fabricated. For easy machinability and further study, we used bare metal for the bipolar plate. It is widely known that bare metals are easily corroded under the typical working condition of a PEMFC. Graphite bipolar plates are generally used for academic purpose because of its high corrosion resistance. In this study, however, we used bare metal not only for the manufacturability of small channels but also for the high potential in various commercial applications. For example, in portable PEMFCs which are used in automobiles, metal bipolar plates are widely chosen for high strength, low volume, easy manufacturability and low cost. AISI stainless steel (type 316L) was used for fabricating bipolar plates; it is widely used in studies for metal bipolar plates because of its high corrosion resistivity. The type of flow channel was chosen to be single serpentine type, which is known for having excellent performance among several flow field designs. Channel width, depth and rib width were designed to be equal. Bipolar plates having the channel size of 1000 μm, 500 μm, 300 μm were fabricated and tested respectively. The activation area of the bipolar plate was 5 cm^2.

2.2 Performance test for a single cell

A single cell with fabricated bipolar plate at cathode was assembled and activated by the current sweep-hold method with a commercial fuel cell test system (WonATech Co. Ltd., WFCTS 1.11) and the detail procedure can be found elsewhere. [7] Polarized test was performed by the current sweep method at a 10 mA/cm^2s^2 rate using the system mentioned above. While fully-humidified H_2 (99%) was input to the anode at a flow rate of 150 ml/min at 70 °C, fully-humidified air (99.9%) as a fuel were input to the cathode at a rate of 800 ml/min. For the test using O_2 as a fuel, fully-humidified O_2 (99%) at a flow rate of 200 ml/min was input while supplying H_2 at the same condition above. The uncertainty of maximum power density in the measurement system is under 3%.

3 Results and discussion

3.1 Fuel cell performance

Figure 1 shows the polarization curves of the tested PEMFC. We observed that overall performance of the tested cells increases as the channel size decreases. As the channel size decreases, maximum current density increased from 1.052 A/cm^2 to 1.311 A/cm^2 and power density increased from 0.37 W/cm^2 to 0.463 W/cm^2. It is worth noting that the rise in the performance from the cell with the channel size of 300 μm to that of 500 μm is much larger than the rise from the cell with the channel size of 1000 μm to that of 500 μm. In the cell with the channel size of 300 μm, a rapid voltage drop near the limit current density was monitored, which is further discussed below.

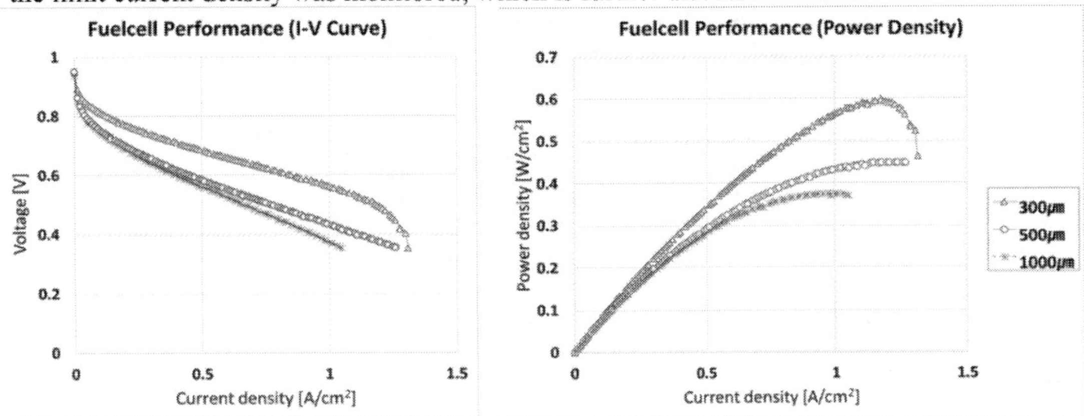

Figure 1. Polarization curves of tested PEMFC (a) I-V curve (b) Power density

3.2 Discussion

Some of the data does not match with the results in preceding numerical studies. [8] It was expected that overall performance tends to converge as the channel size decreases. In our results, however, the rise in the performance increases as the channel size becomes smaller. Further studies should be conducted to explain the difference between the result of preceding studies and our work.

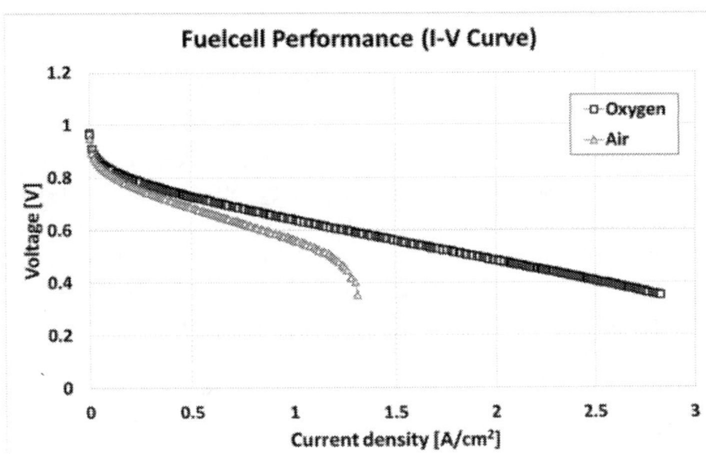

Figure 2. Performance comparison between different inlet gases

Also, In case of the channel size of 300 μm, performance abruptly drops near the limit current density, which was not observed in other cases. To check whether this unpredicted phenomenon occurs due to the water accumulation inside the channel, we changed the input gas from air to pure O_2.

To prevent the performance change due to the total amount of input O_2, pure O_2 was inserted with 200 ml/min which was calculated to contain the same amount of O_2 in case of 800 ml/min of air. As we can see in Figure 2, the overall performance improved without any sudden voltage drop. From this, we draw a conclusion that the rapid voltage drop is not due to the water accumulation because the total amount of produced water between the case of supplying air and O_2 would be the same.

We assumed that the sudden voltage drop are due to the air pressure growth inside the flow channel which may cause an increase in the total amount of supplied O_2 in the reaction site. We will further conduct the performance test with air at a lower flow rate so that we will be able to observed a stable performance of the cell with the channel size of 300 μm. Also, to generalize our results in other types of bipolar plates, additional test with different flow field designs will be conducted. Electrochemical impedance spectroscopy (EIS) analysis will be performed to investigate the major cause of overall performance improvement as the channel size decreases.

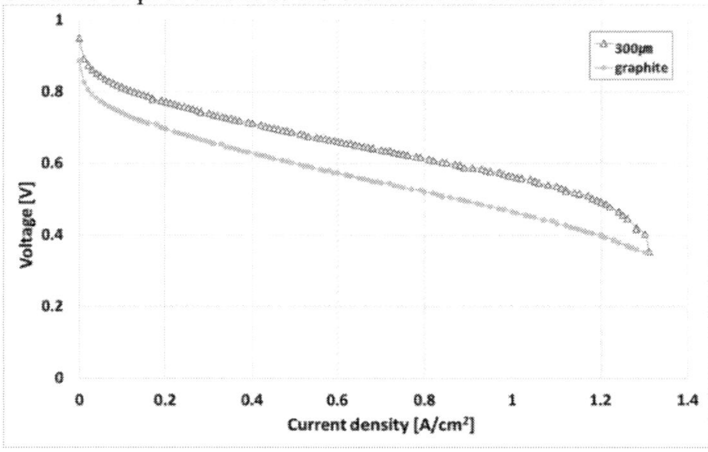

Figure 3. Performance comparison between original graphite bipolar plate and bipolar plate with 300 μm channel

To see whether the performance of PEMFC with metal bipolar plate is comparable to that of PEMFC with graphite bipolar plate, performance of PEMFC with reference graphite bipolar plates is observed under same experimental condition. Reference graphite bipolar plate has same feature with the bipolar plate which has 1000 μm channel. As you can see in Figure 3, overall performance of PEMFC with 300 μm channel bipolar plate is comparable to the PEMFC with graphite bipolar plate. Based on this result, we verified that PEMFC with metal bipolar plates can be compatible with PEMFC with graphite bipolar plates.

4 Conclusions

By demonstrating the performance test of a PEMFC with metal bipolar plates, we conclude that reducing the dimension of the flow channel can be an effective way to minimize the performance deterioration of PEMFC due to the change of material of bipolar plates. We observed continuous increases in the limit current density and the power density as the channel dimension decrease from 1000 μm to 500 and 300 μm, which supports preceding reports claiming the effectiveness of sub-millimetre-sized flow channels. Further study is to be performed in order to elucidate the sudden drop in the voltage plot by modulating the flow rate of supplied fuel.

Acknowledgement

This work was supported by the Brain Korea 21 Plus Project in 2015.

References

1. H. Tawfik, Y. Hung, D. Mahajan, J. Power Sources **163**, 755-767 (2007)
2. S. Karimi, N Fraser, B. Roberts, F.R. Foulkes, Adv. Mater. Sci. Eng. **2012**, 828070 (2012)
3. S. Shimpalee, J.W. Van Zee, Int. J. Hydrog. Energy **32**, 842-856 (2007)

4. E. Mancusi, E. Fontana, A. Augusto, U. Souza, S.M.A. Guelli, U. Souza, Int. J. Hydrog. Energy **39**, 2261-2273 (2014)
5. J. Cui, Y. Fu, J. Bionic. Eng. **9**, 99-109 (2012)
6. A. Aiyejina, M.K.S. Sastry, J. Fuel Cell Sci. Technol. **9** (2012)
7. J.W. Lim, Y.H. Cho, M.J. Ahn, D.Y.Chung, Y.H. Cho, H. Jung, Y.S. Kang, O. Kim, M.J. Lee, M. Kim, Y.E. Sung, J. Electrochem, Soc, **159(4),** B378-B384 (2012)
8. X.D. Wang, W.M. Yan, Y.Y. Duan, F.B. Weng, G.B. Jung, C.Y. Lee, Energy Conv. Manag. **51**, 959-968 (2010)

MATEC Web of Conferences **51**, 03003 (2016)
DOI: 10.1051/matecconf/20165103003

Performance Evaluation of PCD Insert 1600 Grade on Turning of Al 6061 Reinforced with 7.5% ZrB_2 Metal Matrix Composite

Ramanathan. M[1,a], Thanish.P[2], R. Rohinth Raja[3]

1-3, under graduate student, Department of Mechanical Engineering, Sri Venkateswara College of Engineering, Irrungattu kottai post, Pennalur, Sriperumbudur Taluk– 602 117, Tamilnadu, Chennai, India.

Abstract. Aluminum matrix composite is the innovation of high performance material technology and it has superior interfacial integrity and thermodynamic stability between the matrix and reinforcement. Making the engineering components from this composite material require subsequent machining operations. This paper presents the detailed experimental investigation of the machining behaviour in turning of Al 6061-7.5% ZrB_2 Metal Matrix Composite (MMC) by using Poly Crystalline Diamond (PCD) insert of 1600 grade. The effect of ZrB_2 reinforcement particles on machinability behaviour need to be studied. It is concluded that the feed rate has great influence on surface roughness and depth of cut has great influence on cutting force. The confirmation experiment indicates that there is a good agreement between the estimated value and experimental Value. Tool wear study also carried out for time duration of 15 minutes.

1 Introduction

Metal matrix composites have attracted the attention of many researchers for its superior hardness, wear resistant, formability, corrosion resistance and elevated strength with minimal weight. Machinability issues of ex-situ metal matrix composites are highly focused in literatures [1-3]. Coarse, uninformed distribution and agglomeration of reinforcements present in the matrix are obstacles for machinability behavior of metal matrix composites[4]. In-situ composites have overcome these problems because of the formation of reinforcements by exothermal chemical reaction. The 6061 aluminum alloy has been used in automotive industry for the fabrication of parts which includes wheel, panels and vehicle structure [5] ZrB_2 is known to have many superior properties such as high melting temperature, hardness, corrosion resistance, oxidation resistance and electrical conductivity, being used for aircraft / rocket components working at special high temperature, cutting tools and cathode material for aluminum refinement furnace [6]. Production of in-situ metal matrix composite by flux assisted synthesis was widely reported in the literatures [7- 8,10].

Anandakrishnan and Mahamani [9, 15] investigated machinability of the in-situ Al6061-TiB2 composites. They reported the effect of speed, feed and depth of cut on flank wear, cutting force and surface roughness. It was observed that presence of small and fine TiB_2 particles have offered significant influence on machinability.

[a] Corresponding author : ramanathan.ramanathan.muthu@gmail.com

Table 1. Chemical composition of Al-6061

Component	Mg	Si	Fe	Cu	Zn	Ti	Mn	Cr	Al
Wt (%)	0.08	0.65	0.34	0.16	0.12	0.15	0.18	0.09	Balance

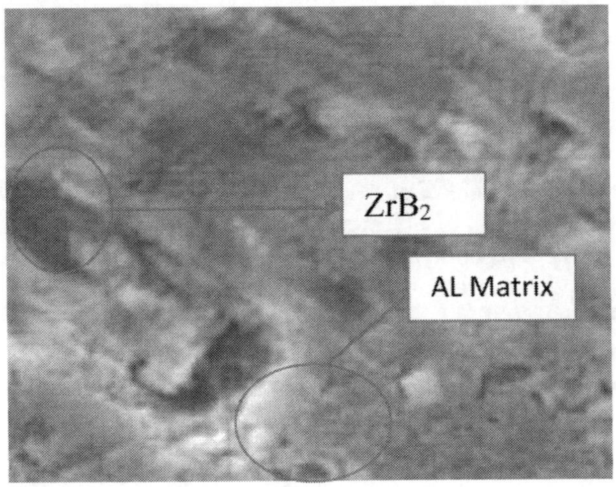

Figure 1. Microstructure of the workpiece (Al6061 – 7.5% ZrB_2)

In this direction an attempt has been made to investigate the turning parameters in machining of Al 6061 -7.5 % ZrB_2 metal matrix composite using PCD 1600 grade insert. Table -1 shows the Chemical composition of Al6061 .Figure- 1 shows the Microstructure of work piece

2 Experimental Procedure

Commercially fabricated cylindrical bars having 7.5 % of Z_rB_2 particles on matrix of Al 6061, using stir casting method of diameter 100 mm and 350 mm long are turned on self centered three jaw chuck, medium duty lathe of spindle power 2 kw. Fig -2 shows the experimental setup with tool dynamometer integral with it. Parameters such as surface roughness of machined component were measured by Mitutoyo surftest (Make-Japan –Model SJ-301) measuring instrument with the cutoff length 2.5 mm. Cutting force was measured by using Unitech lathe tool dynamometer with digital indicator. The cutting tool selected for machining Al-6061/ZrB_2 metal matrix composites was Poly Crystalline Diamond (PCD) insert of 1600 grade. The PCD inserts used were of ISO coding CNMA 120408 and tool holder of ISO coding PCLNR 2020M12.

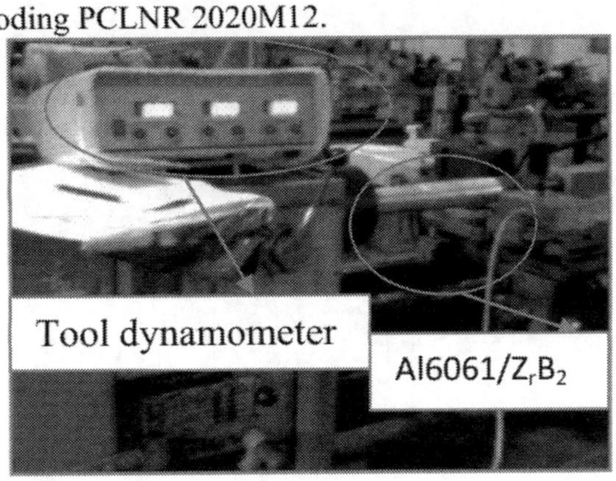

Figure 2. Experimental set up

Table 2. Machining parameter and their levels

Symbol	Machining parameter	Unit	Level 1	Level 2	Level 3
A	Cutting Speed	m/min	40	80	120
B	Feed	mm/rev	0.1	0.2	0.3
C	Depth of cut	mm	0.5	1.0	1.5

3 Results and Discussions

3.1 Surface Roughnss

Surface roughness is an important index of the machinability because the performance and service life of the machined component are often affected by the surface roughness. At high cutting speed, the machining condition follows the adiabatic system. With in the short interval of time the material passed away. As spindle speed increases, the strain rate will also increase, the matrix which difficult to deform, so that the particle will be crushed. Hence pits and cracks are reduced. As spindle speed decreases, the strain rate will also decrease, the matrix will be easy to deform so that particle will be pulled out [11, 12]. Hence pits and cracks formation on the machined surface is increased. Increase in feed rate also increased the chatter and produced incomplete machining at a faster transverse, which leads to higher surface roughness. Increasing in depth of cut, increased the deposition of build- up edge (BUE) over the machined surface. From fig 3and 4, it is understood that, cutting speed increases surface roughness decreases, at depth of cut 0.5 mm, and feed rate 0.3 mm/rev, surface roughness found to be 1.83 microns at cutting speed of 120 m/min. At the same feed and cutting speed at 1.5 mm depth of cut surface roughness found to be approximately two times greater than at depth of cut. It is clearly evident that depth of cut also having influence on the surface roughness [11] similar trend exists in the other depth of cut 1.5 mm.

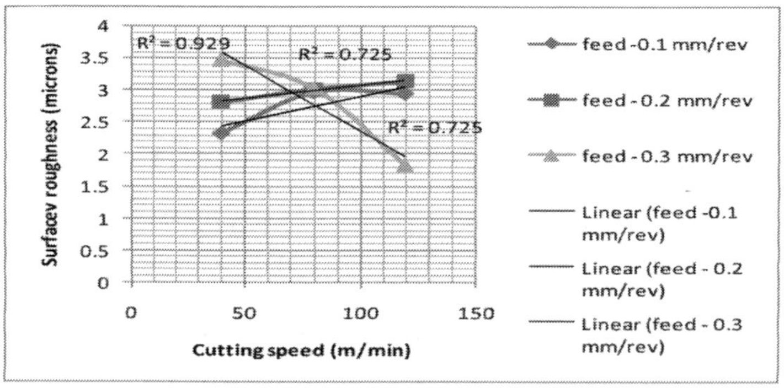

Figure 3. Cutting Speed versus Surface roughness Ra (Depth of cut 0.5 mm)

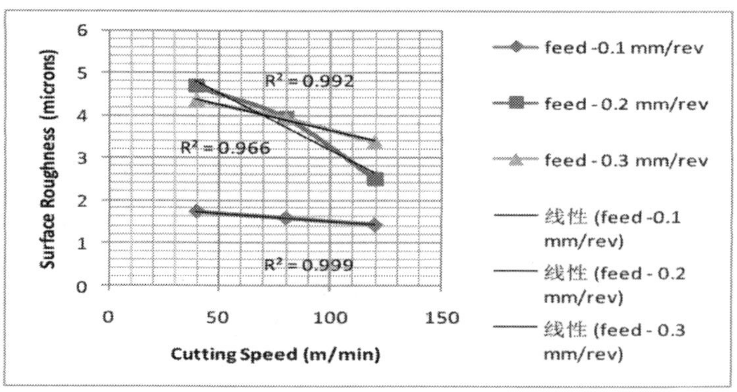

S

Figure 4. Cutting Speed versus Surface roughness Ra (Depth of cut 1.5 mm)

3.2 Cutting Force

Figure 5 and 6 show the trend line for cutting speed versus main cutting force in Newton for depth of cut 0.5 mm and 1.5 mm. Cutting forces exerted by the cutting tool on the work piece during machining action to be identified in order to control the tool wear and occurrence of vibration, thus to improve tool-life [11].The increase in cutting speed will minimize the tool –chip contact length, which reduces the cutting force. Increasing the feed rate increases the cutting force drastically. When machining the work piece with 0.5 mm depth of cut at 0.3 mm/rev feed rate shows the cutting force increases with the increase in cutting speed up to 80 m/min after that it decreases with the increase in cutting speed. This is the fact that at higher cutting speed with more feed rate which softens the matrix metal and pulling the reinforcement particle from the matrix very easy which results in decreasing the cutting force [12, 14]

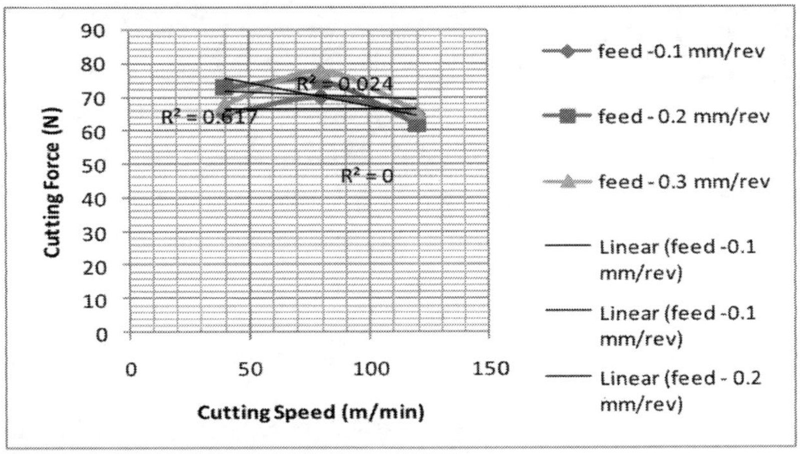

Figure 5. Cutting Speed versus Cutting force (Depth of cut 0.5 mm)

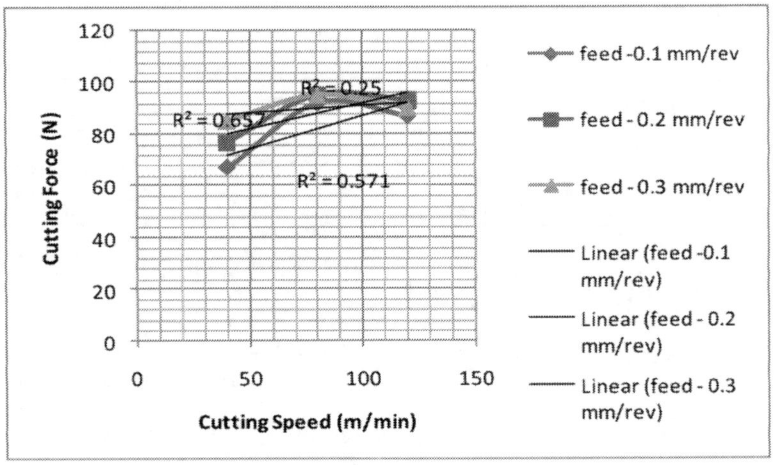

Figure 6. Cutting Speed versus Cutting force (Depth of cut 1.5 mm)

3.3 Analysis of Variance

Analysis of Variance (ANOVA) is a method of apportioning variability of an output to various inputs. Table 3 and 4 show the results of ANOVA analysis for surface roughness and cutting force respectively. This analysis was carried out at a 95% confidence level. The last column of the tables shows the percentage of contribution of each factor on the total variation of the output. This was carried out to identify the factors which most influence the machinability.

Table 3. Analysis of variance for Surface roughness

Source	DF	Seq SS	Adj SS	Adj MS	F_{test}	F_{table} at 95%	% cont.
A	2	1.1293	1.1293	0.5647	1.38	4.459	6.75
B	2	5.2221	5.2221	2.611	6.36	4.459	31.50
C	2	1.0456	1.0456	0.5228	1.27	4.459	6.56
A*B	4	2.0799	2.0799	0.52	1.27	3.8378	12.60
A*C	4	0.6793	0.6793	0.1698	0.41	3.8378	4.21
B*C	4	3.0653	3.0653	0.7663	1.87	3.8378	18.40
Error	8	3.2842	3.2842	0.4105			19.88
Total	26	16.5057					100.00

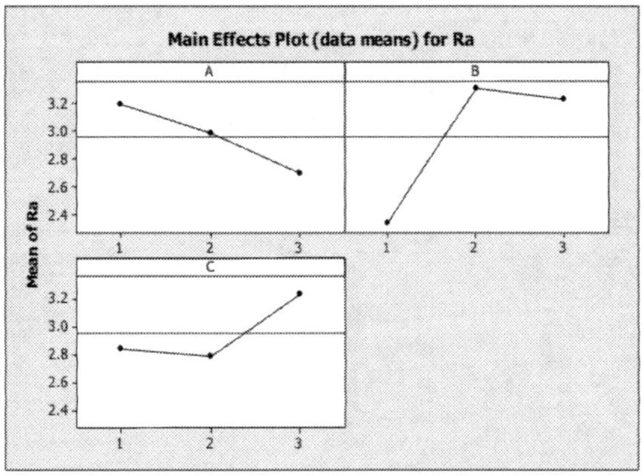

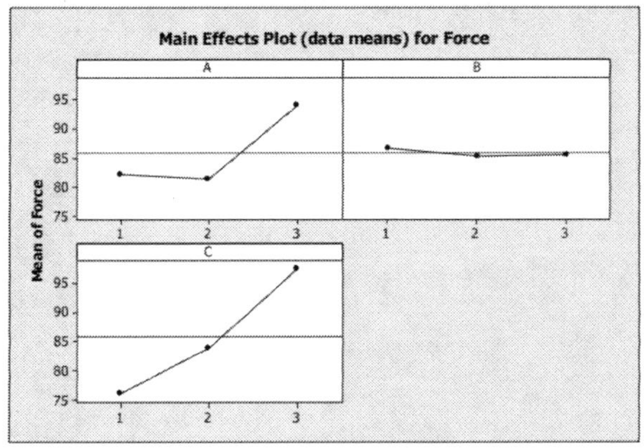

Figure 7. main effects plot for surface roughness **Figure 8.** main effects plot for Cutting force

From .Table 3, it is clear that the surface roughness is mainly influenced by feed rate 31.50% , cutting speed has 6.75%, Interaction effect of cutting speed and feed has also influencing the surface finish by 12.60%. Feed and depth of cut also having some effect on the surface finish (18.40%). From fig -7, it is clearly understood that, to get good surface finish, the optimum parameters are $A_3B_1C_1$

Table 4. Analysis of Variance for Cutting force

Source	DF	Seq SS	Adj SS	Adj MS	F_{test}	F_{table} at 95%	% cont.
A	2	907.19	907.19	453.59	10.22	4.459	16.35
B	2	9.85	9.85	4.93	0.11	4.459	0.18
C	2	2098.07	2098.07	1049.04	23.64	4.459	37.80
A*B	4	1479.7	1479.7	369.93	8.34	3.8378	26.66
A*C	4	62.81	62.81	15.7	0.35	3.8378	1.13
B*C	4	637.48	637.48	159.37	3.59	3.8378	11.49
Error	8	354.96	354.96	44.37			6.40
Total	26	5550.07					100.00

From .Table 4, it is clear that the cutting force is mainly influenced by depth of cut, its contribution is 37.80%, followed by interaction effect of cutting speed and feed which is having 26.66% and cutting speed having 16.35%, other factors also having less influence on cutting force. From fig – 8, it is understood that, to reduce the cutting force, best machining parameters are $A_2B_3C_1$.

3.4 Confirmation Experiment

Once the optimal level of machining parameters is selected the final step is to predict and verify the improvement of the performance characteristics using the optimal level of the machining parameters. Table 5 shows the results of the confirmation experiment

$$Percentage\,of\,error = \left(\frac{\exp erimental value - prediction value}{\exp erimental value}\right) X100 \tag{1}$$

Table 5. Results of confirmation experiment

	Optimal machining parameters		
	Prediction	Experiment	% of error
Setting level	$A_1B_1C_1$	$A_1B_1C_1$	
Suraface roughness (Ra) in µm	2.18	2.54	14.17
Cutting Force in (N)	59	72	18.05

Using the optimal machining parameters, Surface roughness Ra is found as 2.54 µm in experimentation and 2.18 µm in prediction, similarly the Cutting force is found as 72 N in experimentation and 59 N in prediction. It is clearly shown that machining performance of Al6061 /ZrB_2 are greatly improved through this study. From this analysis, it is found that the percentage of error for surface roughness is found (using the equation -1) to be 14.17 %, where as the percentage of error for cutting force is found to be 18.05 %.

3.5 Tool Wear

By setting the optimum parameters $A_1B_1C_1$ as a constant parameter and tool wear study was carried out for time duration of 15 minutes and the tool flank wear study was carried out. Tool flank wear was caused by abrasive nature of the hard particles present in the work piece. At low cutting speed worn flank encourages the adhesion of work piece material on the tool insert and formed Built-Up-Edge [1-2,9,12]. Fig-9 shows the Scanning Electron Microscope (SEM) image of fresh insert. Fig- 10 shows SEM image of PCD 1600 grade insert after machining the work piece for 30 minute duration. It is proved that hard ZrB2 particles which have higher hardness than diamond abrading the cutting tool [1, 9]. It is observed that the tool life of PCD 1600 grade is performing well in the chosen cutting condition

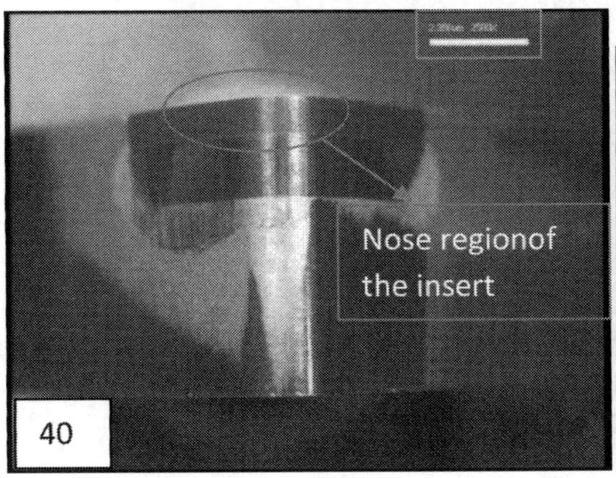

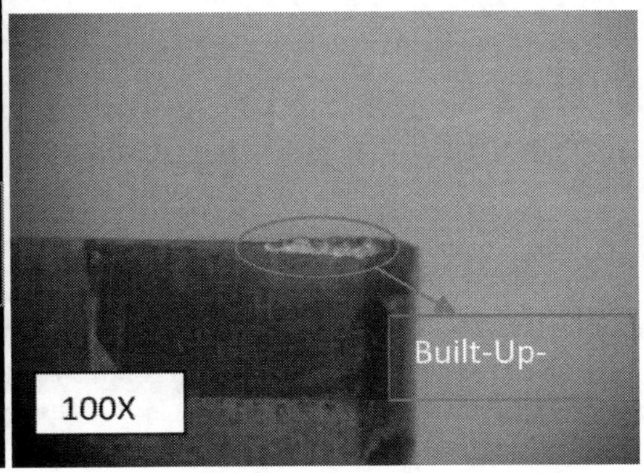

Figure 9. SEM image of fresh PCD 1600 grade Fig-10 SEM image of insert after 15 minute duration

4 Conclusion

a. The use of orthogonal array to optimize the Al-6061/7.5 % ZrB2 Metal Matrix Composites machining process has been reported in this paper.

b. Good surface finish is obtained at higher cutting speeds. Surface roughness obtained at feed rate 0.3 mm/rev at depth of cut 1.5 is nearly doubled the value of surface roughness obtained at depth of cut 0.5 mm.
c. Feed rate has strong influence on the surface finish. Depth of cut having less influence on surface roughness.
d. Percentage of error in predicted and experimental value was found to be lesser than 20 %.
e. Main cause of the tool wear is caused by abrasion of hard particles. At lower cutting speed formation of BUE is observed on the flank face of the insert.

References

1. Muthukrishnan, N., Murugan, M., & Prahladarao, K. (2008). Machinability issues in turning of Al–SiC (10p) metal matrix composites. International Journal of Advanced Manufacturing Technology, Vol. 39, 2008, pp 211-218.
2. Manna, A., & Bhattacharyya, B. (2003). A study on machinability of Al/SiC–MMC. Journal of Materials Processing Technology, 140, No.1, pp 711–716.
3. Ozben, T., Kilickap, E., & Çakır, O. (2008). Investigation of mechanical and machinability properties of SiC particle reinforced Al–MMC. Journal of Materials Processing Technology, 198, 220–225.
4. Daniel, B. S. S., Murthy, V. S. R., & Murty, G. S. (1997). Metal-ceramic composites via in situ methods. Journal of Materials Processing Technology, 68, 132–155.
5. Demir, H., & Gunduz, S. (2009). The effects of aging on machinability of 6061 aluminum alloy. Materials & Design, 30, 1480–1483.
6. Guo, D., Zhang, M., Jin, Z., & Kang, R. (2006). Pulse plating of copper-ZrB2 composite coatings. Journal of Materials Science and Technology, 22, 514–518
7. Naveenkumar, G., Narayanasamy, R., Natarajan, S., Kumareshbabu, S. P., Sivaprasad, K., & Sivasankaran, K. (2010). Dry sliding wear behavior of AA 6351-ZrB2 in situ composite at room temperature. Materials & Design, 31, 1526–1532.
8. Ozcatalbas, Y. (2003a). Chip and built-up edge formation in the machining of in situ Al4C3–Al composite. Materials & Design, 24, 215–221.
9. Anandakrishnan, A., & Mahamani, A. (2010). Investigations of flank wear, cutting force, and surface roughness in the machining of Al-6061–TiB2 in situ metal matrix composites produced by flux-assisted synthesis. International Journal of Advanced Manufacturing Technology, 55(1-4), 65–73.
10. Doniavi, A., Eskandarzade, M., Abdi, A., & Totonchi, A. (2008). Empirical modeling of EDM parameters using Grey relational analysis. Asian Journal of Scientific Research, 1, 502–509.
11. Ganeshbabu, B., Selladurai, V., & Shanmugam, R. (2008). Analytical modeling of cutting forces of end milling operation on aluminum silicon carbide particulate metal matrix composite material using response surface methodology. ARPN Journal of Engineering and Applied Sciences, 3, 5–18.
12. Haq, A. N., Marimuthu, P., & Jeyapaul, R. (2008). Multi response optimization of machining parameters of drilling Al/SiC metal matrix composite using grey relational analysis in the Taguchi method. International Journal of Advanced Manufacturing Technology, 37, 250–255.
13. Huang, J. T., & Liao, Y. S. (2003). Optimization of machining parameters of Wire-EDM based on Grey relational and statistical analyses. International Journal of Production Research, 41, 1707–1720.
14. Kumar, S., Subramaniyasarma, V., & Murty, B. S. (2007). Influence of in-situ formed TiB2 particles on the abrasive wear behavior of Al-4Cu alloy. Materials Science and Engineering A, 465, 160–164.
15. Lu, L., Lai, M. O., & Chen, F. L. (1997). Al-4% Cu composite reinforced with in-situ TiB2 particles. Acta Materialia, 45, 4297–4309. doi:10.1016/S1359-6454(97)00075-X

MATEC Web of Conferences **51**, 03004 (2016)
DOI: 10.1051/matecconf/20165103004

Investigation on the Temporal Surface Thermal Conditions for Thermal Comfort Researches Inside A Vehicle Cabin Under Summer Season Climate

Wencan Zhang[1,a], Jun Liu[1]

[1]*School of Mechanical & Electrical Engineering, Foshan University, Foshan 528000, China*

Abstract. With the proposes of improving occupant's thermal comfort and reducing the air conditioning power consumption, the present research carried out a comprehensive study on the surface thermal conductions and their influence parameters. A numerical model was built considering the transient conduction, convective and radiation heat transfer inside a vehicle cabin. For more accurate simulation of the radiation heat transfer behaviors, the radiation was considered into two spectral bands (short wave and long wave radiation), and the solar radiation was calculated by two solar fluxes (beam and diffuse solar radiation). An experiment was conducted to validate the numerical approach, showing a good agreement with the surface temperature. The surface thermal conditions were numerically simulated. The results show that the solar radiation is the most important factor in determining the internal surface thermal conditions. Effects of the window glass properties and the car body surface conditions were investigated. The numerical calculation results indicate that reducing the transitivity of window glass can effectively reduce the internal surface temperature. And the reflectivity of the vehicle cabin also has an important influence on the surface temperature, however, it's not so obvious as comparison to the window glass.

1 Introduction

Thermal comfort in vehicles cabins is one of the major concerns by car's owners, researchers and automotive industries as it is known to affect the occupant's travel experience as well as the driver's physiological arousal activity level and concentration ability. It has now become an important criterion for consumers to select a new car. Some researches indicate that the principal factors affecting the occupant's thermal comfort depend on the such thermal conditions as air temperature, relative humidity, mean radiant temperature, and air velocity; in addition to the clothing worn on the occupant and the physical activity of the occupant[1]. Most of investigations on occupant's thermal comfort focus on the thermal environment, especially of the air-flow field and the temperature distributions during cooling or heating period in vehicle cabin[2-6]. The surface temperature plays an important role. The thermal conditions of the surface which contact with occupant segment, for example the seat and the steering wheel, have a direct impact on the occupant's thermal comfort. Moreover, low or high surface temperature surrounding the occupant will lead to an unwanted mean radiant temperature. On a hot summer day, the internal surface temperature of a vehicle parked in the

[a] Corresponding author : zhangwencan@live.com

sun can exceed 90℃, it greatly affect the occupant's thermal comfort. Reducing the temperature of the vehicle interior surface can greatly improve passenger and driver's travel experience. However, few studies have been undertaken focusing on the surface conditions in a vehicle cabin.

Today, with the urgent need to reduce the economic and environmental cost of energy consumption, not only thermal comfort but also the energy efficiency has been recognized as essential parameters of vehicle cabin environmental design. Air conditioning is the major ancillary load for a compact vehicle. As a quantitative example, approximately 26 billion liters of fuel is consumed for the HVAC system in vehicle cabin annually in the United States[7]. The HVAC system in vehicle cabin wastes a large amount of energy, which is used to cool the entire vehicle cabin rather than just occupants. In a cooling period, a person can produce only about 100W of heat. However, a typical cooling system in a vehicle provides about 4000W of cooling[8]. The study of Levinson et al shows that most of the heat removed by the air conditioning comes from the cabin surface, rather than the air[9]. Many studies have considered the vehicle cabin as a potential approach to reduce ancillary load and improve fuel economy by decrease the its surface temperature. Turler al et[10] experimental studied the effect of using advanced thermal insulation material and window technologies on reducing vehicle auxiliary loads, the test results indicated that the retrofitted vehicle warm up under soak conditions was delayed compared to a factory vehicle. In Hoke and Greiner's study[11], experiments and numerical simulations were conducted to determine the effect of paint properties on the thermal load present and the resulting temperatures during vehicle soak. They found out that decreasing the solar absorptivity of the painted surface reduced the vehicle soak temperature and the transient air conditioning cool down time to comfort. Levinson al et[9] performed an experimental comparison of otherwise identical black and silver compact sedans, the results indicated that increasing the solar reflectance of the car's body by about 0.5 lowered the soak temperature of breath level air by about 5-6℃. And the air conditioning capacity required to cool the cabin air in the silver car to 25℃ within 30 min is 13% less than that required in the black car.

This current study focuses on the internal surface thermal conditions of a vehicle cabin under the sun. Firstly, a description of the heat transfer modes of a vehicle at daytime and the theoretical support of the simulation is presented. Secondly, an experiment is conducted under the typical climate conditions of South China summer season with the propose to validate the calculation model. Then, the numerical simulations are performed to investigate the effect of the vehicle window and body properties on the interior surface temperature. And this paper ends with a summary of major conclusions.

2 Numerical method

Fig. 1 shows the basic heat transfer modes of a vehicle at daytime. As shown, the heat transfer modes include conduction, convection and radiation. The short wave radiation can penetrate the glass window and goes into the cabin but the long wave radiation emitted by the cabin internal surface is prevented from escaping. Thus, in order to propose an accurate and feasible vehicle cabin heat transfer mode, the radiations considered here are divided into the long wave radiation and the short wave radiation. The short wave radiation includes the solar radiation; while the long wave radiation includes the sky radiation, the environment radiation and the surface radiation. The solar radiation can be calculated by two fluxes: the direct solar radiation and the diffuse solar radiation.

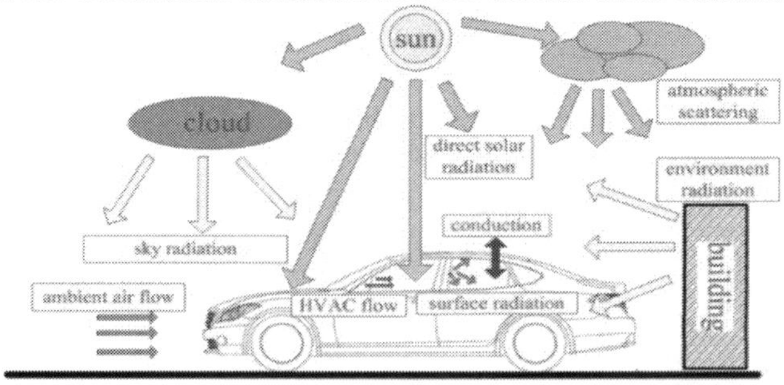

Figure 1. Heat transfer process of vehicle

The nodal network meth was used for numerical calculation, the procedure is shown as Fig. 2: (a) discretize the vehicle cabin into a series of small elements from an original CAD model, and set up the material properties for the vehicle cabin; (b) calculate the view factor; (c) obtain and set up the initial conditions; (d) calculate the heat transfer of the vehicle cabin, obtain the surface and air temperatures; (e) validate the numerical results with experiment data; (f) according to the validation result, repeat (a) to (e), or accept the numerical results.

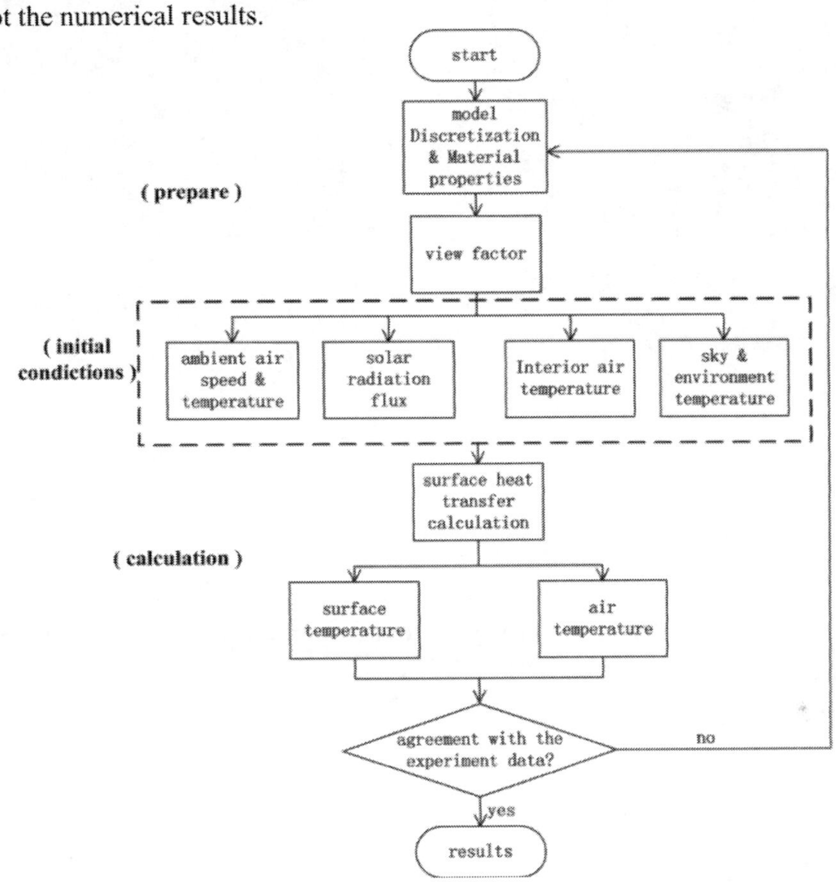

Figure 2. The numerical solution procedure

3 Experiment

For the purpose of validating the heat transfer calculation methods and the numerical model, an experiment was designed and conducted at the State Key Laboratory of Environment Adaptability for Industrial products in the summer season climate of south China. As shown in Fig. 3(a), a sedan parked in the outdoor turf with direct exposuring to sunlight was used in the test. The test vehicle was heated by different solar radiation intensity in the daytime. A small meteorological station shown in Fig. 3(b) was applied to obtain the meteorological conditions. According to the aforementioned heat transfer process of vehicle, the environment conditions including the ambient temperature, the ambient air speed, the direct solar radiation and the diffuse solar radiation flux were monitored during the experiment. The internal surface temperature and the interior air temperature were collected by using a discrete thermocouples network. The surface temperature at the dashboard, the seats and the doors were monitored, these places are expected to have a significant influence on the occupant's thermal comfort. Fig. 3(c) is an illustration of the surface temperature measurement method in the present study. The thermocouple was adhered to the surface by a type of light color material with low radiation absorptance to minimize the absorption of the solar radiation and surface radiation which would lead to an over measurement in surface temperature. Another thermocouple suspended approximately150mm from the ceiling in the middle of the cabin was used to measure the interior air temperature. In order to avoid exposing to direct sunlight, the thermocouple was wrapped in a small cylinder with low solar absorptance and low thermal emittance as shown in Fig. 3(d).

Figure 3. Test vehicle, Meteorological station, surface and air temperature measurement in the test

4 Results and discussion

4.1 Experiment results

The test was carried out from 6:00 to 18:00. The temperature measurement data were collected every 30 seconds by using a data acquisition system integrates with and a laptop computer, while the meteorological conditions were measured in 10 minutes intervals during the test. Fig. 4 gives the radiation flux and temperature measurement results in the experiment. As shown, the diurnal temperature variation is greatly depended on the solar radiation. The peak temperature at each surface always occurs between 12:00 to 14:00 when the solar has its maximum irradiation intensity. During the soaking period, the temperature distribution inside the vehicle cabin is highly non-uniform, the temperature differences among these surfaces can exceed 30℃. Because the sun rises from the east in the summer season, the east-facing surfaces in the test usually have higher temperature. The hottest surface in the present experiment is the dashboard due to it is completely exposed to the solar radiation transmitted through the windshield, its peak temperature is about 90℃, the next hottest surfaces are the seats, which also have large area exposed to the solar radiation. Another observation is that a greenhouse effect presents in the vehicle cabin. The temperature profiles of both interior and outside ambient air are similar. However, the interior air temperature is well above the outside ambient temperature. It is because the solar radiation can pass through the glass window of the vehicle and heats the internal surfaces, while the long wave radiation emitted by the internal surfaces is not able to pass through the window. Thus, all the internal surfaces will reach a considerable high temperature. Then, the interior air is conductively and convectively heated by the hot internal surface. Under the influence of this greenhouse effect, the interior air temperature is up to 60℃ and about 30℃ higher than that of the outside ambient air.

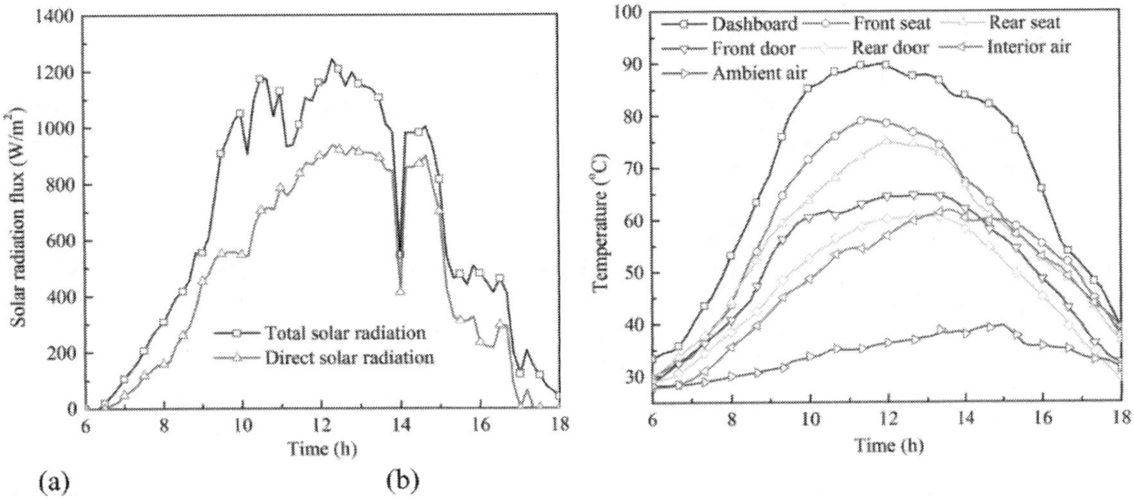

Figure 4. Experiment results: (a) the solar radiation flux; (b) temperature

4.2 Numerical model validation

The experiment results demonstrated above are use to validate the numerical model and methods. Table 1 lists the material properties of the vehicle cabin used in the simulation model.

Table 1. Material properties of the vehicle cabin used in the simulation model

Parts	Material	Density(kg/m^3)	Specific heat(J/kg·K)	Thermal conductivity (W/m·K)
Car body	Steel	7785	449	100
Door	ABS	996	1464	0.166
Dashboard	Lexan	1200	1502	0.3
Seat	Leather	860	1500	0.135
Window	Glass	2530	754	1.1717

Fig. 5 shows the comparisons between the simulation temperature profiles and the experiment data for both internal surfaces of the vehicle cabin. It can be observed that the simulation results are in good agreement with the values of experiment for all the internal surfaces temperature. The maximum surface temperature deviation between the predicted and the measured result is about 5℃ and within 10%. Therefore, this simulation method is able to satisfactorily predict the internal surface temperature variation of the vehicle cabin under the sun.

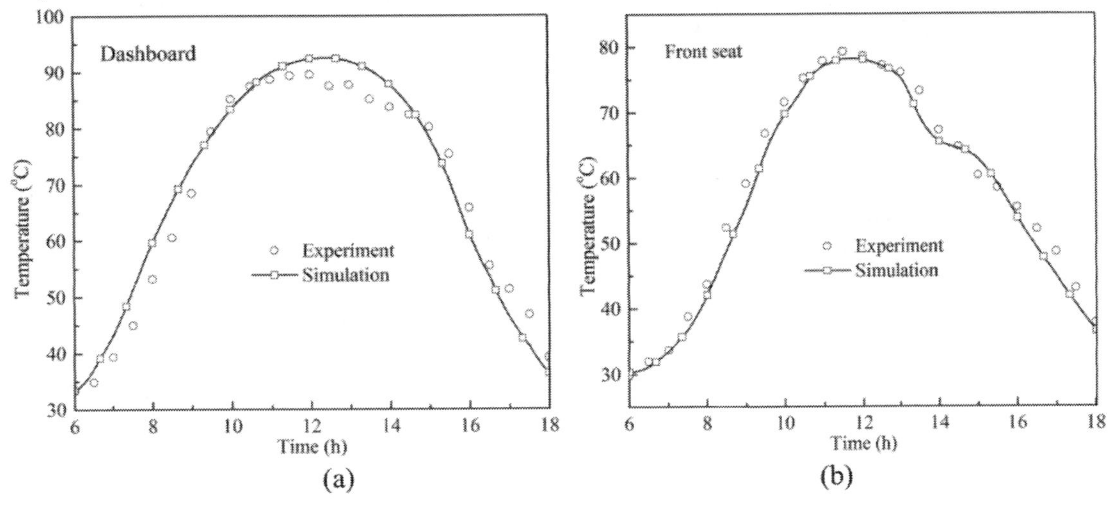

Figure 5. Comparisons between the simulation and experiment temporal temperature variation: (a) Dashboard; (b) Front seat

To investigate the contribution of each type heat transfer mode to the internal surface temperature, Fig. 6 gives the net heat flux of the conduction, convection, surface radiation and solar radiation on each surface. As shown in Fig. 6(a)-(c), the solar radiation received by the dashboard and the seats is very large. It is why the dashboard and the seats have high surface temperature when soaking in the sun as described above. And their heat dissipates conductively, radiatively and convectively to other surfaces and to the air. The heat flux evolution of the seats is not smooth as depicted in Fig. 6(b) and (c), a shake-up is found in the solar radiation heat flux profile. It is due to the shade effect of the car body and the internal surfaces, the solar radiation heat flux sharp increases when the solar irradiation reaches the surface and sharp decreases when the solar irradiation is shaded by the car body and other surfaces. Fig. 6(d) shows the net heat flux of the doors, unlike the dashboard and the seats, the solar radiation heat flux is very small. The influence of the solar radiation to their surface temperature is not the most important as compares to other types of heat transfer. Another important phenomenon revealed in Fig. 6(d) need to be pointed out is the net conduction heat flux. During the soaking period, there is dynamic temperature difference between the internal surface and the external surface of the door. At the first stage, the external surface temperature is higher than that of the internal surface due to it is exposed to the sun illumination and receives more solar radiation, the internal surface of the door is conductively heat by the external surface. At the second stage, the external door surface temperature is lower than that of the internal surface. Because the solar radiation received by the external surface decreases due to the position of the sun has changed and its radiation intensity weakens. Furthermore the external surface is also cooled convectively by the ambient air. Thus, the internal door surface is cooled conductively by the external surface, and its net conduction heat flux decreases and becomes negative.

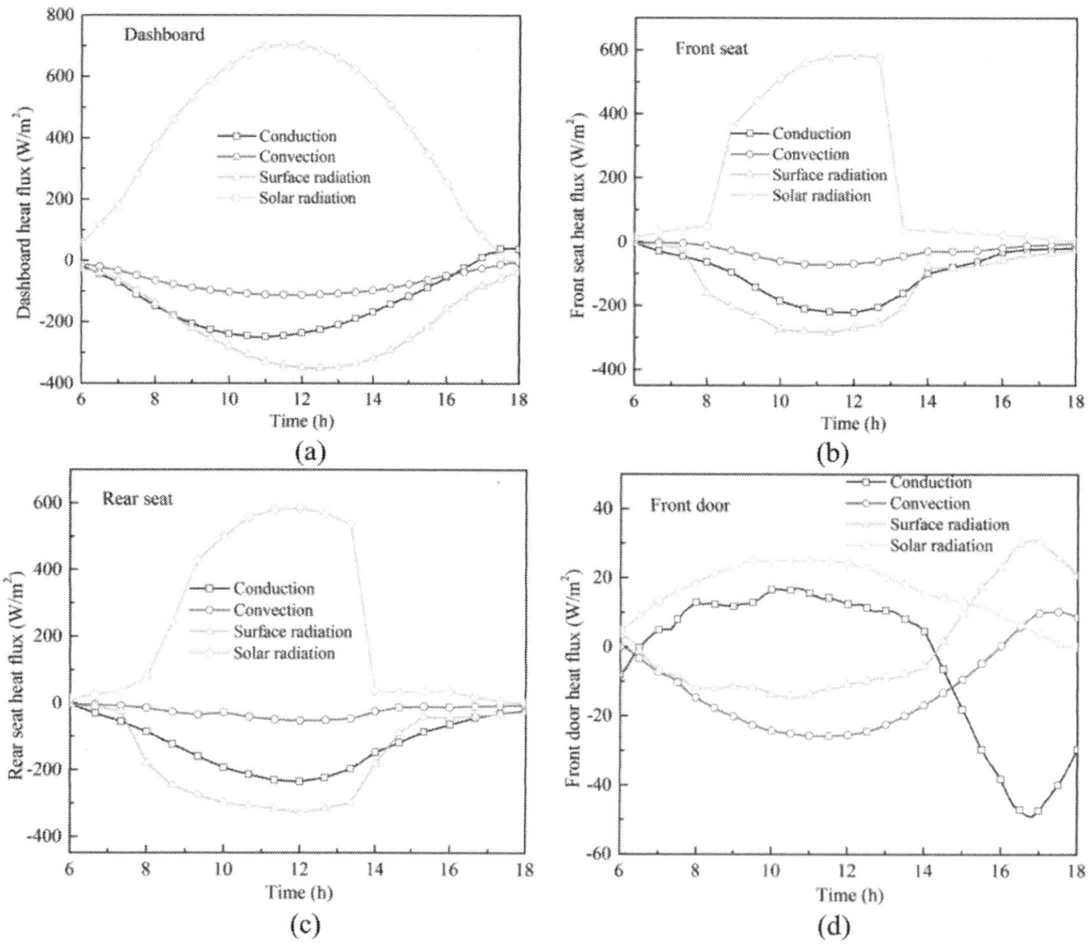

Figure 6. The heat flux of each internal surface

4.3 The effect of window glass properties

As aforementioned, the solar radiation has a decisive influence on the internal surface temperature of a vehicle cabin. Restricting the solar irradiance into the cabin is an effective method to reduce the surface temperature, then to improve the occupant's thermal comfort and reduce the energy consumption of the HVAC system in the cabin. 4 different types of glasses are selected for parametric analysis, in order to quantify the effect that the glass window on the internal surface temperature distribution. The glass optical properties, including the base line and other three types, are given as table 2.

Table 2. Optical properties of the four glass types

Glass type	Transmittivity	Absorptivity	Reflectivity
Base line	0.76	0.16	0.08
Glass type I	0.56	0.16	0.28
Glass type II	0.36	0.16	0.48
Glass type III	0.16	0.16	0.68

Fig. 7 depict how the window glass can affect the internal surface temperature distribution of a vehicle cabin. As can be seen, the glass optical properties have a pronounced influence on the temperature of all the surfaces. However, the sensitivity of different surfaces to the optical properties is quite different. As previously noted, the dashboard is just behind the front windshield and it is totally exposed to the solar illumination, its surface temperature is greatly depended on the solar intensity passed through the windshield. When decrease the transmittivity of the glass, the solar radiation energy obtained by the dashboard is significantly reduced. Therefore, the dashboard is the most sensitivity surface to the glass optical properties. From the geometric of the vehicle cabin and the location of the sun, it's obvious that the next sensitivity surfaces are the seats and then the doors. As revealed in Fig. 7(d), the maximum surface temperature at the dashboard and the seats greatly reduce with the transmittivity of window glass reduces (reflectivity of the glass increases accordingly). When the transmittivity of window glass reduces each 0.2, the maximum surface temperature reduces about 15℃ and 9℃ at the dashboard and the seats correspondingly. As above mentioned, the solar radiation received by the internal door surface is very small and its impact is not the most important as comparison to other factors. Therefore, reducing the transmittivity of window glass will not have significant influence on the internal door surface, it only reduces about 2℃ when the transmittivity of window glass reduces each 0.2.

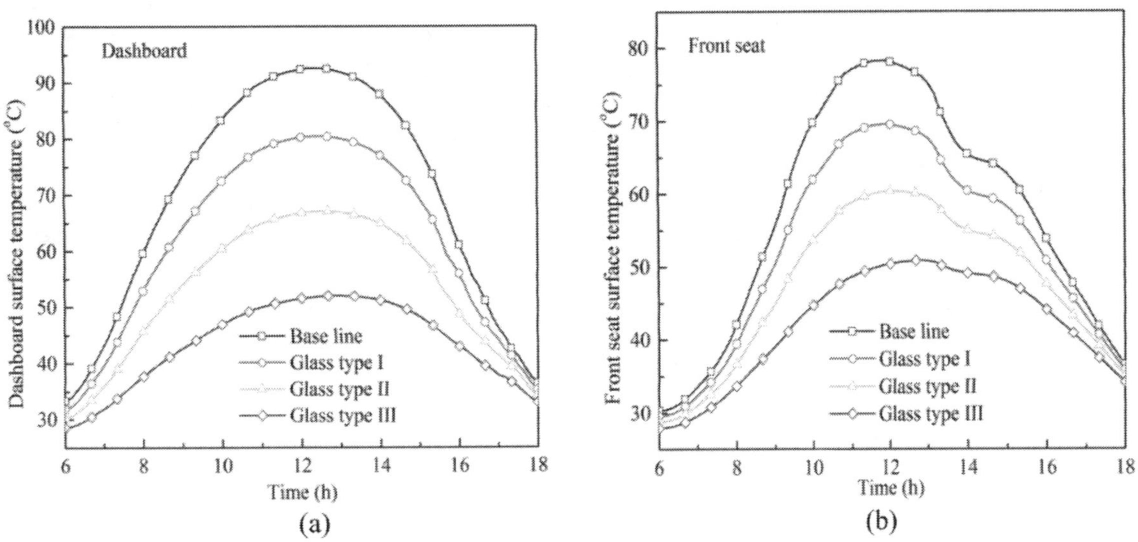

(a) (b)

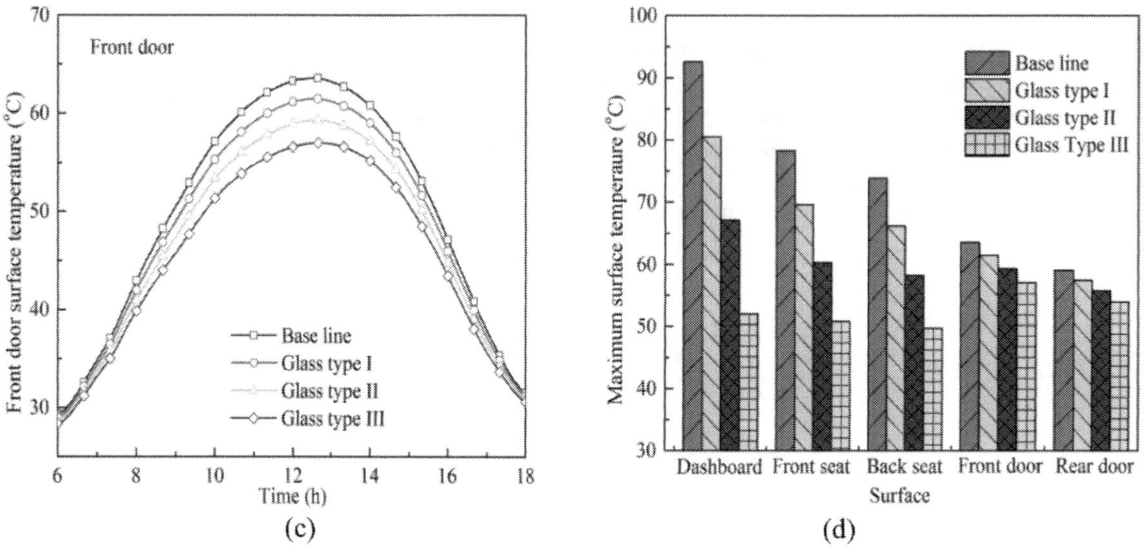

Figure 7. Effect of glass properties on the internal surface temperature

4.4 The effect of car body surface conditions

The car body surface condition is another major parameter that affects the vehicle cabin interior thermal environment, changing the car body surface condition is also an especially effective process to reduce the cabin surface temperature, and then the interior air temperature and the fuel consumption. To quantify the influence of the external surface absorptivity and reflectivity on the internal surface temperature, 4 types of surface conditions listed in table 4 are considered in this study.

Table 3. surface conditions of the four car body types

Surface condition	Absorptivity	Reflectivity
Base line	0.74	0.26
Surface condition I	0.54	0.46
Surface condition II	0.34	0.66
Surface condition III	0.14	0.86

As expected, decreasing the absorptivity or increasing the reflectivity of the external surface results in a decrease of internal surface temperature of the vehicle cabin as shown in Fig. 8. As depicted, the effect of car body surface condition is not so obvious as comparison to the properties of window glass. From Fig. 8(d), it is known that the most sensitive surface to the external surface condition is the door. Once decreases the absorptivity by each 0.2 (increase the reflectivity by 0.2 accordingly), the maximum temperature of the door would decrease about 4℃. Because the internal door surface is directly linked to the external surface, any change of the external surface temperature would increase or decrease the conductively heat transfer to the internal door. And the value for the dashboard and the seat is about 1.5℃ and 2.5℃ correspondingly. Indeed, the dashboard and the seats surface are mainly heated by the solar radiation. Only changing the external surface condition of the car body does not affect the solar illumination intensity on the internal surface. The effect of the external surface condition on the dashboard and the seats is mainly through the surface radiation heat transfer. The change of the surface temperature, for example the temperature at the door and the ceiling, will change the surface radiation heat transfer among internal surfaces.

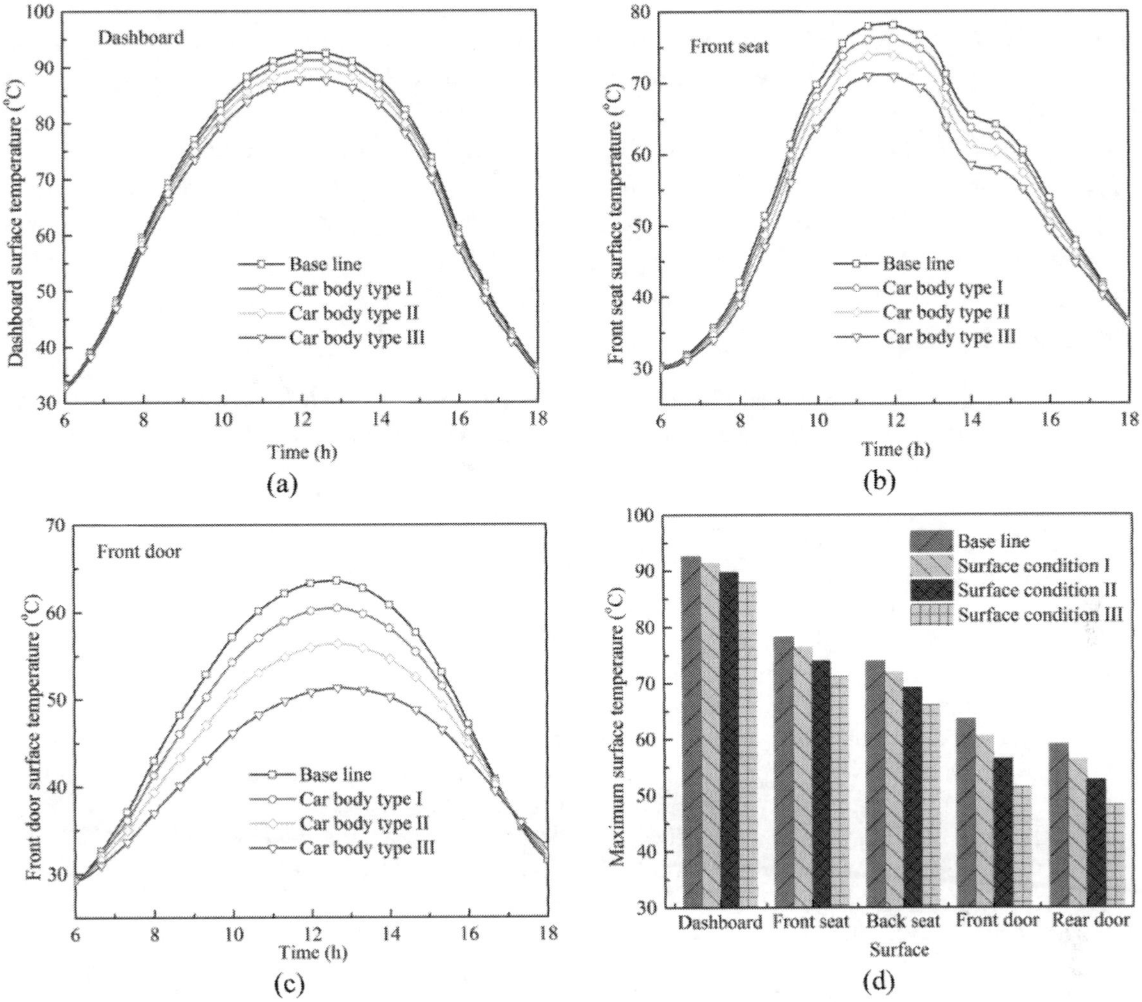

Figure 8. Effect of car body external surface conditions on the internal surface temperature

5 Conclusions

This study aims to find a way to obtain a more accurate surface thermal boundary conditions and decrease the internal surface temperature in a vehicle cabin in summer, then improve occupant's thermal comfort and reduce air conditioning power consumption. In this paper, we described the theoretical support of the numerical approach to study the thermal conditions in a vehicle cabin. Experimental and numerical investigations of temporal surface thermal conditions in summer conditions were carried out. The influence of the window glass and car body properties was quantificationally and systematically investigated. Based on the presented data and discussions, within the investigated parameter ranges, the following conclusions can be drawn:

(1) The surface temperature distribution inside a vehicle cabin is greatly depended on the solar radiation, and it is highly non-uniform through the cabin, the max temperature difference can exceed 30℃. Moreover, there is a green house effect in the cabin, the interior air temperature is well above the outside ambient temperature under the sun.

(2) The numerical method presented in this paper could be reliably applied to the problem of surface thermal conditions inside a vehicle cabin under sun.

(3) The surfaces exposed to the sun, such as dashboard and the seats, are mainly heated by the solar radiation and dissipates by conductively, radiatively and convectively heat transfer.

(4) The glass optical properties have a pronounced influence on the internal surface temperature, especially the surfaces exposed to the sun. With the transmittivity of window glass reducing every 0.2, the maximum surface temperature reduces about 15 ℃ and 9 ℃ at the dashboard and seats correspondingly, but only 2℃ on the door in this study.

(5) Increasing the reflectivity of the external surface can also results in a decrease of internal surface temperature. But, it's not so obvious as comparison to the window glass properties. With the absorptivity reducing by each 0.2, the maximum temperature decrease about1.5℃, 2.5℃ and 4℃on the dashboard, seat and door correspondingly.

Acknowledgements

This project is supported by National Natural Science Foundation of China (Grant No. 51375170) and Guangdong province public monweal research and capacity building project (2014A020216052)

References

1. Y.D. Cheng, J.L. Niu, N.P. Gao, Thermal comfort models: A review and numerical investigation. Build Environ, **47**:13-22(2012)
2. C. Walgama, S. Fackrell, M. Karimi, A. Fartaj, G.W Rankin, Passenger thermal comfort in vehicles–a review, P I Mecheng D-J Aut, **220**: 543-562(2006)
3. J. Currle, Numerical simulation of the flow in a passenger compartment and evaluation of the thermal comfort of the occupants, SAE Paper No. 970529
4. C.H,Chien J.Y Jang, Y.H Chen, S.C Wu, 3-d numerical and experimental analysis for airflow within a passenger compartment, Int J Auto Tech-Kor, **9**: 437-445 (2008)
5. H.J. Zhang, L. Dai, G.Q. Xu, al et, Studies of air-flow and temperature fields inside a passenger compartment for improving thermal comfort and saving energy, Part I: Test/numerical model and validation, Appl Thermeng, **29**: 2022–2027(2009)
6. H.J. Zhang, L. Dai, G.Q. Xu, al et, Studies of air-flow and temperature fields inside a passenger compartment for improving thermal comfort and saving energy, Part II: Simulation results and discussion, Appl Thermeng, **29**: 2028–2036(2009)
7. J. Rugh, V. Hovland, National and world fuel saving and CO2 emission reductions by increasing vehicle air conditioning COP, Proceedings from the 2003 Alternate Refrigerant Systems Symposium in Phoenix, AZ,(2003)
8. R. McGuffin, R. Burke, C. Huizenga, H. Zhang, A. Vlahinos and G. Fu, Human thermal comfort model and manikin, SAE Paper 2002-01-1955
9. R. Levinson, H. Pan, G. Ban-Weiss, P. Rosado, R. Paolini, H. Akbari, Potential benefits of solar reflective car shells: Cooler cabins, fuel savings and emission reductions, Appl Energ. **88**: 4343-4357(2011)
10. D. Turler, D. Hopkins, H. Goudey, Reducing vehicle ancillary loads using advanced thermal insulation and window technologies, SAE Paper No. 2003-01-1076.
11. P.B Hoke, C. Greiner, Vehicle paint radiation properties and affect on vehicle soak temperature, climate control system load, and fuel economy, SAE Paper No. 2005-01-1880 .

MATEC Web of Conferences **51**, 03005 (2016)
DOI: 10.1051/matecconf/20165103005

Analysis of the Effects of Back-stayed Arrangement on the Mechanical Properties of Corbel Shelf

PENG Dou[1], WANG Zhuan[1,a] and SHA Cong[1]

[1]*School of Mechanical Engineering, University of Science and Technology Beijing, 100083 Beijing, China*

Abstract. With AS/RS corbel shelf as research object, the orthogonal experiment was designed, which is composed of 21 groups of schemes and 3 kinds of load conditions. By using the Rack Computer-Aided FEA System based on SAP84, this paper presents the effects of grounding back-stayed and the ground floor height of back-stayed on the strength, rigidity and stability of corbel shelf under normal load, horizontal load in X direction and horizontal load in Y direction. The results show that in all conditions, grounding back-stayed improves the strength and stability of column, and the strength and stability of corbel shelf will be enhanced with the decrease of the ground floor height of back-stayed. This research provides a theoretical basis for the analysis and design of corbel shelf.

1 Introduction

With the rapid development of the logistics industry, AS/RS corbel shelf has become the main storage equipment of modern logistics center. AS/RS corbel shelf is a unit multi shelf which use the cantilever beam as the loading component [1-2]. The strength and stability of corbel shelf has become the core problem in shelf design, owing to its structure and bearing capacity. As an important component of the AS/RS corbel shelf, back-stayed directly affect shelf stiffness and stability. Therefore, how to arrange the back-stayed has become an urgent problem to be solved.

The current research on AS/RS shelf are mainly focus on three aspects. Firstly, the effects of beam-to-column and column-to-ground semi-rigid connection on the stability of the racks. Freitas et al.[3] studied the effects of connection of beam-to-column on the overall stability of rack. Aajori[4] analyzed the Modal analysis of cold-formed pallet rack structures with semi-rigid connections. Secondly, influence of some parameters on the stability of the shelf. Prabha[5] analyzed the effects of the thickness of the column on the behaviour of Cold-formed steel shelves by using ABAQUS software. Thirdly, evaluation the stiffness and strength of the specific shelf model through the finite element analysis. Affolter[6] studied on the reasons for collapse of a high storage rack. Liu and Zu[7] studied collapse of shelf when torsion and compression effect on the end of the column at the same time. Jia et al.[8] studied the strength calculation and design method of the AS/RS shelf under the earthquake generated little alternating load. In summary, finite element analysis of AS/RS corbel shelf are mainly focus on the influence of different components connection methods and some parameters on the overall stability of the shelf. But there are few research on the effects of grounding back-stayed

[a] Corresponding author : wzpaper@126.com

and the ground floor height of back-stayed on the mechanical properties of corbel shelf especially under different calculation conditions and loadings at present.

This work analyses the effects of back-stayed arrangement on the mechanical properties of corbel shelf under normal load, horizontal load in X direction and horizontal load in Y direction, provides a theoretical basis for the analysis and design of corbel shelf.

2 Calculation model

2.1 Structure

In this paper, take common beam pallet rack as the research object, the rack is mainly composed of column, corbel, corbel shelf, continuous beam, back-stayed, tie beam, diagonal bracing and row spacer. The shelf scale is 2 row *7 column *10 layer, the total height is 15.5m, length is 16.8m, width is 4m.

Two pallets are placed in each cargo box, the load of each pallet is 1000kg, and column is divided into two sections. Coordinate system is set as: X direction on behalf of the rack aisles direction, Y direction on behalf of the shelf into the deep direction, Z direction on behalf of shelf height direction. Grounding back-stayed grid structure are shown in Figure 1.

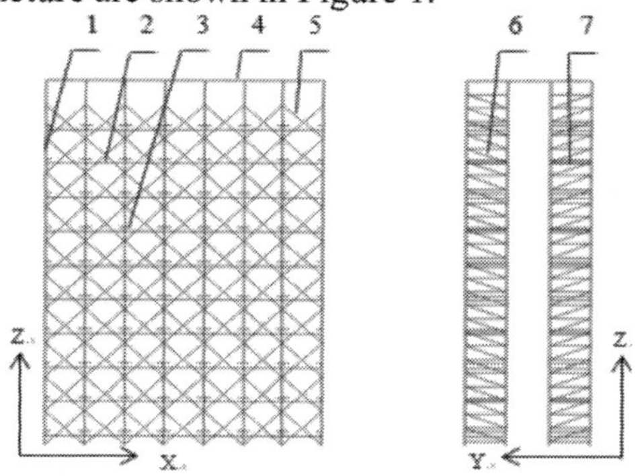

1-Column 2-Continuous beam 3-Corbel 4-Top beam 5-Back-stayed 6-Brace 7-Shelf

Figure 1. The grid structure of grounding back-stayed corbel shelf.

2.2 Finite element model

2.2.1 Element constraint

In finite element model, column, corbel, corbel shelf, continuous beam and diagonal bracing use frame element, back-stayed uses truss element. All elements are connected rigidly.

2.2.2 Node constraints and boundary conditions

All nodes of frame element have six degrees of freedom in SAP84, three translational degrees of freedom and three rotational degrees of freedom. The number 0 means the release of the degrees of freedom and the number 1 means constraints.

All node constraints are released, which can be expressed as R=0, 0, 0, 0, 0, 0.

The constraints of column-to-ground connected nodes can be expressed as R=1, 1, 1, 1, 1, 1.

All three-dimensional truss element nodes only has X, Y, Z direction of the translational degrees of freedom without rotational degrees of freedom, which can be expressed as R=0, 0, 0, 1, 1, 1.

Constrained Y direction of the translational degrees of freedom due to back-stayed is in the XOZ plane, and the constraints of its intersection node can be expressed as R=0, 1, 0, 1, 1, 1.

Constrained Z direction of the translational degrees of freedom due to tie beam is in the XOZ plane, and constraints of its intersection node can be expressed as R=0, 0, 1, 1, 1, 1.

2.3 Calculation conditions and loading

In order to analyze the effect of different load combinations on the mechanical properties of rack, the following three calculation conditions are studied.

Normal condition one: constant load + live load.

Normal condition two: constant load + live load + horizontal load in X direction.

Normal condition three: constant load + live load + horizontal load in Y direction.

In this AS/RS corbel shelf case, its load type, load size and loading position as shown in Table 1.

Table 1. The loads information of corbel shelf.

The number of loads types	Load type	Explanation	Load size(N)	Loading position
1	Constant load	Weight of the rack	1864776	
2	Vertical impact load	50% of the maximum static load of goods unit	4900	The first row of the first column top corbel shelf
3	Stacker force	goods reaction force	822	Rail crane beam
4	horizontal load in X direction	1.5% of all constant loads and live loads	147	All joints
5	horizontal load in Y direction	1.5% of all constant loads and live loads	147	All joints

3 Simulation experiment scheme

3.1 Experimental scheme

According to the calculation model above, the orthogonal experiment was designed, which is composed of 21 groups of connection schemes and 3 kinds of load conditions. Analysis the effects of back-stayed arrangement on the AS/RS corbel shelf stiffness, strength and stability under the different conditions, the simulation experiment schemes are shown in Table 2.

Comparison of scheme A11 and B16, A21 and B26, A31 and B36, we research the effects of grounding back-stayed on the mechanical properties of corbel shelf. In this experiment, the ground floor height of back-stayed is 300mm.

We study the effects of the ground floor height of back-stayed on the mechanical properties of corbel shelf by all B scheme.

Table 2. Simulation experiment scheme

Back-stayed \ calculation conditions		Conditions 1	Conditions 2	Conditions 3
Grounded	0mm	A11	A21	A31
Ungrounded	50mm	B11	B21	B31
	100mm	B12	B22	B32
	150mm	B13	B23	B33
	200mm	B14	B24	B34
	250mm	B15	B25	B35
	300mm	B16	B26	B36

3.2 Mechanical parameters

According to the experimental scheme, we use finite element analysis software to modelling, and extract the following mechanical parameters as evaluation indicators:

1) The maximum displacement of nodes (including X, Y and Z directions).
2) The maximum stress of column elements.
3) The stability reduced stress of column elements.

4 Analysis of experimental results

This paper selects the maximum displacement of nodes, the maximum stress of column elements and the stability reduced stress of column elements as the research objects, analyses the displacement results of shelves, stress results and the stability results statistically.

4.1 The effects of grounding back-stayed on the mechanical properties of corbel shelf

This work analyses the effects of grounding back-stayed on the stiffness, strength and stability of corbel shelf under normal load, horizontal load in X direction and horizontal load in Y direction. The grounding back-stayed experimental results are shown in Table 3.

Table 3. The grounding back-stayed experimental results.

Conditions indicator		The maximum displacement（mm）			The maximum stress（MPa）	The stability（MPa）
		X	Y	Z	column	reduced stress
Grounded	Condition 1	0.501	-0.695	-2.862	-91.839	101.369
	Condition 2	1.402	0.994	-2.936	-107.462	116.649
	Condition 3	0.63	12.46	-3.364	-119.005	133.204
Ungrounded	Condition 1	0.357	0.807	-2.855	-115.621	117.523
	Condition 2	1.438	1.084	-2.928	-151.866	154.424
	Condition 3	0.456	12.461	-3.356	-158.044	156.334
Fluctuation	Condition 1	40.34%	-13.88%	0.25%	-20.57%	-13.75%
	Condition 2	-2.50%	-8.30%	0.27%	-29.24%	-24.46%
	Condition 3	38.16%	-0.01%	0.24%	-24.70%	-14.80%

As can be seen from the data of Table 3:

1)In condition 1, the grounding back-stayed makes X direction displacement increase about 40%, Y direction displacement decreases about 13%, Z direction displacement has no obvious change, column stress decreases about 20%, stability enhances about 13%.

2) In condition 2, the grounding back-stayed makes Y direction displacement decreases about 8%, X and Y direction displacement has no obvious change, column stress decreases about 30%, stability enhances about 25%.

3) In condition 3, the grounding back-stayed makes X direction displacement increase about 40%, Y and Z direction displacement has no obvious change, column stress decreases about 23%, stability enhances about 15%.

4.2 The effects of the ground floor height of back-stayed on the mechanical properties of corbel shelf

In this work, we analyse the effects of the ground floor height of back-stayed on the stiffness, strength and stability of corbel shelf under the above three conditions. The ground floor height of back-stayed experimental results are shown in Table 4 to 6.

Table 4. The ground floor height of back-stayed experimental results of condition 1

Indicator ground floor height of back-stayed(mm)	The maximum displacement(mm)			The maximum stress (MPa)	The stability (MPa)
	X	Y	Z	column	reduced stress
50	0.478	0.713	2.869	96.403	100.236
100	0.45	0.732	2.859	106.009	106.402
150	0.42	0.752	2.857	112.776	113.559
200	0.393	0.771	2.856	116.11	117.299
250	0.378	0.789	2.855	116.792	118.366
300	0.357	0.807	2.855	115.621	117.523
The maximum value	0.478	0.807	2.869	116.792	118.366
The minimum value	0.357	0.713	2.855	96.403	100.236
Fluctuation	33.89%	13.18%	0.49%	21.15%	18.09%

Table 5. The ground floor height of back-stayed experimental results of condition 2

Indicator ground floor height of back-stayed(mm)	The maximum displacement(mm)			The maximum stress (MPa)	The stability (MPa)
	X	Y	Z	column	reduced stress
50	1.397	1.001	2.935	117.34	116.784
100	1.399	1.018	2.933	132.234	132.188
150	1.405	1.035	2.932	142.368	142.96
200	1.413	1.052	2.93	148.496	149.773
250	1.424	1.068	2.929	151.394	153.348
300	1.438	1.084	2.928	151.866	154.424
The maximum value	1.438	1.084	2.935	151.866	154.424
The minimum value	1.397	1.001	2.928	117.34	116.784
Fluctuation	2.93%	8.29%	0.24%	29.42%	32.23%

Table 6. The ground floor height of back-stayed experimental results of condition 3

Indicator ground floor height of back-stayed(mm)	The maximum displacement(mm)			The maximum stress (MPa)	The stability (MPa)
	X	Y	Z	column	reduced stress
50	0.601	12.459	3.363	139.436	137.725
100	0.573	12.459	3.361	149.922	148.248
150	0.548	12.458	3.359	155.887	154.221
200	0.517	12.458	3.358	158.504	156.83
250	0.481	12.458	3.357	158.845	157.155
300	0.456	12.461	3.356	158.044	156.334
The maximum value	0.601	12.461	3.363	158.845	157.155
The minimum value	0.456	12.458	3.356	139.436	137.725
Fluctuation	31.80%	0.02%	0.21%	13.92%	14.11%

The effect of the grounding back-stayed on displacement of nodes can be seen from the Table 4 to 6:

1) In condition 1, X direction displacement decreases with the increase of the ground floor height of back-stayed, while Y direction displacement increases. The ground floor height of back-stayed has high effect on the X direction displacement, the fluctuations is 33.89%, Y direction displacement of the second, the fluctuations is 13.18%, Z direction displacement of the last, the fluctuations is less than 1%.

2) In condition 2, X direction displacement decreases with the increase of the ground floor height of back-stayed, the ground floor height of back-stayed have little effect on the X direction displacement, which fluctuations is 8.29%, Y and Z direction displacement are less than 2%.

3) In condition 3, X direction displacement decreases with the increase of the ground floor height of back-stayed. The ground floor height of back-stayed have high effect on X direction displacement, the fluctuations is 31.8%, but have little effect on Y and Z direction displacement, their fluctuations are less than 0.5%.

The element stress under different conditions are shown in Figure 2 to 4.

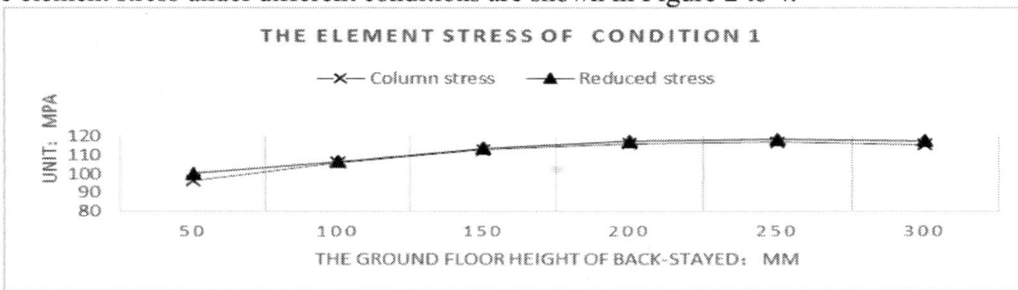

Figure 2. The element stress of condition 1

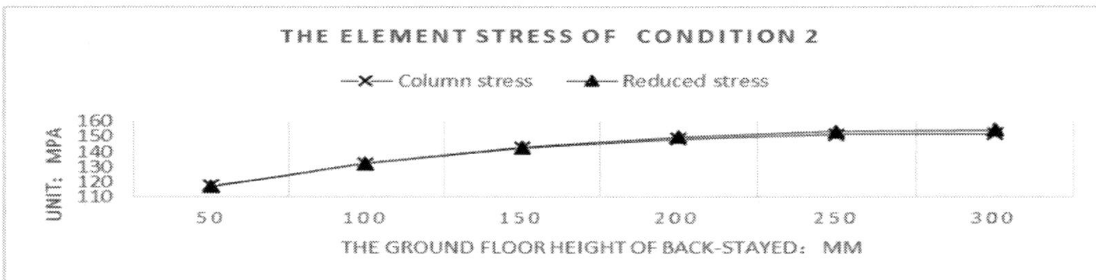

Figure 3. The element stress of condition 2

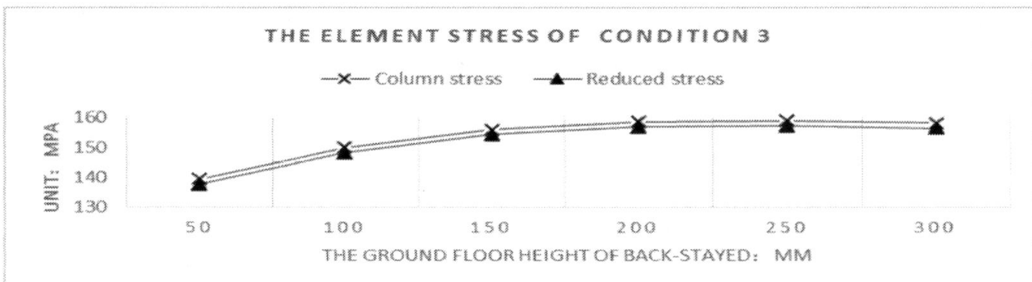

Figure 4. The element stress of condition 3

The effect of the grounding back-stayed on displacement of nodes can be seen from the Figure 2 to 4:

1) In condition 1 and 3, the column strength and stability decreases rapidly with the increase of the ground floor height of back-stayed when the ground floor height of back-stayed is less than or equal to the ground floor height of bracing (200mm).The column strength and stability enhances slightly with

the increase of the ground floor height of back-stayed when the ground floor height of back-stayed is more than the ground floor height of bracing.

2) In condition 2, the column strength and stability enhances with the increase of the ground floor height of back-stayed

5 Conclusion

This paper designs and analyses the AS/RS corbel shelf in different back-stayed arrangement by using the FEA software, get the results of displacement and stress under different condition, and draws the following useful conclusions.

1) The grounding back-stayed makes the column strength increases 20% to 30% and the stability increases 13% to 25%, enhances the mechanical properties of corbel shelf greatly.

2) The column strength and stability decreases with the increase rapidly of the ground floor height of back-stayed when the ground floor height of back-stayed is less than or equal to the ground floor height of bracing (200mm). The ground floor height of back-stayed have little impact on the column strength and stability when the ground floor height of back-stayed is more than the ground floor height of bracing.

3) The grounding back-stayed and increase of the ground floor height of back-stayed make X-displacement increased under the condition 1 and 3, while other conditions have no obvious change.

The above experimental results can provide a theoretical basis for structure optimization and design of this kind goods shelves.

References

1. GA50018-2002 Technical Code of Cold-Formed Thin-Wall Steel Structures[S].China Association for Engineering Construction Standardization. (2002).
2. Chen Ming, Ma Xiaofei, Zhao Gentian. Seismic behavior of T-shaped joint using cold-formed compound section steel [J]. Engineering Mechanics, **32**,184—191 (2015).
3. Arlene M.S. Freitas. Analysis and behavior of steel storage drive-in racks[J]. Thin-Walled Structures ,**48**,110 - 117 (2010)
4. Kamal M. Bajoria. Modal analysis of cold-formed pallet rack structures with semi-rigid connections [J]. Journal of Constructional Steel Research,,**66**,428–441 (2010)
5. P. Prabha. Evaluation of connection flexibility in cold formed steel racks[J]. Journal of Constructional Steel Research, **66**, 863 - 872 (2010)
6. Ch. Affolter. Collapse of a high storage rack [J]. Engineering Failure Analysis, **16**,1846—1855 (2009)
7. Liu,Y.Z.&Zu. Stability and bifurcation of helical equilibrium of a thin elastic rod [J]. Acta Mechanica, **167**,29—39 (2004)
8. Jia Zhengxian, Cao Xijing, Liu Changqi. The strength calculation and design of AS/RS rack [J]. Facilities and Equipment, **12**,29—54 (2003)

MATEC Web of Conferences **51**, 03006 (2016)
DOI: 10.1051/matecconf/20165103006

Component Analysis of Deposits in Selective Catalytic Reduction System for Automotive Diesel Engine

Neng Zhu[1,a], Lin Lv[1] and Chengwei Ye[1]

[1]*Key Laboratory of High Performance Ship of Ministry of Education, Wuhan University of Technology, Wuhan, China*

Abstract. In this paper, deposits in exhaust pipes for automotive diesel engines were studied by various chemical analysis methods and a kind of analysis process to determine the compositions of organic matter was proposed. Firstly, the elements of the deposits were determined through the element analysis method. Then using characteristic absorption properties of organic functional groups to the infrared spectrum, the functional groups in the deposits were determined. Finally, by GC-MS (gas chromatography - mass spectrometry) test, the content of each main component was determined quantitatively. Element analysis results indicated that the deposits adsorbed metal impurities from fuel oil, lubricating oil, mechanical wear and urea water solution. The result of GC-MS test showed that the area percentage of cyanuric acid was the biggest (about 85%), the second was urea (about 4%), and the content of biuret and biurea was scarce.

1 Introduction

Diesel engines are widely used in heavy vehicles because of its superior performance on the economy and reliability. Selective catalytic reduction (SCR) which uses urea water solution (UWS) as the reductant is widely regarded as one of the most promising technology to reduce NO_X emissions of diesel engine and to meet increasingly stringent emission regulation standards [1-4]. As for SCR system in practical engineering application, however, there are still many problems to be solved, especially the deposit problem, which has brought serious challenges for the application of the SCR system.

In the ideal situation, UWS generates reductant NH_3 after evaporation, pyrolysis and hydrolysis [5-7]. Schaber, etc. [8] studied the thermal decomposition process of urea by thermo-gravimetric analysis method in the lab. The results showed that the urea pyrolysis and the generation of related by-products are related to the temperature. When the temperature was above 152℃, urea started to break down, and at the same time biuret would form. When the temperature was above 190℃, a lot of cyanuric acid generated. Thagard [9] dealt with the chemical analysis of the by-products formed by urea decomposition at about 150 ℃ - 200 ℃ in a dielectric barrier discharge (DBD). Recent researches [10-14] showed that we could get hydrogen by urea pyrolysis reaction, which was considered as a potential source of hydrogen/fuel cell power. In addition, other researchers [15-17] found urea related deposits formed at different places in vehicles with SCR system, such as the exhaust pipe wall, the surface of catalyst entrance, the inside of urea nozzle and urea pump. We also observed a large

[a] Corresponding author : zhuneng555@163.com

number of deposits existing on the exhaust pipe wall through road test for a certain type of commercial vehicle. The exhaust pipe was serious congestion, which affected the performance of diesel engine seriously. What's more, the deposits formed in diesel engine exhaust pipes after continuous injection of urea aqueous solution for 5 h caused exhaust back pressure increased by 1.4 kPa and NO_X conversion efficiency reduced by 3%.

In terms of SCR system for automotive diesel engine, the operation condition of the given engine and exhaust flow field is very complex, the formation of deposits is an extremely complicated physical chemistry process. In order to avoid or reduce the formation of deposits, it is necessary to study its composition. In this paper, the deposits from a vehicle diesel engine exhaust pipe wall were analyzed by various chemical analysis methods. Not only its compositions were determined, but also the content of each component in the deposits was analyzed quantitatively.

2 Experimental

2.1 Reagents

A road test was conducted using a certain type of commercial vehicle with SCR system. Technical parameters of the engine and the test vehicle are shown in Table 1. After 1000 km of road test, a large amount of deposits were found on the exhaust pipe wall, as shown in Figure 1. The vehicle driving route included suburban road, plain expressway and mountain road. The collected deposits weighted 131.67g,which was the sample for the following chemical analysis.

Table 1. Technical Parameters.

Parameter Name		Parameter Value
Vehicle	The Maximum Allowable Total Mass	31000 kg
	Unladen Mass	10200 kg
Engine	Displacement	11.12 L
	Quantity of cylinders	6
	Air Admission Mode	Supercharged Inter-cooling
	Fuel System	High Pressure Common Rail
	Rated Rotate Speed	1900 r/min
	Rated Power	257 kW

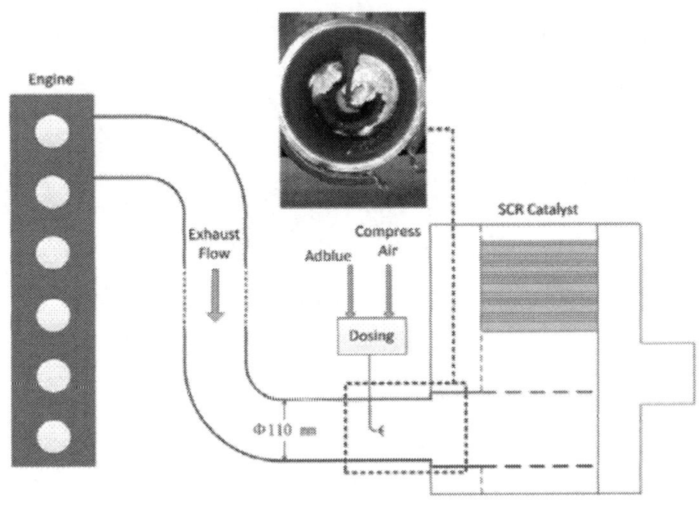

Figure 1. The deposits on the exhaust pipe wall.

2.2 Apparatus

Based on the chemical and physical properties of deposits, its components were researched by chromatographic analysis, spectral analysis, spectrum analysis and thermal analysis. Chemical analysis apparatus are shown in **Table 2**.

Table 2. Test apparatus.

Apparatus name	Specifications	Manufacturer
X-ray Fluorescence Spectrometer	AXIOS	PANalytical.B.V
Plasma Emission Spectrometer	Optima4300DV	PerkinElmer
CHNS/O elemental analyzer	Vario EL cube	Elementar
Fourier transform infrared spectrometer	NEXUS	Therno Nicolet
GC-MS	Agilent 7890/ 5975	Agilent

2.3 Procedures

Firstly, the elements of the deposits were determined through the element analysis method. According to JY/T 016-1996 type of general principle of wavelength dispersion X-ray fluorescence spectrometer method, the contents of inorganic elements in the sample were determined using Axios advanced X-ray fluorescence spectrometer. In order to study the sources of the inorganic elements in the deposits, the inorganic elements contents of UWS and deposits were compared and analyzed by Optima4300DV plasma emission spectrometer. In addition, the UWS for SCR system used in the test process was provided by KeLanSu Environmental Protection Technology Co., Ltd. The deposits sample was dissolved by aqua regia and diluted with hydrochloric acid of 10% concentration. What's more, carbon, hydrogen, nitrogen, oxygen and etc. are the common elements in the organic matter. In this paper, organic elements in the sample were determined by Vario EL cube model CHNSO element analyzer manufactured by Elementar in German.

Then the deposits were analyzed using potassium bromide (KBr) pressed-disk technique by Fourier transform infrared spectrum and the functional groups in the deposits were determined as the characteristic absorption properties of organic functional groups to the infrared spectrum.

Gas chromatography-mass spectrometry (GC-MS) technology has high detection sensitivity and good separation effect, which has become one of the most effective means to analyze complex unknown material. In this paper, the deposits were dissolved using methanol as solvent and the supernatant was collected to be analyzed by GC-MS. As a result, the content of each main component was determined quantitatively.

3 Results and Discussion

3.1 Elemental analysis

The results of x-ray fluorescence spectrum test are shown in **Table 3**. The loss on ignition of the sample was as high as 99.8% after burning under the high temperature of 1000 ℃ for 2 h, which indicated that the main ingredients in the sample were organic matters.

Table 3. Results of x-ray fluorescence spectrum test.

Elements	Si	Al	S	Mo	Zr	Ba	Pb	Br	Loss on ignition
Contents (%)[b]	0.042	0.033	0.007	0.015	0.003	0.016	0.046	0.038	99.8

b. Calculated by oxide content.

In order to study the sources of the inorganic elements in the deposits, the results of plasma emission spectrum test are shown in **Table 4**. It is observed that the mass fraction of metal elements detected in the samples are higher than that in the UWS. Hence, we can infer that the deposits have absorbed metal impurities from fuel oil, lubricating oil, mechanical wear and urea water solution.

Table 4. Results of plasma emission spectrum test.

Elements	Wave length /nm	Urea water solution /10^{-6}	Deposits/10^{-6}
Al	396.153	69	110
Cr	267.716	3	10
Fe	238.204	2	50
Mg	285.213	0	50
Mn	257.610	1	10
Pb	220.353	0	10
Zn	206.200	0	100
K	766.490	8	70
Na	589.592	63	120
Ca	317.933	0	850
Si	251.611	0	20

The results of organic elemental analysis are shown in **Table 5**. At the same time, **Table 5** lists the theoretical value of the content of each element in urea and related by-products .It indicates that the sample is composed of a variety of organic compounds by comparing the organic elements in the sample, urea and by-products.

Table 5. Results of organic elemental analysis.

Elements	C (%)	H (%)	N (%)	O (%)
Deposits	27.26	2.74	36.33	32.1
Urea	20	6.67	46.67	26.67
Biuret	23.3	4.85	40.78	31.07
Cyanuric acid	27.91	2.33	32.56	37.21

3.2 Qualitative analysis

Figure 2 shows the Fourier infrared spectra of deposits. As the characteristic absorption peaks shown in Figure 2, two absorption peaks at 3466 and 3334 cm^{-1} respectively are the absorption bands of antisymmetric stretching vibration $\nu(NH_2, as)$ and symmetric stretching vibration $\nu(NH_2, s)$. Strong absorption peaks at 1778, 1752 and 1721 cm^{-1} are the absorption bands of stretching vibration $\nu(C{=}O)$. Absorption peak at 1605 cm^{-1} is the absorption band of deformation vibration $\delta(NH_2)$. Strong absorption peak at 1053 cm^{-1} is the absorption band of stretching vibration $\nu(C\text{-}N)$. Absorption peaks across 900 ~ 700 cm^{-1}, whose peak type is wide and strong, is the characteristic absorption bands of wagging vibration $\omega(NH_2)$. These characteristic absorption peaks appeared at the same time, which showed that the composition of the sample contained aliphatic amide. Its structure formula is

$$\begin{array}{c} R \\ | \\ O = C - NH_2 \end{array}$$

Strong absorption peak at 3334 cm^{-1} is the absorption band of stretching vibration $\nu(N\text{-}H)$ with hydrogen-bonding association. Strong absorption peaks at 1778, 1752 and 1721 cm^{-1} are the absorption bands of stretching vibration $\nu(C{=}O)$. Absorption peak at 1461 cm^{-1} is the absorption band of deformation vibration $\delta(N\text{-}H)$. Strong absorption peak at 1053 cm^{-1} is the absorption band of stretching vibration $\nu(C\text{-}N)$. These characteristic absorption peaks appeared at the same time, which showed that the composition of the sample contained aliphatic secondary amide. Its structure formula is

$$O = \overset{\overset{R'}{|}}{C} -NH-R$$

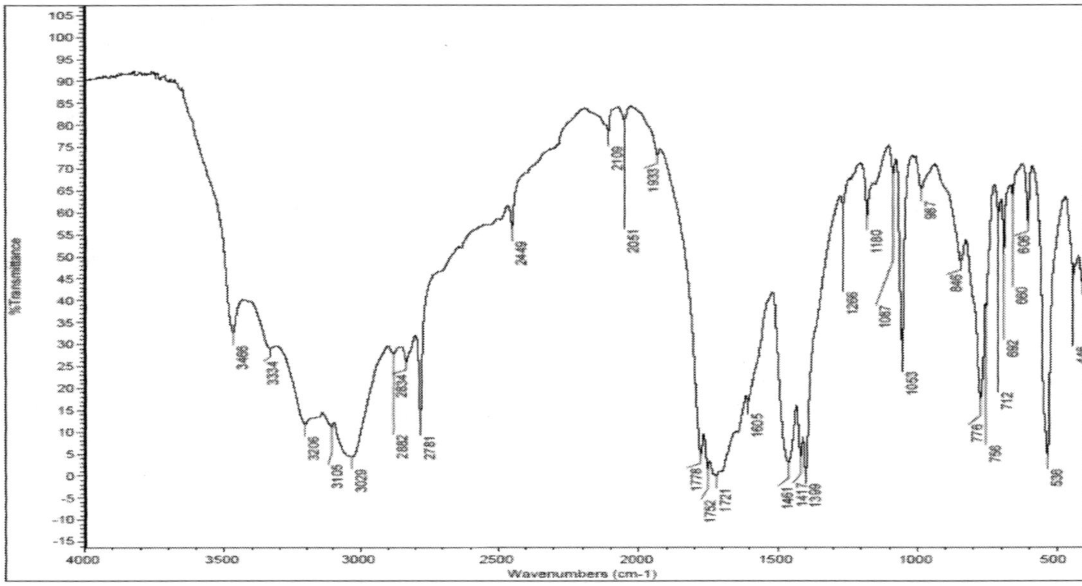

Figure 2. Fourier infrared spectra of deposits.

The infrared spectrograms of the deposits, biuret and cyanuric acid are compared in **Figure 3**. The results show that the matching degree of deposits and cyanuric acid is as high as 70% above, so we can primarily infer that the main ingredient in deposits is cyanuric acid.

3.3 Quantitative analysis

Figure 4 shows the chromatograms of deposits dissolved in methanol. Mass spectrometry results show that, the deposits contain biuret, urea and cyanuric acid. What's more, biurea, which haven't been observed in deposits ever, is also detected here. Hence, the reaction related biurea should be taken into account on the research of deposits in the future.

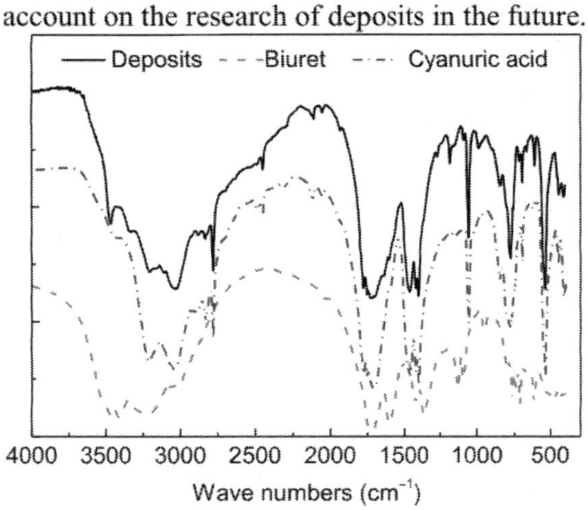

Figure 3. Infrared spectrum comparison results.

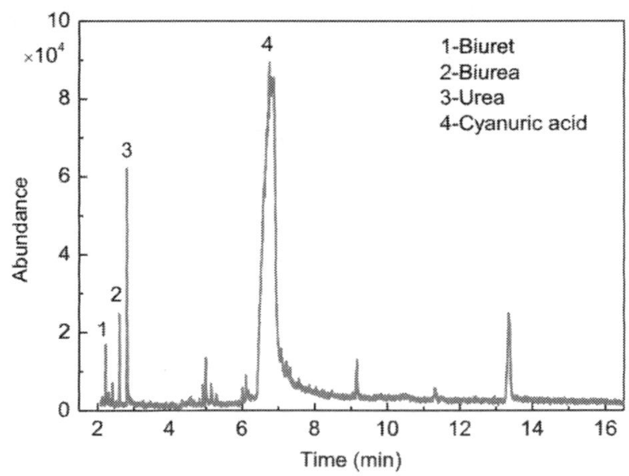

Figure 4. Chromatograms of the deposits.

According to the results of the area percentage, cyanuric acid has the largest peak area (about 85%), followed by urea (about 4% peak area), and the content of biuret and biurea is scarce. Similarly,

Paul Way [18] has analyzed multiple deposits ever by HPLC method and found that the content of cyanuric acid in the deposits was about 70 ~ 95%, urea was about 5 ~ 15%, which is consistent with the results in this paper. But the small amount of biurea was not detected in Paul Way's research.

4 Conclusions

This paper puts forward a kind of analysis process to determine the composition of organic matter: element analysis is to determine the content of each element in the sample, infrared spectrum analysis is to determine the characteristic functional groups of the sample, and the GC-MS analysis is to determine the content of each component quantitatively. From the results shown above, the following items can be concluded:

(1) The deposits are mainly composed of urea and its by-products. But beyond that, the deposits also adsorbed metal impurities from fuel oil, lubricating oil, mechanical wear and urea water solution;

(2) The organic matter of the deposits contains aliphatic amide ($\begin{matrix} & R & \\ & 1 & \\ O= & C & -NH_2 \end{matrix}$) and aliphatic secondary amide ($\begin{matrix} & R' & \\ & 1 & \\ O= & C & -NH-R \end{matrix}$);

(3) According to the area percentage results of GC-MS analysis, cyanuric acid has the largest peak area (about 85%), followed by urea (about 4%), and the content of biuret and biurea is less.

Acknowledgements

The authors thank the support and grants by the National Natural Science Foundation of China (No. 51379165) and the National Special Foundation of China (No. L1422029). And we also appreciate the support of Key Laboratory of High Performance Ship Technology of Ministry of Education (Wuhan University of Technology).

References

1. Tan Feng, Wuhan: Wuhan University of Technology, 5 (2013)
2. Haoyu Wang, Wuhan: Wuhan University of Technology, 5 (2013)
3. Tim Johnson, SAE Paper, 130105 (2013)
4. M. Koebel and M. Elsener, Catalysis Today, 59, 335~345 (2000)
5. C. Scott Sluder and John M. E. Storey, SAE paper, (2004)
6. Gaia Piazzesi, Martin Elsener and Oliver Krocher, Applied Catalysis B: Environmental, 65, 169~174 (2006)
7. Noel W. Cant and Dean C. Chambers, Topics in Catalysis, 10, 13~20 (2000)
8. Peter M. Schaber, James Colson and Steven Higgins, Thermochimica Acta, 424, 131~142 (2004)
9. S.M. Thagard, A. Mihalcioiu and K. Takashima, IEEE Trans. Plasma Sci, 37, 444~448 (2009)
10. B.K. Boggs, R.L. King and G.G. Botte, Chem. Commun, 32, 4859~4861 (2009)
11. R. Lan, S. Tao and T.S. Irvine, Energy Environ. Sci, 3, 438~441 (2010)
12. A.N. Rollinson, G.L. Rickett and V. A Lea-Langton, Appl. Catal. B, 106, 304~315 (2011)
13. A.N. Rollinson, J. Jones and V. Dupont, Energy Environ. Sci, 4, 1216~1244 (2011)
14. Jenny M. Jonesa and Andrew N. Rollinson, Thermochimica Acta, 565, 39~45 (2013)
15. Saito S, Shinozaki R and Suzuki A, SAE Paper, 01, 3248 (2003)
16. Seo J, SAE Paper, 01, 1135 (2011)
17. Chen Yue, Lu Lin, Shen Jie, Chinese Internal Combustion Engine Engineering, 2, 36 (2015)
18. Paul Way, Karthik Viswanathan and Phebe Preethi, SAE Paper, 01, 0633 (2009)

MATEC Web of Conferences **51**, 03007 (2016)
DOI: 10.1051/matecconf/20165103007

Varying Load Detection in A Gearbox System Based on Adaptive Threshold Estimation

A. Kouadri [1,a], G. R. Ibrahim [2] and A. Albarbar [3]

[1] *Signals and Systems Laboratory, Institute of Electrical and Electronics Engineering, University of Boumerdes,*

[2] *Department of Mechanical Engineering, College of Engineering, University of Anbar, Iraq.*

[3] *School of Engineering, Manchester Metropolitan University, ManchesterM1 5GD, UK.*

Abstract. This article presents an adaptive threshold computation for varying load detection in gearbox system. This method was applied on the vibration signal which is captured from gearbox housing. The experimental work was carried out using a three-stage gearbox connected with a motor and a generator. The developed adaptive threshold is obtained through several repeated experiments in the healthy mode and under the same operating conditions. An appropriate statistical test is used to examine the validity of the adaptive threshold estimation approach. Besides and for all experiments, a confidence interval is obtained which is closely linked to the distribution frequencies of the gearbox vibration signal as a random variable. In addition, several significance levels are considered to show the performances of the proposed adaptive thresholding technique compared to the limitations of the fixed threshold through the rate of false detection alarms. It is demonstrated from various experimental varying load mode of a gearbox system the effectiveness and accuracy of the adaptive threshold in term of evaluating gearbox operating conditions.

1 Introduction

Gearboxes fault diagnosis based on vibration measurements have grown significantly due to the progress in digital signal processing algorithms and the implementation techniques. A great deal of effort has been put into the development of monitoring approaches for such faults within mechanical transmission systems. For instance, Belsak *et al*[1] studied fatigue crack in tooth roots of the spur gear, using it to conduct the finite element method (FEM) to simulate fatigue crack growth and short- time fourier transformation (STFT). Kara *et al* applied advanced signal-processing techniques, such as the discrete wavelet transform (DWT) and a corrected multi-resolution Fourier transform (MFT), to investigate the vibration and current transients in a multistage gearbox under transient load [2]. Wu *et al* [3] studied the effect of tooth crack on the vibration response of a one-stage gearbox with spur gear. The vibration response of a pair of meshing gears was simulated by a lumped parameter model. They used several statistical indicators to reflect the change in the vibration response caused by the tooth crack. Ibrahim and Albarbar [4] presented a comparison between empirical mode decomposition (EMD) and smoothed pseudo-Wigner–Ville distribution (SPWVD) methods based on vibration signature and energy calculation procedure for monitoring gearbox systems. This article illustrates

[a] Corresponding author : a_kouadri@hotmail.com

that the calculation of energy using EMD techniques offers a more effective way to detect early faults than the SPWVD method, and that the computation of energy using the EMD technique is faster than the SPWVD method.

In this work, the gearbox vibration signal is used as a varying load detection index. This detection is a test of the adherence of the gearbox vibration signal to the confidence interval which is obtained after several independent experiments which are conducted in the same functioning conditions. The accuracy and effectiveness of the vibration signal–based approach for varying load detection are based on the powerful adaptive confidence interval which quantifies the noises in and on the considered system. More specifically, an appropriate test statistics based on the statistical characteristics of the analysed signal is used to define the best random phenomena associated with it and examine the validity of the proposed fault detection approach.

This paper is organized as follows: the exploitable experimental database is firstly presented in section 2. Then, a qualitative statistic analysis of the available database is carried out by giving some statistical quantities. In section 4, the proposed procedures are presented in a general manner to estimate the adaptive threshold. The use of these procedures to overcome the problem of uncertainties in measurement and detection with accuracy of varying loads in the gearbox is reported in section 5 in which the best results achieved are discussed and, finally, some conclusions are drawn in section 6.

2 Test rig and baseline data

The experimental work was carried out using a three-stage helical gearbox connected with a motor and a generator as shown in figure 1. Tests were carried out for a healthy pair of gears under varying loads (0%, 20% and 80% of the total load). The drive pinion at the first stage had 40 teeth meshing with a 20-tooth wheel. The pinion gear at the second stage had 80 teeth meshing with 20-tooth wheel. The pinion gear at the third stage had 80 teeth meshing with 32-tooth wheel. The vibration signals were collected using an accelerometer mounted vertically on the gearbox housing. The accelerometers were B&K type 4371 with sensitivity of 10mV/g and suitable for vibration measurements within a range of 1Hz to 12 kHz. The sampling frequency was set to 5 kHz. Before the accelerometer signals were fed to the analogue-to-digital converter NI USB 9233 card, they passed through a B&K type 2635 charge amplifier to condition the signal. The charge amplifier converts the accelerometer high impedance, low charge (in the range of Pico-coulomb) signal into low impedance and high voltage (in the range of mV) and the cut-off frequency for filter initialisation was set to 10 kHz.

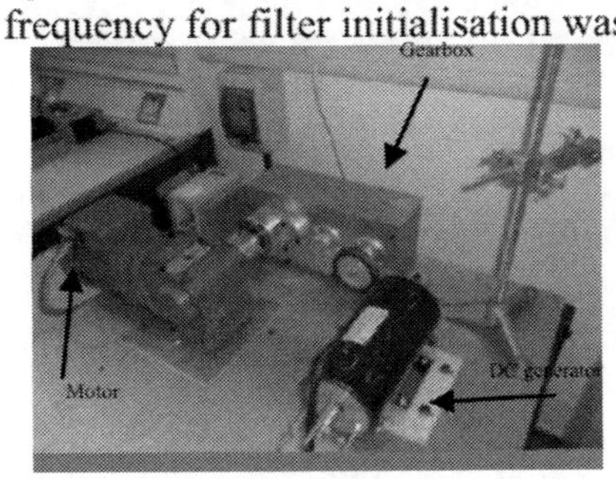

Figure 1. Gearbox test rig

In this paper, the database was divided into N=66 parts, recorded from N different repeated experiments. These experiments were conducted in a healthy mode and under the same operating conditions. Each database part includes T=400 samples. Corresponding to 0.08 seconds with reading frequency equals 20 milliseconds. These database parts are organized in a matrix $\boldsymbol{S}$ of 66 rows and 400 columns.

3 A statistical and qualitative analysis of vibration signals

In the present section, a preliminary statistical analysis of the database is carried out. Figure 2 reports the instantaneous statistical parameters of the database. High-order statistics are also illustrated and can be used as a reference value for adaptive threshold estimation performance [5] (figure 2). At each sample time, we can determine the mean of the variable

$$M(t)=\frac{1}{N}\sum_{i=1}^{N}S_i(t) \tag{1}$$

Also, we can compute the other instantaneous statistical parameters as follow:

$$V(t)=\frac{1}{N}\sum_{i=1}^{N}(S_i(t)-M(t))^2 \tag{2}$$

$$Sk(t)=\frac{1}{N}\sum_{i=1}^{N}\left[\frac{(S_i(t)-M(t))}{\sqrt{V(t)}}\right]^3 \tag{3}$$

$$Ku(t)=\frac{1}{N}\sum_{i=1}^{N}\left[\frac{(S_i(t)-M(t))}{\sqrt{V(t)}}\right]^4 \tag{4}$$

where $V(t)$, $Sk(t)$ and $Ku(t)$ represent respectively the variance, skewness and kurtosis at sample time t.

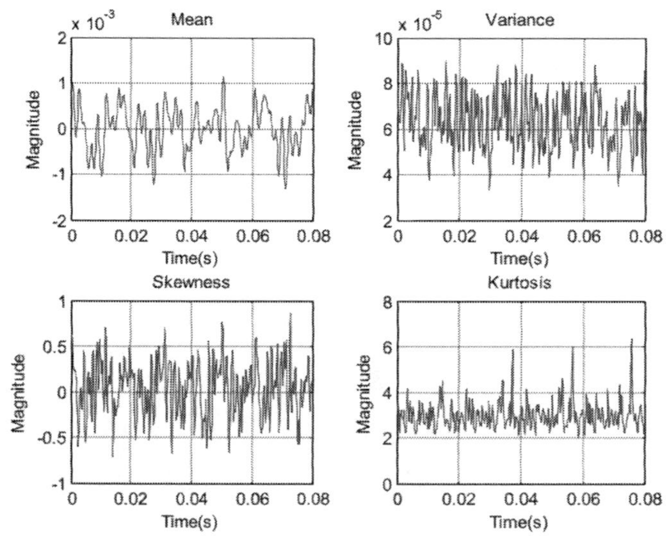

Figure 2. Different instantaneous statistical parameters of a gearbox vibration signal for repeated experiments

We focused on the two first statistical parameters, because it can be seen a large variation in the mean which indicates the non homogeneity of the vibration signal. The variance confirms that it is strongly related to the noise on the measurement device. Therefore, the vibration signal is a non stationary process. In addition, $S(j)$ at the sample time j is considered as a random variable for which we can find its probability distribution law. By looking at the histograms make randomly for several sample time and their corresponding skewness and kurtosis parameters, it seems that a normal distribution well describes its distribution [6] (figure 3).

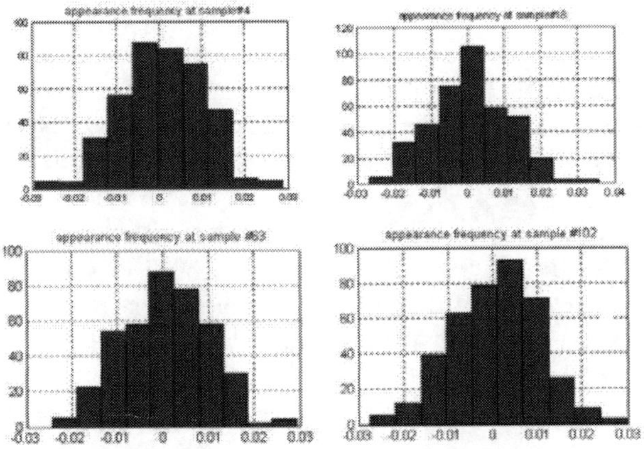

Figure 3. Different histograms of vibration signal at different sample time.

4 Adaptive threshold computation

The measurements are corrupted by noise, and, more importantly the characterization of signal's dispersion in the presence of faults, random factors, and disturbances on the system is difficult to interpret within a quantitatively perfect threshold [7]. Based on the fact that the vibration signal of the gearbox changes from one experiment to another under the same operating conditions, the sampled vibration signal is classified as a random variable $\Sigma(t)$. Thus, for each experiment, an instantaneous vibration signal $S_i(t), i=1,\ldots,N$, is evaluated. To increase the sensitivity of the fault detection index based on vibration signal $S(t)$ at each sample time t, an adaptive threshold is adopted in this study. Specifically, it should be constructed, from a given deduction, a random interval containing $S_i(t)$ with an acceptable probability. This probability will define a confidence level. In practice, the confidence level $(1-\alpha)$ is typically selected to be 95-99%. Considering the centred interval in $S_i(t)$, i.e. $\left]S_i(t)-\varepsilon(t), S_i(t)+\varepsilon(t)\right[$, $\varepsilon(t)$ is determined as:

$$P\left(S_i(t)-\varepsilon(t) \prec \hat{S}_i(t) \prec S_i(t)+\varepsilon(t)\right)=1-\alpha \tag{5}$$

where α represents the error risk incurred that $S_i(t)$ is located in the considered interval. The parameter $\varepsilon(t)$ depends on the value of the error risk α and the distribution law associated to the random variable $\Sigma(t)$, so:

$$\varepsilon(t)=F^{-1}(1-\alpha)\sqrt{V(t)} \tag{6}$$

where $F^{-1}(.)$ represents the inverse of the probability distribution function value. Therefore, the adaptive threshold limits at a confidence level $(1-\alpha)$ for which the confidential condition of the equation (5) verifies, are given by the confidential interval, so:

$$CI_\alpha(t)=\left]M(t)-F^{-1}(1-\alpha)\sqrt{V(t)}, M(t)+F^{-1}(1-\alpha)\sqrt{V(t)}\right[\tag{7}$$

For the purpose of adaptive thresholding, it is required to evaluate the confidence interval at each sample time t using equation (7). A fault in a system can be detected, if a sampled vibration signal is outside the corresponding confidence interval.

The probabilistic computation of the fault detection index based on the gearbox vibration signal $S(t)$ and its associated adaptive threshold given by the confidence interval is based on the fact that the random variable $\Sigma(t)$ follows a specified distribution law of mean $M(t)$ and variance $V(t)$ [8]. This hypothesis should apply to all sequences of the vibration signal and at each sample time. The skewness and kurtosis indicate automatically the normality distribution of the phenomena under study (See figure 2 and 3). In the normal distribution case, the skewness and kurtosis are close to 0 and 3, respectively. However, the skewness is an asymmetric distribution measure of the random variable $\Sigma(t)$ around its mean $M(t)$, the skewness defined by equation 3 is normally distributed, such that:

$$Sk(t) \propto \mathcal{N}\left(0, \sqrt{\frac{6}{N}}\right) \tag{8}$$

The kurtosis measures the pointed or flat characteristics of the $\Sigma(t)$ distribution law. It follows:

$$Ku(t) \propto \mathcal{N}\left(3, \sqrt{\frac{24}{N}}\right) \tag{9}$$

At a given sample time t, the random variable $\Sigma(t)$ obeying a normal distribution law $\mathcal{N}\left(M(t),\ \sqrt{V(t)}\right)$ if the two quantitative statistics measurement $Sk(t)$ and $Ku(t)$ follow jointly a centered and reduced normality distribution $\mathcal{N}(0,\ 1)$, respectively.

5 The adaptive thresholding performances

The adaptive threshold for the vibration signal of the gearbox is calculated from 66 trajectories under the hypothesis that the given sampled vibration signal $S(t) \propto \mathcal{N}\left(M(t), \sqrt{V(t)}\right)$ with different significant values. The efficiency of the detection algorithm is based on the computation of the confidence intervals which provide an estimate of the amount of error involved in the considered data and characterize the precision of the computed statistical estimates. The Skewness-Kurtosis non-parametric test validates the normal distribution law, for a typical confidence level of 95%, of the vibration signal with acceptable probabilities (p-value equals 0.857).

The probability, of observing given sampled gearbox vibration signal of all experiments under the assumption that this later is normally distributed, is computed and evaluated. Furthermore, the normally distribution is justified to an appropriate satisfaction probability degree. However, the chosen significance level is related to the degree of certainty you require in order to reject that the sampled gearbox vibration signal follows normal law. This is true if the probability of observing sampled result is less than the significance level. Therefore, the vibration signal of the gearbox is fully satisfied to be normally distributed with fairly acceptable jointly probabilities of skewness and kurtosis. These obtained probabilities are at almost samples closed to 1.

To evaluate the performance of the detection scheme based upon the adaptive threshold; several cases, a healthy pair of gears under varying loads (20% and 80% of the total load), are examined. Figure 6 illustrates the vibration signal variation at different loads with the confidence interval limits which are developed at the confidence level 99%. It can be seen clearly that the gearbox vibration signal is outside the confidence interval limits. Therefore, the varying load is accurately detected using the adaptive thresholding technique. Regarding the amplitude of the gearbox vibration signal which becomes quite small at high load, the adaptive threshold indicates well the occurrence of the varying load in spite of the confidence interval is quite large at the confidence level 99%. Consequently, the results confirm the high sensitivity and accuracy of the adaptive threshold in detecting varying load.

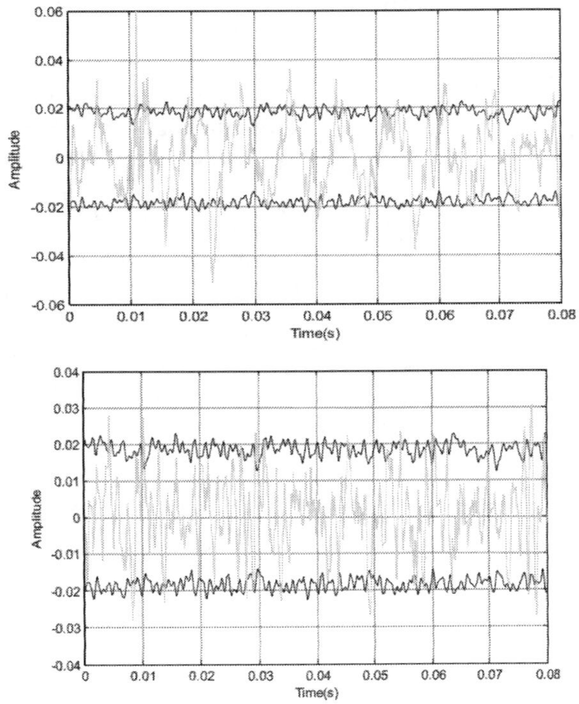

Figure 4. Load detection (20% and 80% load, respectively)

6 Conclusion

The present paper deals with one of the major problems that arise when searching for a reliable, accurate, and easily exploitable, adaptive threshold, which can indicate with accuracy the presence of a varying load in the gearbox. The developed adaptive threshold is based on the vibration signal measurement for which it is able to overcome the problem of false detection and non detection of a fault by using the fixed thresholding techniques. The obtained results also illustrate that under different situations the adaptive threshold can accurately detect changes in the experimental, three-stage gearbox connected with motor and generator, variables. Furthermore, the principle of calculation of confidence interval at different confidence level, after testing the distribution law of the database using Skewness-Kurtosis method, is sufficient to constitute an interesting and helpful tool to overcome the difficulties of parametric uncertainties quantification and devices measurement inaccuracies which increase the number of false alarms. It is also important to note the high sensitivity of the adaptive threshold in detecting varying loads with fairly low amplitudes in gearbox vibration signal.

References

1. Belsak, A. and J. Flasker. Detecting cracks in the tooth root of gears. Engineering Failure Analysis 14 (2007) 1466–1475.
2. Kara, C. and A. Mohanty. Vibration and current transient monitoring for gearbox fault detection using multiresolution Fourier transform. Journal of Sound and Vibration 311 (2008) 109–132.
3. Wu, S., M. Zuo and A. Parey. Simulation of spur gear dynamics and estimation of fault growth. Journal of Sound and Vibration 317 (2008) 608–624 .
4. G. R., Ibrahim and A., Albarbar. Comparison between Wigner–Ville distribution- and empirical mode decomposition vibration-based techniques for helical gearbox monitoring. Proc. IMechE

Part C: J. Mechanical Engineering Science. Published online 16 June 2011. DOI: 10.1177/0954406211403571

5. Zhijian Wang, Zhennan Han, Fengshou Gu, James Xi Gu and Shaohui Ning, A novel procedure for diagnosing multiple faults in rotating machinery, ISA Transactions 55 (2015) 208–218,
6. Lotfi Saidi, Jaouher Ben Ali and Farhat Fnaiech, Application of higher order spectral features and support vector machines for bearing faults classification, ISA Transactions 54 (2015) 193–206,
7. Vicenç Puig, Saúl Montes de Oca and Joaquim Blesa. Adaptive threshold generation in robust fault detection using interval models: time-domain and frequency-domain approaches, International Journal of Adaptive Control and Signal Processing, 27 (10) (2013) 873–901.
8. Joanes DN, Gill CA. Comparing measures of sample skewness and kurtosis. Journal of the Royal Statistical Society, (Series D): The Statistician 47 (1) (1998) 183–189.

MATEC Web of Conferences **51**, 03008 (2016)
DOI: 10.1051/matecconf/20165103008

Micro-Satellite Attitude Determination with Only A Single Horizon Sensor

Gaoxiang Ouyang [a], Xin Dong, Xin Li

Academy of Opto-Electronics, Chinese Academy of Sciences, Beijing, China

Abstract. Through using measurement from only a single horizon sensor, this paper presented a quaternion-based 3-axis attitude determination method, which can be implemented on board micro-satellites and applied over a whole orbital period. Firstly, a description of attitude representation on the quaternion is given. Secondly, a detailed modeling formulation with nadir vector and measurement equations on attitude estimation system is demonstrated. Afterwards, a correction is made to eliminate the estimation error resulted from Earth's oblateness, and able to further improve the accuracy of the attitude determination algorithm. Finally, a six degree-of-freedom closed-loop simulation is used to validate the accuracy of the attitude determination method given in this paper.

1 Introduction

At present, micro-satellites are increasingly applied to different space missions 1, at the same time also need to face the drastic restrictions imposed by the Cubesat standard 2. Taking into account power consumption, weight and other constraints on micro-satellites, the adoption of only a single attitude sensor has become a practical requirement of the micro-satellites such as Cubesat. If the final target is to maintain the overall attitude accuracy, then it will be accompanied by huge challenges. In fact, the attitude determination based on single-sensor essentially belongs to a class of the under-measurement problem, the biggest difficulty lies in the lack of information dimension measurement, which maybe result in weak observability of the dynamic system 3.

In attitude information acquisition process, the attitude measurement is often achieved by devices such as sun sensors, star sensors and magnetometers. However, these sensors have clear limitations: sun sensors will lose their functionalities in periods of eclipse in orbit, magnetometers cannot acquire high accuracy attitude measurements due to the constantly changing Earth magnetic field, star sensors are too expensive for micro-satellites. Compared with other sensors, Earth horizon sensors (EHS) have emerged as efficient and relatively inexpensive means of more precise attitude determination, capable of satisfying attitude knowledge requirements of micro-satellites in low Earth orbit (LEO), especially for missions with Earth-specific science objectives. Accordingly, this paper proposes a 3-axis attitude determination method based on only a single scanning horizon sensor.

To reduce the effects of noise on the measurement, it is more practical for micro-satellites to introduce dynamic attitude estimation and filtering techniques, relying on body measurements of known vectors in the reference frame, and use the dynamical equations of the system under

[a] Corresponding author : ouyanggx@aoe.ac.cn

consideration. In this paper, a quaternion-based extend Kalman filter (EKF) is developed to estimate the attitude history of LEO satellites using EHS and gyros, meanwhile the 3-axis attitude pointing accuracy always remains within the error range of 4 deg. Remarkably, the pointing accuracy of the roll-pitch orientation, which mainly impacts on Earth observation missions 1, stays within the 1 deg of margin.

Here are some coordinate system definitions, the Local Vertical Local Horizontal (LVLH) frame describes the current orbit frame of the satellite, and has its origin at the center of mass of the satellite. The coordinate axis z_L points towards the center of the earth (direction of the nadir), y_L points opposite to the satellite's angular momentum, and x_L completes the right-handed triad. The instantaneous LVLH frame is used as reference to measure the local attitude of the satellite. Body frame is an orthogonal coordinate system fixed to the satellite body with origin at its center of mass. Measurement frame is known in advance, and all measurements of the sensor are expressed in the measurement frame. The symbols $\hat{x}$ and $x^{\times}$ denote the estimate and the cross-product matrix associated with x, respectively.

The paper is organized as follows. Section 2 provides the modeling on system state equation and measurement equation, and a full expression of the nadir vector in the satellite's body frame. A simple analytical method for Earth's oblateness correction is given in Section 3. Section 4 presents a six degree-of-freedom closed-loop simulation, which is used to observe and compare the performance of the attitude determination algorithm developed by different type of measurement errors and estimators. Finally, Section 5 summarizes this paper.

2 Quaternion-Based EKF Estimator

2.1 System Model

A seven-element state vector, x, consisting of the quaternion $\mathbf{q}$ and the gyro-bias $\mathbf{b}$, is defined as

$$x = [\mathbf{q} \quad \mathbf{b}] \tag{1}$$

For simply, the angular rate $\boldsymbol{\omega}_{m,}$, as the gyro outputs, is directly used as an alternative to attitude dynamics, so the system of differential equations governing the state is defined as

$$\dot{\mathbf{q}} = \frac{1}{2}\Omega(\boldsymbol{\omega}_m - \mathbf{b}) \cdot \mathbf{q} \tag{2}$$

where the state equation Ω is consistent with definition given in 4, the gyro bias $\mathbf{b}$ is assumed to be Gaussian white noise, and other noise is omitted. For implementation of the discrete time Kalman filter (KF) equations, we need to discretize the above state propagation model (2) as follows

$$\hat{q}_{k+1}^{-} = \boldsymbol{\Theta}\ \hat{q}_k^{+} \tag{3}$$

where the matrix $\boldsymbol{\Theta}$ is given as

$$\boldsymbol{\Theta} = \mathrm{c}|\Delta\hat{\boldsymbol{\omega}}|\mathbf{I}_{3\times3} - \frac{\hat{\boldsymbol{\omega}}^{\times}}{|\hat{\boldsymbol{\omega}}|}\mathrm{s}|\Delta\hat{\boldsymbol{\omega}}| + \frac{\hat{\boldsymbol{\omega}}\hat{\boldsymbol{\omega}}^{\mathrm{T}}}{|\hat{\boldsymbol{\omega}}|^2}(1 - \mathrm{c}|\Delta\hat{\boldsymbol{\omega}}|) \tag{4}$$

where the common practice of writing s· for sin(·) and c· for cos(·) is adopted. The expression $\Delta\hat{\boldsymbol{\omega}}$ is represented as $\Delta\hat{\boldsymbol{\omega}} = (\boldsymbol{\omega}_m - \mathbf{b})\ \Delta t$, and Δt is the sampling period. The symbol $\mathbf{I}_{3\times3}$ is identity matrix with three dimensions.

Because of singularity caused by the unit norm constrain of the quaternion 4, a six-state error model is used for propagation of covariance. Moreover, a multiplicative error model is employed, then the discrete process model for the error state is obtained as

$$\Delta x = \begin{bmatrix} \delta\dot{\mathbf{q}} \\ \Delta\dot{\mathbf{b}} \end{bmatrix} = \begin{bmatrix} \boldsymbol{\Theta} & \boldsymbol{\Psi} \\ \mathbf{0}_{3\times3} & \mathbf{I}_{3\times3} \end{bmatrix} \begin{bmatrix} \delta\mathbf{q} \\ \Delta\mathbf{b} \end{bmatrix} \tag{5}$$

where the matrix $\boldsymbol{\Theta}$ is the same as (4), and the matrix $\boldsymbol{\Psi}$ is defined as

$$\mathbf{\Psi} = -\Delta t \cdot \mathbf{I}_{3\times3} + \frac{\hat{\boldsymbol{\omega}}^{\times}}{|\hat{\boldsymbol{\omega}}|^2}(1 - \mathrm{c}|\Delta\hat{\boldsymbol{\omega}}|) - \frac{1}{|\hat{\boldsymbol{\omega}}|^3}(\Delta t|\Delta\hat{\boldsymbol{\omega}}| - \mathrm{s}|\Delta\hat{\boldsymbol{\omega}}|)(\hat{\boldsymbol{\omega}}^{\times})^2 \tag{6}$$

The covariance matrix Q_k of the noise in the discrete time system can be approximated by two-order Taylor series, according to reference [5].

$$Q_k = \frac{\Delta t}{2}(\begin{bmatrix} \mathbf{\Theta} & \mathbf{\Psi} \\ \mathbf{0}_{3\times3} & \mathbf{I}_{3\times3} \end{bmatrix} \cdot Q + Q \cdot \begin{bmatrix} \mathbf{\Theta} & \mathbf{\Psi} \\ \mathbf{0}_{3\times3} & \mathbf{I}_{3\times3} \end{bmatrix}^{\mathrm{T}}) \tag{7}$$

where the Q is the spectral density of the white noise process of system dynamics.

2.2 Roll/Yaw Measurements

In the measurement configuration given by this article, the single EHS fixed on body y axis directly provides pitch axis knowledge, besides, based on the geometrical relationship, the EHS can provide roll axis attitude information [6], which is calculated as follows

$$\begin{aligned} \theta &= \frac{1}{2}(\omega_{so} - \omega_{si}) \\ \varphi &= \frac{\pi}{2} - \arcsin(\rho / \sqrt{\mathrm{c}^2\gamma + \mathrm{s}^2\gamma \times \mathrm{c}0.5\omega_{so} \times \mathrm{c}0.5\omega_{si}}) \\ &\quad - \arctan(\mathrm{c}\gamma / \mathrm{s}\gamma \times \mathrm{c}0.5\omega_{si}) \end{aligned} \tag{8}$$

where the scan angle width ω_{so} and ω_{si} are defined the arc length of the EHS's scanning axis which crosses from the Earth-in to Earth-out and from Earth-in to baseline respectively. Semi-scan angles γ is a known hardware parameter. The Earth disk visible from the satellite at an altitude h is of angular radius ρ is defined by

$$\rho = \arcsin(\frac{R_e}{R_e + h}) \tag{9}$$

where the variable R_e denotes the radius of the Earth.

Although the EHS fixed on pitch axis is unable to directly measure the yaw information, the attitude kinematics equation implies that the roll angle φ couples with yaw attitude ψ. If the 3-axis angles are small, the attitude coupled kinematics equation is given by

$$\begin{bmatrix} \dot{\varphi} \\ \dot{\theta} \\ \dot{\psi} \end{bmatrix} = \begin{bmatrix} \omega_x \\ \omega_y \\ \omega_z \end{bmatrix} + \begin{bmatrix} 1 & \psi & -\theta \\ -\psi & 1 & \varphi \\ \theta & -\varphi & 1 \end{bmatrix} \begin{bmatrix} 0 \\ \omega_0 \\ 0 \end{bmatrix} \tag{10}$$

where orbital angle velocity ω_0 is always assumed to be constant. When ω_x=ω_z=0, then the trajectory of roll and yaw angle follow after solving differential Equation(10) as

$$\begin{aligned} \varphi(t) &= \varphi(t_0)\cos\omega_0(t - t_0) + \psi(t_0)\sin\omega_0(t - t_0) \\ \psi(t) &= \varphi(t_0)\sin\omega_0(t - t_0) + \psi(t_0)\cos\omega_0(t - t_0) \end{aligned} \tag{11}$$

As obviously seen from Equation(11), every 1/4 orbital period, the roll and yaw angle values exchange, which is also known as the orbital gyrocompass in reference 7-8. So it makes full use of the dynamical coupling effect to estimate the yaw angle through roll axis measurements indirectly.

2.3 Vector Measurements

We now will give a detailed derivation of the unit vector in the coordinate system of the horizon sensor, which points from the center of the pinhole camera to the center of the Earth. This vector will be referred to as the "nadir vector". The horizon sensor is always looking in the -z direction, thus the pointing vector $\mathbf{E}_o$ of a nadir-pointing EHS should be $[0\ 0\ \text{-}1]^{\mathrm{T}}$. Since the EHS is not looking directly at the Earth center in the presence of attitude angles, a sequence of two rotations can be defined to describe the alignment of the EHS coordinate system with that of a nadir-pointing EHS. These consist of (i) rotating the EHS again about itself z-axis by an angle $-\varphi$; (ii) rotating a nadir-pointing EHS about itself x-axis by an angle θ, then. These angles have already been measured with EHS: they are the pitch and negative roll angles given in Equation(8) respectively. To compute the nadir vector in the coordinate system of the rotated EHS, we simply apply the rotation matrices to the nadir vector $\mathbf{E}_o$ as follows:

$$\mathbf{E}_s = \mathbf{R}_x(\theta)\ \mathbf{R}_z(-\varphi)\ \mathbf{E}_o = \begin{bmatrix} -\sin(\theta) \\ -\sin(-\varphi)\cos(\theta) \\ \cos(-\varphi)\cos(\theta) \end{bmatrix} \tag{12}$$

Afterwards, combining with the known of EHS setup matrix, the above vector $\mathbf{E}_s$ is transferred and expressed in the body frame as follows,

$$\mathbf{E}_b = \begin{bmatrix} 0 & -1 & 0 \\ 1 & 0 & 0 \\ 0 & 0 & 1 \end{bmatrix} \begin{bmatrix} -\sin(\theta) \\ -\sin(-\varphi)\cos(\theta) \\ \cos(-\varphi)\cos(\theta) \end{bmatrix} = \begin{bmatrix} \sin(-\varphi)\cos(\theta) \\ -\sin(\theta) \\ \cos(-\varphi)\cos(\theta) \end{bmatrix} \tag{13}$$

Now, the nadir vector $\mathbf{E}_b$ is used as the measurement in body frame to yield filtering residual.

2.4 Measurement Model

As shown in the previous section, the EHS measurements have been used to construct the nadir vector in the body frame. Hence, the measurement equation can be written as,

$$z_k = \mathbf{E}_b + v_k \tag{14}$$

Which results in the measurement residual,

$$z_k - \hat{z}_k = \mathbf{E}_b - C(\hat{\mathbf{q}})\mathbf{E}_o + v_k \tag{15}$$

The measurement noise covariance matrix is given as

$$R_k = E\left[v_k^{\mathrm{T}} v_k\right] = \begin{bmatrix} \sigma_\varphi^2 & 0 \\ 0 & \sigma_\theta^2 \end{bmatrix} \tag{16}$$

Using an approximation to attitude direction matrix $C(\hat{\mathbf{q}}) \approx (\mathbf{I}_{3\times3} - \delta\boldsymbol{\theta}^\times)C(\mathbf{q})$ as shown in 4, so the required sensitivity matrix for our measurement model can be derived as

$$H_k = \frac{\partial z_k}{\partial \mathbf{E}_b}\frac{\partial \mathbf{E}_b}{\partial \boldsymbol{\theta}}\frac{\partial \boldsymbol{\theta}}{\partial x} = \begin{bmatrix} 2(r_x^{\times}\hat{z}_k)^{\mathrm{T}} & \mathbf{0}_{1\times3} \\ 2(r_y^{\times}\hat{z}_k)^{\mathrm{T}} & \mathbf{0}_{1\times3} \end{bmatrix} \tag{17}$$

where the vector r_x and r_y are the first and second row of the transformation matrix from the body to the sensor frame, respectively. In order to reduce the amount of computations in the filtering process, we draw lessons from reference 9 to transfer the above measurement model (14)-(17) to equivalent form as below

$$z_k = \mathbf{E}_b^{\times}(C(\hat{\mathbf{q}})\cdot\mathbf{E}_o) + v_k$$
$$R_k = \begin{bmatrix} \sigma_\varphi^2 & 0 \\ 0 & \sigma_\theta^2 \end{bmatrix},\ H_k = \begin{bmatrix} \mathbf{I}_{2\times2} & \mathbf{0}_{2\times4} \end{bmatrix} \tag{18}$$

The advantage of the measurement model (18) is that both the sensitivity matrix H_k and the discrete process model (5) are constant, so a steady-state KF is applied by solving the corresponding Riccati equation 10. The resulting filter calculation cost will be greatly reduced compared with the situation using the measurement model (14)-(17), meanwhile the attitude estimation precision obtained by the steady-state KF is not decreased, as shown in following simulation.

2.5 Correction Equations

In the case, Murrell's approach to sequentially process the vector measurements is implemented 4. The recursive calculation form of KF gain is given as follows

$$K_k = P_k^- H_k^{\mathrm{T}} \left[H_k P_k^- H_k^- + R_k \right]^{-1} \tag{19}$$

For covariance propagation, Joseph's form is used as follows

$$P_k^+ = (\mathbf{I}_{6\times6} - K_k H_k) P_k^- (\mathbf{I}_{6\times6} - K_k H_k)^{\mathrm{T}} + K_k R_k K_k^{\mathrm{T}} \tag{20}$$

Then, the state vector x is updated as follows

$$\hat{\mathbf{q}}_k^+ = \delta\hat{\mathbf{q}}_k \otimes \hat{\mathbf{q}}_k^- = \frac{1}{\sqrt{1 + \|\delta\hat{\mathbf{q}}_k\|^2}} \begin{bmatrix} \delta\hat{\mathbf{q}}_k \\ 1 \end{bmatrix} \otimes \hat{\mathbf{q}}_k^- \tag{21}$$

$$\hat{\mathbf{b}}_k^+ = \hat{\mathbf{b}}_k^- + \delta\hat{\mathbf{b}}_k$$

3 Earth'S Oblateness Corrections

Actually, the Earth's oblateness contributes the most of error source of attitude determination based on horizon sensor 11. Instead of treating the total measurement errors as white noise in this paper, an effort is made to develop a realistic systemic error model.

Assuming the spacecraft's position known, the error duo to the Earth's oblateness can be eliminated by considering variation of Earth's radius with latitude. The correction based on earth model for the Earth's oblateness is given as

$$\cos\eta = \cos\eta_0 \left[1 + f \cdot F(\phi,\sigma) \right] \tag{22}$$

where η_0 is the nadir angle ignoring the effect of flattening, the f is Earth ellipsoid, $F(\varphi,\sigma)$ is the function of error modification for Earth's oblateness, as follows,

$$F(\phi,\sigma) = \frac{R_e^4}{r^2(r^2 - R_e^2)} \left[\mathrm{s}^2\phi + \frac{r^2 - R_e^2}{R_e^2} \mathrm{c}^2\phi \mathrm{s}^2\sigma + \frac{2\sqrt{(r^2 - R_e^2)}}{R_e} \mathrm{c}\phi \mathrm{s}\phi \mathrm{s}\sigma \right] \tag{23}$$

where r represents the distance between the satellite and the Earth center, φ is the latitude of the sub-satellite point, σ is the azimuth of horizon point with respect to the local eastward direction. Then the correction for Earth's oblateness is given by

$$\delta\varphi = \frac{f \cdot c\eta_0}{2} \left[\frac{c\alpha}{s\gamma \cdot s\beta_0} (F(\phi,\sigma_1) - F(\phi,\sigma_2)) - \frac{s\alpha}{c\gamma} (F(\phi,\sigma_1) + F(\phi,\sigma_2)) \right] \tag{24}$$

$$\delta\theta = \frac{f \cdot c\eta_0}{2} \left[\frac{s\alpha}{s\gamma \cdot s\beta_0} (F(\phi,\sigma_1) - F(\phi,\sigma_2)) + \frac{c\alpha}{c\gamma} (F(\phi,\sigma_1) + F(\phi,\sigma_2)) \right]$$

where the variables σ_1, σ_2 and χ are calculated respectively as follows,

$$\sigma_1 = \alpha - \chi - \arcsin \frac{\sqrt{R_e^2 - r^2 \cdot c^2\gamma}}{R_e}$$

$$\sigma_2 = \alpha - \chi + \arcsin \frac{\sqrt{R_e^2 - r^2 \cdot c^2\gamma}}{R_e}$$

$$\chi = \mathrm{arcctg}(\mathrm{tg}(i)\cos(u)), \quad s\beta_0 = \frac{\sqrt{R_e^2 - r^2 \cdot c^2\gamma}}{r \cdot s\gamma}$$

Finally, the correct calculation of the roll and pitch attitude angles are shown as follows,

$$\hat{\varphi} = \hat{\varphi}_0 - \delta\varphi, \quad \hat{\theta} = \hat{\theta}_0 - \delta\theta \tag{25}$$

4 Simulation Results

The initial conditions considered for our simulation were chosen as shown in Table 1:

Table 1 Parameter setting for simulation

Parameter	Value	Unit
a[a]	6921.7990	km
e[b]	0	
i^c	98.0475	deg
$\mathbf{q}_0$	[-0.000207, 0.000233, 0.000281, 1.0]	
$\boldsymbol{\omega}_{angle}{}^{d}$	[0,-0.004,0.002]	rad/sec
γ	55	deg
$\boldsymbol{b}_0{}^{e}$	0.05	deg/hr
Δt	0.5	sec
$\mathbf{I}_s{}^{f}$	diag{57,150,152}	kg·m^2
σ_φ	10^{-2}	
σ_θ	10^{-2}	

[a]The variable a is orbital semi-major.
[b]The variable e is orbital eccentricity.
[c]The variable i is orbital inclination.
[d]The ω_{angle} is initial angle rate.
[e]The $\mathbf{b}_0$ is initial gyro drift bias.
[f]The I_s is the moment of inertia matrix ,which is diagonal matrix.

We assume that no Sun obtrusions are present (although this is not truly realistic). At each time instant all available LVLH nadir vector and body measurements are used to form the KF gain matrix in Equation(19). Then, the quaternion estimate is found using Equation(21).

It is clearly shown in the Figure 1 that the yaw errors are much larger than other axis. This is since the knowledge of yaw axis is not directly acquired by the EHS. Furthermore, as expected, the accuracy changes as the orbital period changes. In contrast, the roll and pitch attitude estimation are all kept within a small margin of error.

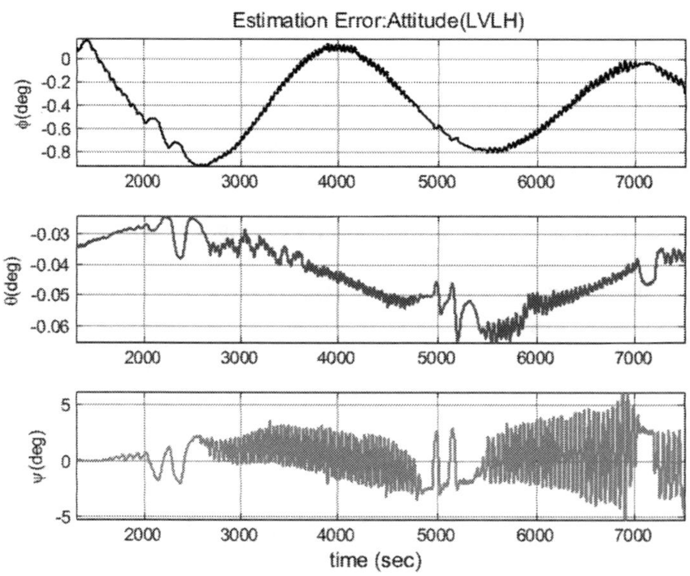

Figure 1. Estimation Errors of recursive EKF without Correction

Through correcting the error caused by the Earth's oblateness, the roll axis pointing accuracy has been further improved, but for improvement of pitch axis was not obvious, as shown in Figure 2. The estimate accuracy of the roll-pitch axis met the pointing accuracy requirement of 1 deg, in spite of the relatively larger estimation error in yaw direction.

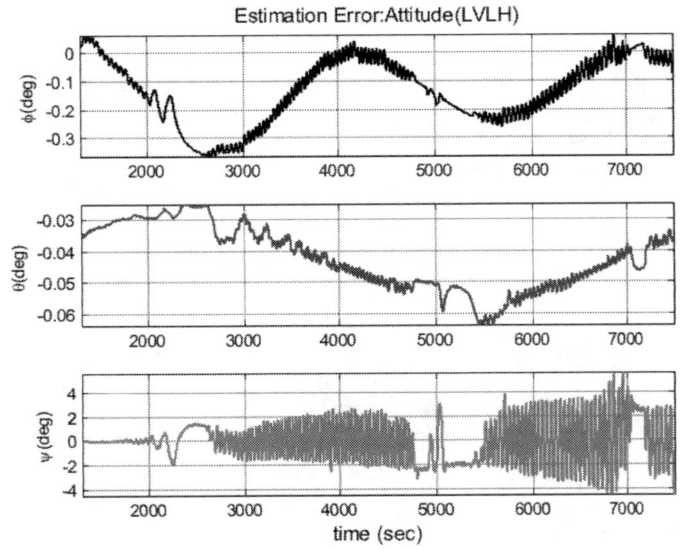

Figure 2. Estimation Errors of recursive EKF with Correction

As mentioned earlier, the recursive KF gain filter can be modified to its steady-state implementation, to do away with its long transients and reduce computing cost. The constant gain K is [0.8697, 389.1524, 0.0, -0.4095, 0.8692, -0.4097]^T, based on the measurement equation (18). The performance with the recursive gain filter for a sample run is illustrated in Figure 3.

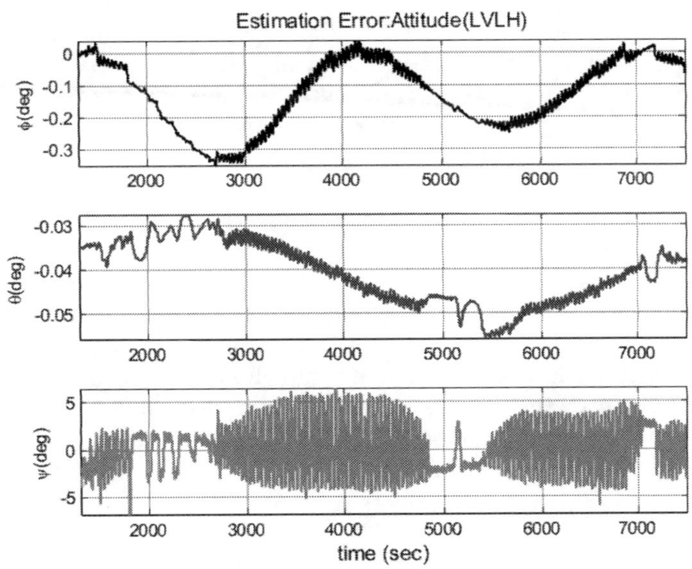

Figure 3. Attitude Errors of Steady-state EKF with Correction

As seen from Figure 3, the estimates of 3-axis attitude angles quickly tend to convergence after the transition process. This indicates that the steady-state KF can receive the same estimation accuracy as the KF with recursive gain, but contrast it will greatly reduce the amount of calculations.

The existence of the attitude leads to the projected component of orbital speed in three axial directions. So the gyro output includes the disturbance resulted from orbital speed, which causes the gyro drift estimates degraded. It is clearly illustrated in Figure 4 that the estimation errors of gyro drift in the roll-yaw direction are relatively larger. In addition, the coupling effect of close loop control will also increase the complexity of the drift estimation process.

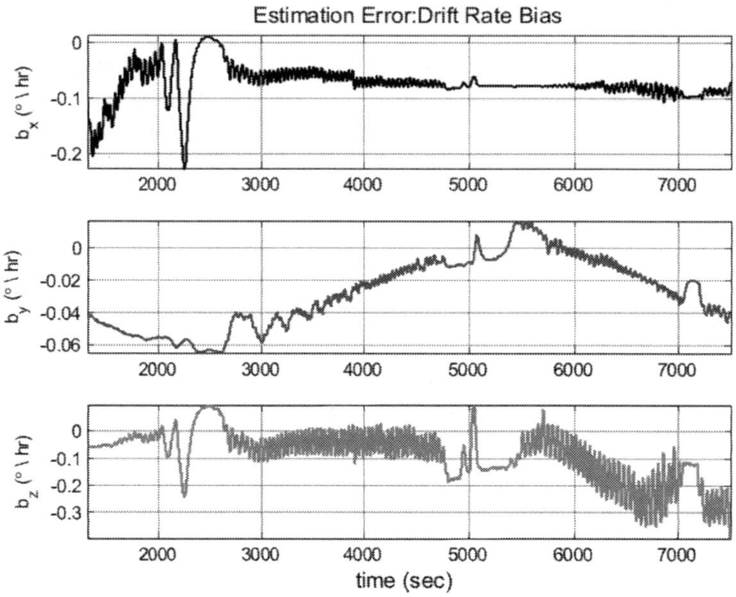

Figure 4. Gyro Drift Estimation Error

5 Conclusion

In this paper, a quaternion-based EKF attitude estimation algorithm with only a single horizon sensor has been studied, analyzed and simulated. Although the EHS can't provide directly yaw axis information, but based on the rate-gyro and satellite attitude kinematics equation, the orbital gyrocompass can give the yaw angle estimation and gradually tracks the real yaw attitude. On the other hand, because of no direct access to the yaw measurement, the accuracy of yaw attitude estimation is relatively poor. In spite of this, for earth observation satellite, its normal mission will be not affected, as long as roll and pitch attitude can maintain a certain level of angle estimation accuracy. To further enhance the performance of the estimation algorithm, an effort is made to consider detailed models of sensor measurements and to mend the algorithms in presence of realistic sensor errors, which are resulted from Earth's oblateness. The simulation reveals that the estimation precision of the roll axis obtained further improvement. It is another important superiority that the attitude determination based on a single EHS can be suitable for a whole orbital period. Therefore, the attitude estimation method given by this paper will be an economic and reliable measurement configuration, especially for a class of earth observation micro-satellites.

References

1. Sandau, R., Roser, H. P., and Valenzuela, A. 2008. *Small Satellites for Earth Observation*. Berlin: Springer.
2. CubeSat Design Specification Rev. 12. California State Polytechnic University. Retrieved August 2015 from: http://org.ntnu.no/studsat/docs/proposal_1/Cubesat.Design.Specification.pdf
3. Trentelman, H. L., Stoorvogel, A. A., and Hautus, M. 2001. *Control Theory for Linear Systems*. Lond: Springer.
4. Markley, F. L. and Crassidis, J. L. 2014. *Fundamentals of Spacecraft Attitude Determination and* Control. New York: Springer.
5. Farrel, J. A. 2008. Aided Navigation GPS with High Rate Sensors. New York: McGraw-Hill Companies.
6. Wertz, J. R.,etc. 1978. *Spacecraft Attitude Determination and Control*. Nethrelands: D.Reidel Publishing Company.

7. Bowers, A. E., Rodden, J. J., and Scott. E. D., et al. Orbital Gyrocompassing Heading Reference. 1968. *Journal of Spacecraft and Rockets*. 5: 901–910.
8. Bryson, A. E. and Kortum, W. 1971. Estimation of the Local Attitude Orbiting Spacecraft. *Automatica*. 7(2): 163–180.
9. Shuster, M.D. 1990. Kalman filtering of spacecraft attitude and the QUEST model. *Journal of the Astronautical Sciences*. 38(3):377–393.
10. Crassidis, J. L., and Junkins, J. L. 2012. *Optimal Estimation of Dynamic Systems*. Boca Raton: CRC Press
11. Tekawy, J. A., Wang, P., and Gray, C. W. 1996. Scanning Horizon Sensor Attitude Correction for Earth Oblateness. *Journal of Guidance, Control, and Dynamics*. 19(3): 706-708.

MATEC Web of Conferences **51**, 03009 (2016)
DOI: 10.1051/matecconf/20165103009

Investigation on Thermocompression Bonding Using Lead Free Sinterable Paste and High Lead Solder Paste for Power QFN Application

Gunaseelan Chandrakasan[1] and Mark Ovinis[2]

[1,2]*Universiti Teknologi PETRONAS, Malaysia*

Abstract. Persistently growing Power QFN packages are used in various fields especially micro-electronics, aerospace, oil and gas as well. However, the particular industries is pushing forward to reduce the use of hazardous materials in the process of manufacturing and assemblies. Thermo-compression die-attach layer is perceived to be the most critical element in power QFN packages as the increase in operating temperature requires new materials with suitable thermo-chemical properties also with suitable melting points of next generation lead free die attachment material. In this situation, Hi-lead solder (RM218: Pb92.5Sn5Ag2.5) which known as high temperature material is widely being used in most semiconductor assembly for die attach, yet it deduce few reliability challenges like solder voids, the clip tilt performance and also solder splash which has been considered as major quality issue in assembly of Power QFN packages (FET die, IC die and clip attach). As a solution, sintering epoxy paste (SPC073-3: Sn96.5/Ag3/Cu0.5) is being considered as a replacement. In this case, sintering epoxy paste demonstrating excellent electrical and thermal performance for Power QFN packages which is known to be demanded in market. Thus, this study investigates the differential pastes sintering paste and also solder paste, in order to identify best die attachment material to be used in thermo-compression bonding method. Therefore, the shear strength was resulting good indication where the sintering paste was recorded 2.4 Kg/mm meanwhile the solder paste was recorded 0Kg/mm at peak temperature of 260°C. Besides of that, the pot life seems promising as the sintering paste seems to have constant viscosity of 100Pa*s throughout the 48 hours tested while, high lead solder paste records viscosity from 100Pa*s marginally increase as the time increase which effects the inconsistency of pot life. Last but not least, the voids mechanisms proves sintering epoxy paste has the same pinhole voids as its individual, but the solder paste's pinhole voids are not same as individuals which easily can fail when the particular shear force was applied. Hence, sintering epoxy paste could resolve the quality issue by using thermo-compression bonding method and produce the better reliability than the solder paste.

1 Introduction

Power Quad Flat No-Lead (PQFN) package is also known as plastic package surface mounted with lead pads mounted on the bottom surface of the package [1]. PQFN packages is a unique as it will be in first choice because of its high speed application where high thermal performance is demanded in various industry. The assembly processes are simplified from the standard QFN assembly with only

FET die, IC die and clip attach hence deduces an addition heat dissipation path to the top surface. In thermo-compression bonding the element that perceived to have high importance, ensure fixation of the die on its substrate as well as dissipation of the heat generated in the die [2]. There are essentially, several criteria that a material needs to have in order to be used for die-attach. The material should demonstrate high thermal conductivity, low coefficient of thermal expansion (CTE) mismatches between the die and the substrate, good wettability as well as adhesion to the die also the substrate, good mechanical properties with stress relaxation behaviours, good fatigue resistance, good corrosion resistance, good rework ability, high electrical conductivity and good reliability [3].

Currently, numerous studies were carried out in order to identify the most suitable material and techniques for die-attach. However, studies on identifying the materials with least disadvantages as well as most durable, with high melting point need to be done. Indeed, existing work and studies have disclosed the processes that includes solder alloys i.e. AuSn, PbSnAg, AuGe and AuSi as well as nano- and micro- particle sintering (regularly with Ag) apart from transit liquid phase bonding [3]. The former is divided into two categories, i.e. Transient Liquid Phase diffusion bonding (TLP) and sintered nanoparticles. TLP bonding using Ag-In can be done either as thin film layers or the mixture of particles. The ratio of Ag- to In- fundamentally, is chosen in such way so the end product will be in Ag- rich region. Meanwhile, the melting point of In- being as low as 156.6°C enables lower temperature processing. Sintered nanoparticle Ag die attachment is done with Ag nanoparticles in a paste with an average silver particle size of 30nm [4]. On one hand, sintering process could be done at 285°C and the end product could withstand 400°C. This is because the uniqueness in sintering process where it is an atomic diffusion process which avoids the liquid phase during the intermediate stage and thus the quality problems such as non-wetting, voids and many more can be eliminated. Besides, sintering particles basically are applied to delay the instruction of mass diffusion and eliminates densification diffusion at high temperature and melting point which able to stabilize according to process condition which leads to improved properties to support high temperature packages. On the other hand, being a single metal system, sintered Nano-silver will not have the intermetallic formation that leads to potential failure mechanism for multiple metal systems. Nevertheless, the occurrence of electrical migration and dry migration which cause Ag to build up at cathode in dendrites form that will reach the anode that will later cause failure of the device [5].

2 Experimental

The purpose of this experiment is to introduce new generation lead free die attachment material using thermo-compressive bonding method for Power Quad Flat No-Lead (PQFN) package. Suitable material and techniques for die-attach are quantified. This experiment exhibits the FET die, IC die and clip attach and lead frame. To analyze this experiment, an analysis phase will be conducted where die attach process tests including die shear test, die tilt test and SEM evaluation. The possible failure mechanism will be identified through these tests. Figure 1 shows the bonding diagram used in this study.

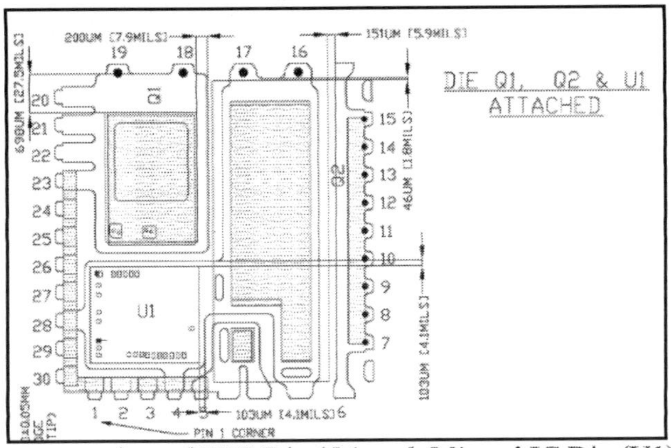

Figure 1. Bonding Diagram consists of FET Die (Q1 and Q2) and IC Die (U1)

A. Die Attach Materials

Two different die attach materials were used, as described in below table:

Table 1. Die Attach Materials

Sample	Type	Thermal Conductivity (W/mK)	Powder	Composition
RM218 (Hi lead Solder)	Water clean	25	Type 3	Pb92.5/Sn5/Ag2.5
SPC073-3 (Sintering Paste)	No-Clean	30	Type 3	Sn96.5/Ag3/Cu0.5

B. Process

i. Screen Print

A stencil with a 2mm x 2 mm opening and a thickness of 7 mils for bottom print and 4 mils for top print was used for manually printing paste onto the DEK machine. Only single print deposit was employed. For each paste, five strips were prepared. Below Table describes the parameter used:

Table 2. Screen Print Parameter

Metal Squeegee Thickness	Blade Angle	Force on Blade per inch of Blade	Blade Speed	Work Life on Stencil
10 mils = 250um	45-60 degrees	0.5 – 1.0 kg	2-4 in./sec	16 hours

ii. Die Attach

The process of joining metal to metal which brought into atomic contact by applying force and heat simultaneously is known as thermos-compression bonding. The test dies and clips was kept on Nitrogen ($N2_)$ contained cabinet in order to avoid oxidization. For this experiment the test dies have been made in dices rather than full wafer formation. Then, the lead frame was placed on the hot plate with vacuum chuck. After that, the collet will pick up die with vacuum and proper alignment. The picked die will be brought into metal to metal contact when the desired pressure was achieved. However, as the pressure determined, the temperature will increase rapidly. Table 2 below presents the bonding conditions:

Table 3. Bonding conditions

Die Attach Material	Bonding Time (ms)	Bond Force (g)	Peak Temperature(C)	Pressure (MPa)
RM218	100	100	260	0.2
SPC073-3	100	100	260	0.2

Machine ESEC 2000 had been used to perform pick and place process for the selected materials. 10 samples had been built using thermo-compression bonding. Thermo-compression bonding was experimented by varying the bonding time, temperature and pressure in order to achieve the quality of bonding. Yet suitable bonding time, peak temperature and pressure had been identified.

iii. Reflow

The strips were reflowed under air atmosphere via a MALCOM infrared oven. The profile with fixed. Peak temperature and soaking temperature were used. The Soaking temperature was 200°C, while the peak temperature was 260°C with total time 0f 90sec, respectively, as shown in below Figure (2):

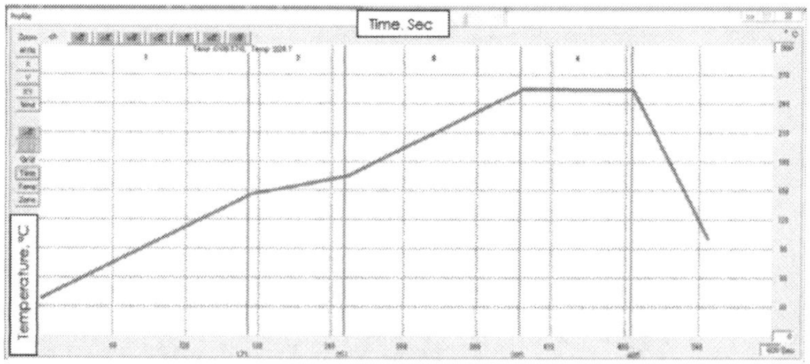

Figure 2. Reflow Profile

3 Results and Discussion

Sinterable epoxy die attachment paste seems to be the best choice in building the Power QFN packages, as it is compatible with silver and gold die back, bare copper lead frames and also the PPF lead frames [6]. Besides that, TLPS materials can also basically issues system in package which having issues with solder such as solder voids, tilting, and others. Even though sinterable epoxy paste does not re-melt, it deduce some good criteria such as reliability and re-workability improvements.

3.1 Hi lead Solder Paste Vs Sinterable Epoxy Paste

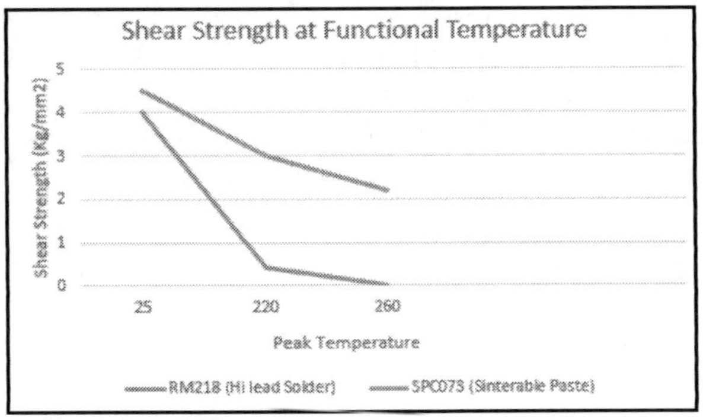

Figure 3. Shear Strength at Functional Temperature

Generally, shear strength defines the ability of die attaching material to withstand the separation between die and lead frame [6]. Hence, the functional temperature had been analyzed on above figure by varying from minimum to maximum peak temperature. For both die attaching material, it is known that minimum peak temperature is 25°C and the maximum is 260C. It is also known that the peak temperature contributes to the shear strength of PQFN packages.

At minimum peak temperature of 25°C, the sinterable paste was found to have high shear strength than the solder paste. At 220°C, the shear strength seems to be dropped drastically, yet sinterable paste seems to be the lower drop than the solder paste which relies on huge gap. When reaching the maximum peak temperature at 260°C, the solder paste seems to fail the shear strength as it records 0 Kg/mm while sinterable paste records, 2.4 Kg/mm. By means, sufficient force was applied to test the strength of the die. Subsequent strength explains an approximate indication of how well the die was attached on substrate. This clearly results the sinterable paste has the good wettability which joins the particles and melting with good condition.

3.2 Pot Life at Room Temperature

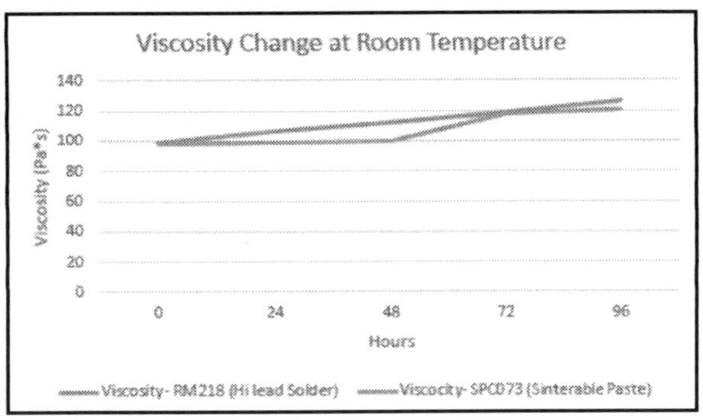

Figure 4. Viscosity Change at Room Temperature

Viscosity change does affects the reliability of a Power QFN package, as the particles become non-reactive as the pot life at room temperature was not controlled. Thus, the SPC073-3 (sintering) paste storage need to be critically studied as it should be stored on laboratory bench top. However, the viscosity was measured at 25°C which known as room temperature and placed it 5mins on a parallel plate on Rheometer to stabilize the paste. Meanwhile, for RM218 (Hi lead) solder paste, the thawing time will be one hour as per the specifications and the stabilization will be placed on plate which rotates clockwise and anti-clockwise at 25°C.

In this case, sinterable paste seems to have lower viscosity which was 120Pa*s at maximum reading of 96 hours, meanwhile, high lead solder paste records 129Pa*s at maximum reading of 96 hours. However, the graph illustrates, the viscosity is directly proportional to the hours of staging for both materials tested. This is a best characteristic in PQFN packages, as it relies lesser the viscosity, better the performance of the paste [6]. Thus, by comparison of both die attaching material, sinterable paste has stable viscosity change while the solder paste increase rapidly. It has been determined to maintain 24 hour for sinterable paste pot life, while solder paste may be the shorter pot life which also considering the cost of usage of die attaching material.

3.3 Mechanisms behind Voiding

Voids from volatiles is something common when the void structures defined where it's can be visualized in "wormy" condition. However, it can be eliminated or minimized with the adjustments with the certain conditions on reflow profile optimization [6]. Generally, when discuss about the reflow profile, the best profile is defined at ramp to peak temperatures. Considering the thermal mass of Power QFN, ramp to peak is often not the best option. Balance soak temperature to thermal mass and minimizing the total time to the peak temperature will results voids from volatiles can be eliminated or minimized drastically [7].Besides that, pinhole voids can be defined as tiny round voids. It may be the cause of time taken to melt the alloy particles in paste's was fast, as the fast melting time of alloy can reacts with the copper particles and preforms solidification condition. However, the late melting time of alloy particles joins the molecules but cannot lead to collapsing. In this case, these pinhole voids are generally non-contiguous and measured in the parameter of 15-25 micron in diameter.

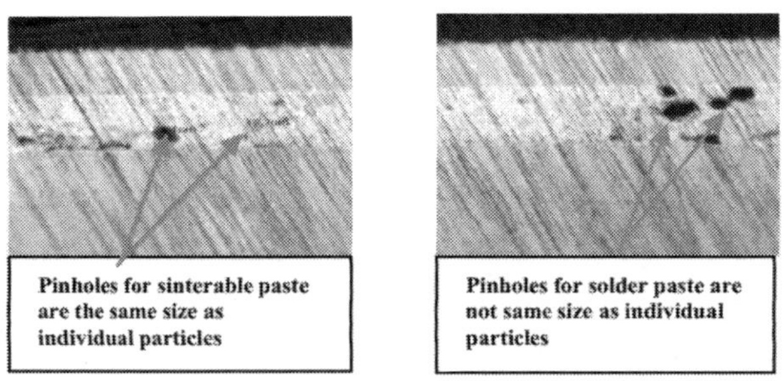

Figure 5. Pinhole voids

When discuss about the way of eliminate this void as it affects the joint strength, yet this cannot be eliminated through the process optimization as they are a normal feature of sinterable paste, as its alloy particles performance criteria need to be consider. At the same time, it is also known that the pinhole voids of sinterable paste were produced in the same size of holes throughout the unit. Meanwhile the pinhole voids hole size for solder paste was inconsistent, which can affect the joint strength.

4 Conclusion

SPC073-3 performed better than the RM218 solder paste at both room temperature and at 260°C for die shear test on bare die and also clip attach. However, sinterable paste is a high lead solder paste replacement as it consist of few advantages. For special case application like Power QFN packages solder paste deduce a lot of quality challenges as discussed above such as step soldering, no solder mask and others. Besides that, sinterable paste does not remelt or change footprint in reflow, which promising the stencil printing quality. Furthermore, the storage, shelf life, pot life and work life are comparable as well with the solder pastes [8]. Apart from that, mechanical, electrical and thermal performance are also seems promising and performs better than the solder joint [9]. Therefore, Sinterable paste is more reliable and can be classified as best die attaching material using the thermos compression bonding method. Therefore, other factors rather than die shear test, die tilt test should be considered in order to deduce a conclusion in this experiment. The major contributing factor to shear strength was the amount of die surface area that actually bonded on the substrate [10]. However, the pattern designed for this research seems successful as there was no die crack was occurred. The bond force applied on this research shall be quantified with different forces, to identify the failures of thermo-compression bonding occurs at which force. Besides that, the possible failures shall be researched in deep by varying more testing methods.

References

1. Bai, J. G., Zhang, Z. Z., Calata, J. N., & Luo, G. Q. (2006). Low- temperature sintered nanoscale silver as a novel semiconductor device-metallized substrate interconnected material.*IEEE transactions on components and packaging technologies, 29*(3), 589- 593.
2. Buttay, C., Masson, A., Li, J., Johnson, M., Lazar, M., Raynaud, C., & Morel, H. (2012). Die attach of power devices using silver sintering- bonding process optimization and characterization. http://hal.archives-ouvertes.fr/docs/00/67/26/19/PDF/article.pdf
3. Hui Shun Chin. "A Review on Die Attach Materials for SiC-Based High-Temperature Power Devices", Metallurgical and Materials Transactions B, 04/06/2010.
4. Edwards, C. & Ho, B. (1999). *Advanced packaging's step by step*. San Diego: HIS Publishing Group. http://www.acoustech-inc.com/pdfs/Step%20by%20Step.pdf
5. H. Reichl, V. Grosser, "Overview and development trends in the field of MEMS packaging", *MicroElectro Mechanical Systems, 2001. MEMS 2001,The 14th IEEE International Conference*, pp. 1–5, 2001.
6. Harris, J. & Matthews, M. (2009). Selecting die attach technology for high-power applications. http://www.ormetcircuits.com/d/bindocs/Selecting_Die_Attach_Technology_for_High_Power_Applications.pdf
   http://www.ormetcircuits.com/d/parts/parts.php
7. Johnson, R. W., Zheng, P., Wiggins, A., Rubin, S., &Peltz, I. (2007, September). *High temperature electronics packaging proceedings*. Presented at HITEN International Conference on High Temperature Electronics, St. Catherine's College Oxford, England.
8. Sabrina C., D.A. Vasilache, S. Girardi, P. Conci, and B. Margesin, "Thermocompression Bonding for 3D RF MEMS Devices Using Gold and Silver As Intermediate Layer", *Proceeding of IEEE*, 2(183-186).
9. Ohme, B., Johnson, B. J., & Larson, M. R. (2007). *SOI CMOS for extreme temperature application, Honeywell aerospace, defense& space.* Retrieved July 25th, 2014 from http://www51.honeywell.com/aero/common/documents/myaerospacecatalog-documents/Defense_Brochures-documents/High_Temperature/SOI_CMOS_for_Extreme_Temperature_Applications.pdf
10. Henson, P. (2010). Considerations and options for high temperature die attach. https://etd.auburn.edu/bitstream/handle/10415/2249/final%20Thesis.pdf

AUTHOR INDEX

EDP Sciences – Web of Conferences
17, Avenue du Hoggar
Parc d'Activité de Courtabouf
BP 112
F-91944 Les Ulis Cedex A
France

ISBN 978-1-5108-2270-2

CURRAN ASSOCIATES INC.
proceedings.com